endorsed for
Edexcel

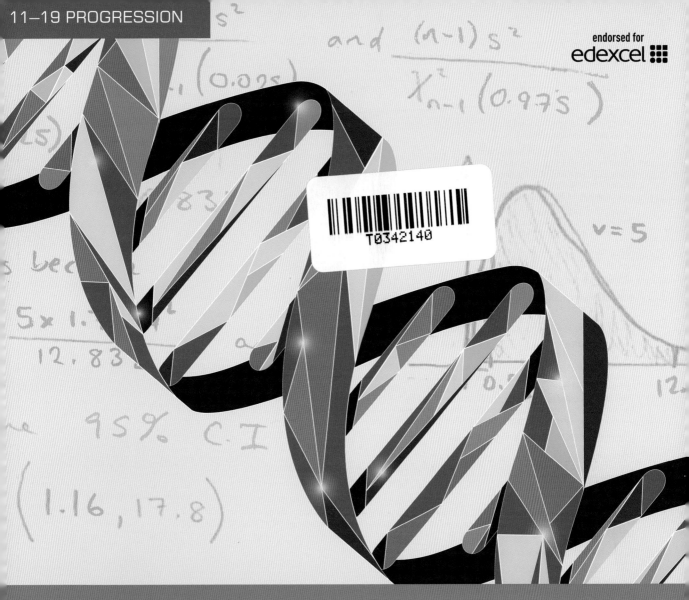

Edexcel AS and A level Further Mathematics

Further Statistics 2

FS2

Series Editor: Harry Smith
Authors: Greg Attwood, Ian Bettison, Alan Clegg, Gill Dyer, Jane Dyer, John Kinoulty,
Keith Pledger, Harry Smith

Pearson

Published by Pearson Education Limited, 80 Strand, London WC2R 0RL.

www.pearsonschoolsandfecolleges.co.uk

Copies of official specifications for all Pearson qualifications may be found on the website:
qualifications.pearson.com

Text © Pearson Education Limited 2018
Edited by Tech-Set Ltd, Gateshead
Typeset by Tech-Set Ltd, Gateshead
Original illustrations © Pearson Education Limited 2018
Cover illustration Marcus@kja-artists

The rights of Greg Attwood, Ian Bettison, Alan Clegg, Gill Dyer, Jane Dyer, John Kinoulty, Keith
Pledger and Harry Smith to be identified as authors of this work have been asserted by them in
accordance with the Copyright, Designs and Patents Act 1988.

First published 2018

24
10 9 8 7 6 5

British Library Cataloguing in Publication Data
A catalogue record for this book is available from the British Library

ISBN 978 1 292 18338 1

Printed in the UK by Bell and Bain Ltd, Glasgow

Acknowledgements
The authors and publisher would like to thank the following for their kind permission to repro-
duce their photographs:

(Key: b-bottom; c-centre; l-left; r-right; t-top)

123 RF: Viparat Kluengsuwanchai 44, 99cr. **Shutterstock:** Gertan 21, 99cl; MPVAN 109, 196l;
Acceptphoto 141, 196c. **Alamy Stock Photo:** Vincent Starr Photography/Cultura Creative (RF) 1,
99l; PCN Photography 92, 99r. **Getty Images:** Ben Queenborough/ Photodisc 163, 196r.

All other images © Pearson Education

A note from the publisher
In order to ensure that this resource offers high-quality support for the associated Pearson
qualification, it has been through a review process by the awarding body. This process confirms
that this resource fully covers the teaching and learning content of the specification or part
of a specification at which it is aimed. It also confirms that it demonstrates an appropriate
balance between the development of subject skills, knowledge and understanding, in addition
to preparation for assessment.

Endorsement does not cover any guidance on assessment activities or processes (e.g. practice
questions or advice on how to answer assessment questions), included in the resource nor does
it prescribe any particular approach to the teaching or delivery of a related course.

While the publishers have made every attempt to ensure that advice on the qualification
and its assessment is accurate, the official specification and associated assessment guidance
materials are the only authoritative source of information and should always be referred to for
definitive guidance.

Pearson examiners have not contributed to any sections in this resource relevant to
examination papers for which they have responsibility.

Examiners will not use endorsed resources as a source of material for any assessment set by
Pearson.

Endorsement of a resource does not mean that the resource is required to achieve this Pearson
qualification, nor does it mean that it is the only suitable material available to support the
qualification, and any resource lists produced by the awarding body shall include this and
other appropriate resources.

Pearson has robust editorial processes, including answer and fact checks, to ensure the
accuracy of the content in this publication, and every effort is made to ensure this publication
is free of errors. We are, however, only human, and occasionally errors do occur. Pearson is not
liable for any misunderstandings that arise as a result of errors in this publication, but it is
our priority to ensure that the content is accurate. If you spot an error, please do contact us at
resourcescorrections@pearson.com so we can make sure it is corrected.

Contents

Overarching themes

The following three overarching themes have been fully integrated throughout the Pearson Edexcel AS and A level Mathematics series, so they can be applied alongside your learning and practice.

1. Mathematical argument, language and proof

- Rigorous and consistent approach throughout
- Notation boxes explain key mathematical language and symbols
- Dedicated sections on mathematical proof explain key principles and strategies
- Opportunities to critique arguments and justify methods

2. Mathematical problem solving

- Hundreds of problem-solving questions, fully integrated into the main exercises
- Problem-solving boxes provide tips and strategies
- Structured and unstructured questions to build confidence
- Challenge boxes provide extra stretch

The Mathematical Problem-solving cycle

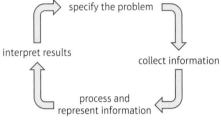

specify the problem

collect information

process and represent information

interpret results

3. Mathematical modelling

- Dedicated modelling sections in relevant topics provide plenty of practice where you need it
- Examples and exercises include qualitative questions that allow you to interpret answers in the context of the model
- Dedicated chapter in Statistics & Mechanics Year 1/AS explains the principles of modelling in mechanics

Finding your way around the book

Access an online digital edition using the code at the front of the book.

Each chapter starts with a list of objectives

Confidence intervals and tests using the *t*-distribution 7

Objectives
After completing this chapter you should be able to:
- Find a confidence interval for the mean of a normal distribution with unknown variance → pages 164–170
- Conduct a hypothesis test for the mean of a normal distribution with unknown variance → pages 170–174
- Carry out a paired *t*-test → pages 174–179
- Find a confidence interval for the difference between means from two independent normal distributions with equal but unknown variances → pages 180–184
- Conduct a hypothesis test for the difference between means from two independent normal distributions with equal but unknown variances → pages 185–189

The real world applications of the maths you are about to learn are highlighted at the start of the chapter with links to relevant questions in the chapter

Farmers often try out different diets to see which is the most effective at producing high-yield animals. They can compare the effectiveness of two diets using a paired *t*-test. → Mixed exercise Q13

Prior knowledge check
1 A random sample of size 20 is taken from a normally distributed population with a standard deviation of 2. The mean of the sample was 16. Find a 95% confidence interval for the mean μ. ← Section 3.1
2 A researcher is comparing the heights of children in two towns. A random sample of 100 children from town A is taken and the sample mean and standard deviation are 145 cm and 4 cm respectively. An independent random sample of 120 children from town B is taken and the sample mean and standard deviation are 146 cm and 3.5 cm respectively. Test, at the 5% level of significance, whether there is evidence of a difference in the mean heights of the children in the two towns. ← Section 3.3

163

The *Prior knowledge check* helps make sure you are ready to start the chapter

Exercise questions are carefully graded so they increase in difficulty and gradually bring you up to exam standard

Exercises are packed with exam-style questions to ensure you are ready for the exams

A level content is clearly flagged

Challenge boxes give you a chance to tackle some more difficult questions

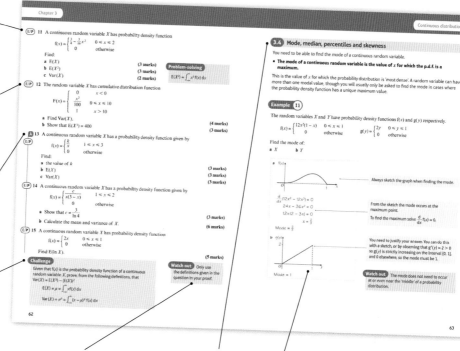

Exam-style questions are flagged with $\mathbf{E}$

Problem-solving questions are flagged with $\mathbf{P}$

Problem-solving boxes provide hints, tips and strategies, and *Watch out* boxes highlight areas where students often lose marks in their exams

Each section begins with explanation and key learning points

Step-by-step worked examples focus on the key types of questions you'll need to tackle

Each chapter ends with a *Mixed exercise* and a *Summary of key points*

Every few chapters a *Review exercise* helps you consolidate your learning with lots of exam-style questions

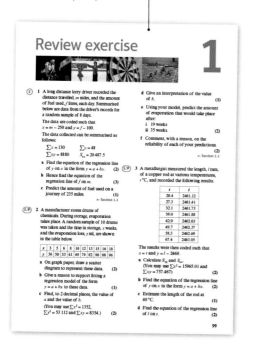

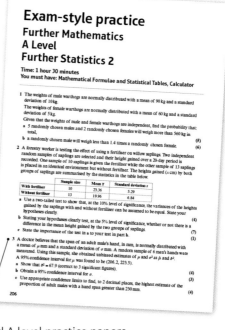

AS and A level practice papers at the back of the book help you prepare for the real thing.

Extra online content

Whenever you see an *Online* box, it means that there is extra online content available to support you.

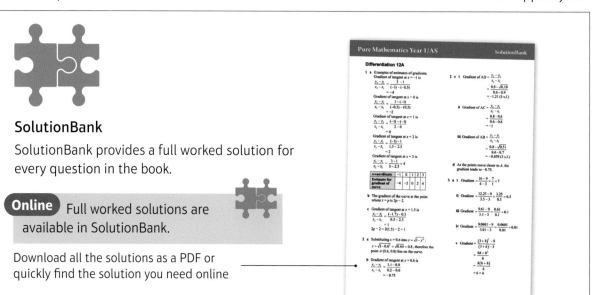

SolutionBank

SolutionBank provides a full worked solution for every question in the book.

Online Full worked solutions are available in SolutionBank.

Download all the solutions as a PDF or quickly find the solution you need online

Use of technology

Explore topics in more detail, visualise problems and consolidate your understanding using pre-made GeoGebra activities.

Online Find the point of intersection graphically using technology.

GeoGebra

GeoGebra-powered interactives

Interact with the maths you are learning using GeoGebra's easy-to-use tools

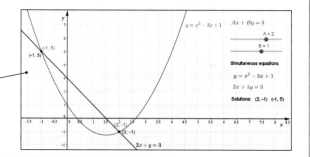

Access all the extra online content for free at:

www.pearsonschools.co.uk/fs2maths

You can also access the extra online content by scanning this QR code:

Linear regression

Objectives

After completing this chapter, you should be able to:

* Calculate the equation of a regression line using raw data or summary statistics → **pages 2-8**
* Use coding to find the equation of a regression line → **pages 8-10**
* Calculate residuals and use them to test for linear fit and identify outliers → **pages 10-15**
* Calculate the residual sum of squares (RSS) → **pages 13-15**

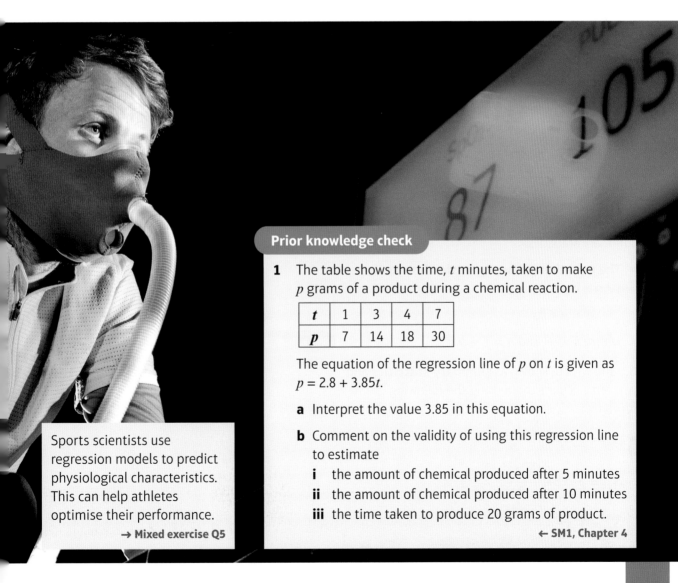

Sports scientists use regression models to predict physiological characteristics. This can help athletes optimise their performance.

→ **Mixed exercise Q5**

Prior knowledge check

1 The table shows the time, t minutes, taken to make p grams of a product during a chemical reaction.

t	1	3	4	7
p	7	14	18	30

The equation of the regression line of p on t is given as $p = 2.8 + 3.85t$.

a Interpret the value 3.85 in this equation.

b Comment on the validity of using this regression line to estimate

  i the amount of chemical produced after 5 minutes

  ii the amount of chemical produced after 10 minutes

  iii the time taken to produce 20 grams of product.

← **SM1, Chapter 4**

1.1 Least squares linear regression

When you are analysing bivariate data, you can use a **least squares regression line** to predict values of the dependent (response) variable for given values of the independent (explanatory) variable. If the response variable is y and the explanatory variable is x, you should use the regression line of **y on x**, which can be written in the form $y = a + bx$.

> **Links** You should only use the regression line to make predictions for values of the dependent variable that are within the range of the given data. This is called **interpolation**. Making predictions for values outside of the range of the given data is called **extrapolation** and produces a less reliable prediction. ← SM1, Section 4.2

The least squares regression line is the line that minimises the **sum of the squares of the residuals** of each data point.

- **The residual of a given data point is the difference between the observed value of the dependent variable and the predicted value of the dependent variable.**

> **Notation** The Greek letter epsilon (ε) is sometimes used to denote a residual.

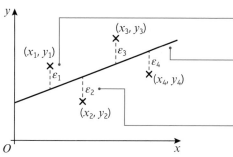

ε_1 is the residual of the data point (x_1, y_1)

The least squares regression line of y on x is the straight line that minimises the value of $\varepsilon_1^2 + \varepsilon_2^2 + \varepsilon_3^2 + \varepsilon_4^2$. In general, if each data point has residual ε_i, the regression line minimises the value of $\sum \varepsilon_i^2$.

The observed value of the dependent variable, y_2, is **less** than the predicted value, so the residual of (x_2, y_2) will be **negative**.

You need to be able to find the equation of a **least squares regression line** using raw data or summary statistics.

- **The equation of the regression line of y on x is:**

$$y = a + bx$$

where $b = \dfrac{S_{xy}}{S_{xx}}$ and $a = \overline{y} - b\overline{x}$

> **Watch out** You can calculate a and b directly from raw data using your calculator. However, you might be given summary statistics in the exam so you need to be familiar with these formulae.

S_{xy} and S_{xx} are known as **summary statistics** and you can calculate them using the following formulae:

- $S_{xy} = \sum xy - \dfrac{\sum x \sum y}{n}$

$S_{xx} = \sum x^2 - \dfrac{(\sum x)^2}{n}$

$S_{yy} = \sum y^2 - \dfrac{(\sum y)^2}{n}$

Example 1

The results from an experiment in which different masses were placed on a spring and the resulting length of the spring measured, are shown below.

Mass, x (kg)	20	40	60	80	100
Length, y (cm)	48	55.1	56.3	61.2	68

a Calculate S_{xx} and S_{xy}.
   (You may use $\sum x = 300$ $\sum x^2 = 22\,000$ $\overline{x} = 60$ $\sum xy = 18\,238$ $\sum y^2 = 16\,879.14$
   $\sum y = 288.6$ $\overline{y} = 57.72$)

b Calculate the regression line of y on x.

c Use your equation to predict the length of the spring when the applied mass is:
   i 58 kg
   ii 130 kg

d Comment on the reliability of your
   predictions.

Online Explore the calculation of a least squares regression line using GeoGebra.

a $S_{xx} = \sum x^2 - \dfrac{(\sum x)^2}{n}$

   $= 22\,000 - \dfrac{300^2}{5}$

   $= 4000$

   $S_{xy} = \sum xy - \dfrac{\sum x \sum y}{n}$

   $= 18\,238 - \dfrac{300 \times 288.6}{5}$

   $= 922$

b $b = \dfrac{S_{xy}}{S_{xx}} = \dfrac{922}{4000} = 0.2305$

   $a = \overline{y} - b\overline{x}$

   $= 57.72 - 0.2305 \times 60$

   $= 43.89$

   $y = 43.89 + 0.2305x$

c i $y = 43.89 + 0.2305 \times 58$

   $= 57.3$ cm (3 s.f.)

   ii $y = 43.89 + 0.2305 \times 130$

   $= 73.9$ cm (3 s.f.)

d Assuming the model is reasonable, the prediction when the mass is 58 kg is reliable since this is within the range of the data.
   The prediction when the mass is 130 kg is less reliable since this is outside the range of the data.

Use the standard formulae to calculate S_{xx} and S_{xy}. Write down any formulae you are using before you substitute.

Use the formulae to calculate a and b. If you want to check your answer using your calculator, make sure you use the correct mode for linear regression with bivariate data. On some calculators this mode is labelled $y = a + bx$.

Remember to write the equation at the end. The numbers should be given to a suitable degree of accuracy.

Substitute the given values into the equation of the regression line.

This is called **interpolation**.

This is called **extrapolation**.

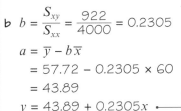

3

Example 2

A scientist working in agricultural research believes that there is a linear relationship between the amount of a certain food supplement given to hens and the hardness of the shells of the eggs they lay. As an experiment, controlled quantities of the supplement were added to the hens' normal diet for a period of two weeks and the hardness of the shells of the eggs laid at the end of this period was then measured on a scale from 1 to 10, with the following results:

Food supplement, f (g/day)	2	4	6	8	10	12	14
Hardness of shells, h	3.2	5.2	5.5	6.4	7.2	8.5	9.8

a Find the equation of the regression line of h on f.

(You may use $\sum f = 56$ $\quad \sum h = 45.8$ $\quad \bar{f} = 8$ $\quad \bar{h} = 6.543$ $\quad \sum f^2 = 560$ $\quad \sum fh = 422.6$)

b Interpret what the values of a and b tell you.

a $S_{fh} = \sum fh - \dfrac{\sum f \sum h}{n}$

$\qquad = 422.6 - \dfrac{56 \times 45.8}{7} = 56.2$

$S_{ff} = \sum f^2 - \dfrac{(\sum f)^2}{n}$

$\qquad = 560 - \dfrac{56^2}{7} = 112$

$b = \dfrac{S_{fh}}{S_{ff}} = \dfrac{56.2}{112}$

$\qquad = 0.5017\dots$ hardness units per g per day

$a = \bar{h} - b\bar{f}$

$\qquad = 6.543 - 0.5017\dots \times 8$

$\qquad = 2.5287\dots$ hardness units

$h = 2.53 + 0.502f$

b a estimates the shell strength when no supplement is given (i.e. when $f = 0$). Zero is only just outside the range of f so it is reasonable to use this value.

b estimates the rate at which the hardness increases with increased food supplement; in this case for every extra one gram of food supplement per day the hardness increases by 0.502 (3 s.f.) hardness units.

Watch out The variables given might not be x and y. Be careful that you use the correct values when you substitute into the formulae. It can sometimes help to write x next to the explanatory variable in the table (f) and y next to the response variable (h).

When dealing with a real problem do not forget to put the units of measurement for the two constants.

Make sure you give your answer in the context of the question. Don't just say that one value increases as the other increases – you need to comment on the **rate** of increase of hardness.

Example 3

A repair workshop finds it is having a problem with a pressure gauge it uses. It decides to have it checked by a specialist firm. The following data were obtained.

Gauge reading, x (bars)	1.0	1.4	1.8	2.2	2.6	3.0	3.4	3.8
Correct reading, y (bars)	0.96	1.33	1.75	2.14	2.58	2.97	3.38	3.75

(You may use $\sum x = 19.2$ $\sum x^2 = 52.8$ $\sum y = 18.86$ $\sum y^2 = 51.30$ $\sum xy = 52.04$)

a Show that $S_{xy} = 6.776$ and find S_{xx}.

It is thought that a linear relationship of the form $y = a + bx$ could be used to describe these data.

b Use linear regression to find the values of a and b giving your answers to 3 significant figures.

c Draw a scatter diagram to represent these data and draw the regression line on your diagram.

d The gauge shows a reading of 2 bars. Using the regression equation, work out what the correct reading should be.

a $S_{xy} = \sum xy - \dfrac{\sum x \sum y}{n}$

$= 52.04 - \dfrac{19.2 \times 18.86}{8} = 6.776$

> Quote the formula you are going to use before substituting the values.

$S_{xx} = \sum x^2 - \dfrac{(\sum x)^2}{n}$

$= 52.8 - \dfrac{(19.2)^2}{8} = 6.72$

b $b = \dfrac{S_{xy}}{S_{xx}} = \dfrac{6.776}{6.72} = 1.0083\ldots$

$a = \bar{y} - b\bar{x} = \dfrac{18.86}{8} - 1.0083\ldots \times \dfrac{19.2}{8}$

$= -0.0625$

Regression line is: $y = -0.0625 + 1.008x$

or $y = 1.008x - 0.0625$

c

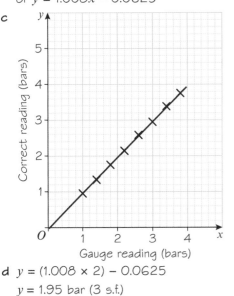

> To draw the regression line either plot the point $(0, a)$ and use the gradient or find two points on the line.
>
> In this case using $x = 1$ gives $y = 0.95$ and using $x = 3$ gives $y = 2.96$.

d $y = (1.008 \times 2) - 0.0625$

$y = 1.95$ bar (3 s.f.)

Exercise 1A

1 The equation of a regression line in the form $y = a + bx$ is to be found. Given that $S_{xx} = 15$, $S_{xy} = 90$, $\overline{x} = 3$ and $\overline{y} = 15$, work out the values of a and b.

2 Given that $S_{xx} = 30$, $S_{xy} = 165$, $\overline{x} = 4$ and $\overline{y} = 8$, find the equation of the regression line of y on x.

3 The equation of a regression line is to be found. The following summary data is given:

$$S_{xx} = 40 \qquad S_{xy} = 80 \qquad \overline{x} = 6 \qquad \overline{y} = 12$$

Find the equation of the regression line in the form $y = a + bx$.

4 Data is collected and summarised as follows:

$$\sum x = 10 \qquad \sum x^2 = 30 \qquad \sum y = 48 \qquad \sum xy = 140 \qquad n = 4$$

a Work out $\overline{x}$, $\overline{y}$, S_{xx} and S_{xy}.

b Find the equation of the regression line of y on x in the form $y = a + bx$.

5 For the data in the table,

x	2	4	5	8	10
y	3	7	8	13	17

Hint Check your answer using the statistical functions on your calculator.

a calculate S_{xx} and S_{xy}

b find the equation of the regression line of y on x in the form $y = a + bx$.

(P) 6 Research was done to see if there is a relationship between finger dexterity and the ability to do work on a production line. The data is shown in the table.

Dexterity score, x	2.5	3	3.5	4	5	5	5.5	6.5	7	8
Productivity, y	80	130	100	220	190	210	270	290	350	400

The equation of the regression line for these data is $y = -59 + 57x$.

a Use the equation to estimate the productivity of someone with a dexterity of 6.

b Give an interpretation of the value of 57 in the equation of the regression line.

c State, giving in each case a reason, whether or not it would be reasonable to use this equation to work out the productivity of someone with dexterity of:

i 2   ii 14

7 A field was divided into 12 plots of equal area. Each plot was fertilised with a different amount of fertilizer (h). The yield of grain (g) was measured for each plot. Find the equation of the regression line of g on h in the form $g = a + bh$ given the following summary data.

$$\sum h = 22.09 \qquad \sum g = 49.7 \qquad \sum h^2 = 45.04 \qquad \sum g^2 = 244.83 \qquad \sum hg = 97.778 \qquad n = 12$$

P 8 Research was done to see if there was a relationship between the number of hours in the working week (w) and productivity (p). The data are shown in the two scatter graphs below.

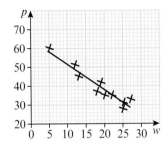 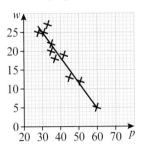

(You may use $\sum p = 397$ $\sum p^2 = 16\,643$ $\sum w = 186$ $\sum w^2 = 3886$ $\sum pw = 6797$)

a Calculate the equation of the regression line of p on w, giving your answer in the form $p = a - bw$.

b Rearrange this equation into the form $w = c + dp$.

The equation of the regression line of w on p is $w = 45.0 - 0.666p$.

c Comment on the fact that your answer to part **b** is different to this equation.

d Which equation should you use to predict:

  i the productivity for a 23-hour working week

  ii the number of hours in a working week that achieves a productivity score of 40.

P 9 In a chemistry experiment, the mass of chemical produced, y and the temperature, x are recorded.

x (°C)	100	110	120	130	140	150	160	170	180	190	200
y (mg)	34	39	41	45	48	47	41	35	26	15	3

Maya thinks that the data can be modelled using a linear regression line.

a Calculate the equation of the regression line of y on x. Give your answer in the form $y = a + bx$.

b Draw a scatter graph for these data.

c Comment on the validity of Maya's model.

E/P 10 An accountant monitors the number of items produced per month by a company (n) together with the total production costs (p). The table shows these data.

Number of items, n (1000s)	21	39	48	24	72	75	15	35	62	81	12	56
Production costs, p (£1000s)	40	58	67	45	89	96	37	53	83	102	35	75

(You may use $\sum n = 540$ $\sum n^2 = 30\,786$ $\sum p = 780$ $\sum p^2 = 56\,936$ $\sum np = 41\,444$)

Watch out The numbers of items are given in 1000s. Be careful to choose the correct value to substitute into your regression equation.

a Calculate S_{nn} and S_{np}. **(2 marks)**

b Find the equation of the regression line of p on n in the form $p = a + bn$. **(3 marks)**

c Use your equation to estimate the production costs of 40 000 items. **(2 marks)**

d Comment on the reliability of your estimate. **(1 mark)**

E/P **11** A printing company produces leaflets for different advertisers. The number of leaflets, n, measured in 100s and printing costs £p are recorded for a random sample of 10 advertisers. The table shows these data.

n (100s)	1	3	4	6	8	12	15	18	20	25
p (pounds)	22.5	27.5	30	35	40	50	57.5	65	70	82.5

(You may use $\sum n = 112$ $\sum n^2 = 1844$ $\sum p = 480$ $\sum p^2 = 26\,725$ $\sum np = 6850$)

a Calculate S_{nn} and S_{np}. **(2 marks)**

b Find the equation of the regression line of p on n in the form $p = a + bn$. **(3 marks)**

c Give an interpretation of the value of b. **(1 mark)**

An advertiser is planning to print t hundred leaflets. A rival printing company charges 5p per leaflet.

d Find the range of values of t for which the first printing company is cheaper than the rival. **(2 marks)**

E/P **12** The relationship between the number of coats of paint applied to a boat and the resulting weather resistance was tested in a laboratory. The data collected are shown in the table.

Coats of paint, x	1	2	3	4	5
Protection, y (years)	1.4	2.9	4.1	5.8	7.2

a Use your calculator to find an equation of the regression line of y on x as a model for these results, giving your answer in the form $y = a + bx$. **(2 marks)**

b Interpret the value b in your model. **(1 mark)**

c Explain why this model would not be suitable for predicting the number of coats of paint that had been applied to a boat that had remained weather resistant for 7 years. **(1 mark)**

d Use your answer to part **a** to predict the number of years of protection when 7 coats of paint are applied. **(2 marks)**

In order to improve the reliability of its results, the laboratory made two further observations:

Coats of paint, x	6	8
Protection, y (years)	8.2	9.9

e Using all 7 data points:

i produce a refined model

ii use your new model to predict the number of years of protection when 7 coats of paint are applied

iii give two reasons why your new prediction might be more accurate than your original prediction. **(5 marks)**

Sometimes the original data is coded to make it easier to manage. You can calculate the equation of the original regression line from the coded one by substituting the coding formula into the equation of the coded regression line.

Example 4

Eight samples of carbon steel were produced with a different percentages, $c\%$, of carbon in them.

Each sample was heated in a furnace until it melted and the temperature, m in °C, at which it melted was recorded.

The results were coded such that $x = 10c$ and $y = \dfrac{m - 700}{5}$

The coded results are shown in the table.

Percentage of carbon, x	1	2	3	4	5	6	7	8
Melting point, y	35	28	24	16	15	12	8	6

a Calculate S_{xy} and S_{xx}.
   (You may use $\sum x^2 = 204$ and $\sum xy = 478$.)

b Find the regression line of y on x.

c Estimate the melting point of carbon steel which contains 0.25% carbon.

a $S_{xy} = \sum xy - \dfrac{\sum x \sum y}{n}$

$\quad = 478 - \dfrac{36 \times 144}{8} = -170$

$S_{xx} = \sum x^2 - \dfrac{(\sum x)^2}{n} = 204 - \dfrac{36^2}{8} = 42$

$\sum x = 1 + 2 + 3 + 4 + 5 + 6 + 7 + 8 = 36$
$\sum y = 35 + 28 + 24 + 16 + 15 + 12 + 8 + 6 = 144$

b $b = \dfrac{S_{xy}}{S_{xx}} = \dfrac{-170}{42} = -4.047\ldots$

$a = \overline{y} - b\overline{x}$

$\quad = \dfrac{144}{8} + 4.047\ldots \times \dfrac{36}{8} = 36.214\ldots$

$y = 36.2 - 4.05x$

$\overline{y} = \dfrac{\sum y}{n}$ and $\overline{x} = \dfrac{\sum x}{n}$

c Method 1

If $c = 0.25$, then $x = 10 \times 0.25 = 2.5$

$y = 36.214\ldots - 4.047\ldots \times 2.5 = 26.095\ldots$

$y = \dfrac{m - 700}{5}$

$m = 5y + 700$

$\quad = 5 \times 26.095\ldots + 700 = 830$ (3 s.f.)

Watch out y and x are coded values. You can either code the given value of c, then reverse the coding for the resulting value of y (method 1). Or you can convert your regression equation in y and x to an equation in m and c (method 2).

Method 2

$y = 36.214\ldots - (4.047\ldots)x$

$\dfrac{m - 700}{5} = 36.214\ldots - 4.047\ldots \times 10c$

$m - 700 = 181.07\ldots - (202.38\ldots)c$

$m = 881.07\ldots - (202.38\ldots)c$

$\quad = 881.07\ldots - (202.38\ldots) \times 0.25$

$\quad = 830$ (3 s.f.)

The estimate for the melting point is 830 °C (3 s.f.)

You can find an equation for the regression line of m on c by substituting $y = \dfrac{m - 700}{5}$ and $x = 10c$ into the regression line of y on x, then rearranging into the form $m = p + qc$

Write a conclusion in the context of the question and give units. If possible, you should check that your answer makes sense. If you substituted $x = 0.25$ into the regression line of y on x you would get a melting point of 35 °C, which is clearly wrong.

Exercise 1B

1 Given that the coding $p = x + 2$ and $q = y - 3$ has been used to get the regression equation $p + q = 5$, find the equation of the regression line of y on x in the form $y = a + bx$.

2 Given the coding $x = p - 10$ and $y = s - 100$ and the regression equation $x = y + 2$, work out the equation of the regression line of s on p.

3 Given that the coding $g = \dfrac{x}{3}$ and $h = \dfrac{y}{4} - 2$ has been used to get the regression equation $h = 6 - 4g$, find the equation of the regression line of y on x.

4 The regression line of t on s is found by using the coding $x = s - 5$ and $y = t - 10$.
The regression equation of y on x is $y = 14 + 3x$.
Work out the regression line of t on s.

5 A regression line of c on d is worked out using the coding $x = \dfrac{c}{2}$ and $y = \dfrac{d}{10}$
  a Given that $S_{xy} = 120$, $S_{xx} = 240$, the mean of $x (\overline{x})$ is 5 and the mean of $y (\overline{y})$ is 6, calculate the regression line of y on x.
  b Find the regression line of d on c.

(E/P) 6 Some data on the coverage area, $a\,\text{m}^2$, and cost, £c, of five boxes of flooring were collected.

The results were coded such that $x = \dfrac{a - 8}{2}$ and $y = \dfrac{c}{5}$

The coded results are shown in the table.

x	1	5	10	16	17
y	9	12	16	21	23

  a Calculate S_{xy} and S_{xx} and use them to find the equation of the regression line of y on x. **(4 marks)**
  b Find the equation of the regression line of c on a. **(2 marks)**
  c Estimate the cost of a box of flooring which covers an area of $32\,\text{m}^2$. **(2 marks)**

(E/P) 7 A farmer collected data on the annual rainfall, $x\,\text{cm}$, and the annual yield of potatoes, p tonnes per acre.

The data for annual rainfall was coded using $v = \dfrac{x - 4}{8}$ and the following statistics were found:

$S_{vv} = 10.21$ $S_{pv} = 15.26$ $S_{pp} = 23.39$ $\overline{p} = 9.88$ $\overline{v} = 4.58$

  a Find the equation of the regression line of p on v in the form $p = a + bv$. **(3 marks)**
  b Using your regression line, estimate the annual yield of potatoes per acre when the annual rainfall is $42\,\text{cm}$. **(2 marks)**

1.2 Residuals

You can use residuals to check the reasonableness of a linear fit and to find possible outliers.

- **If a set of bivariate data has regression equation $y = a + bx$, then the residual of the data point (x_i, y_i) is given by $y_i - (a + bx_i)$. The sum of the residuals of all data points is 0.**

Consider the following data set:

x	1	2	4	6	7
y	1.2	1.7	3.1	5.2	5.8

The equation of the regression line of y on x is $y = 0.2 + 0.8x$.

You can calculate the residuals for each data point and record them in a table:

x	y	$y = 0.2 + 0.8x$	ε
1	1.2	1.0	0.2
2	1.7	1.8	−0.1
4	3.1	3.4	−0.3
6	5.2	5.0	0.2
7	5.8	5.8	0

Use ε for the residual column. Remember that if the observed value is **less** than the predicted value then the residual will be negative.

Notice that the sum of the residuals adds up to zero. → **Mixed exercise, Challenge**

The residuals can be plotted on a **residual plot** to show the trend:

The distribution of the residuals around zero is a good indicator of linear fit. You would expect the residuals to be randomly scattered about zero. If you see a trend in the residuals, you would question the appropriateness of the linear model.

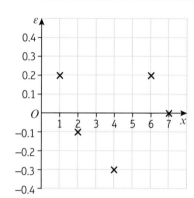

Non-random residuals might follow an increasing pattern, a decreasing pattern or an obviously curved pattern. Here are three examples of residual patterns which might indicate that a linear model is not suitable:

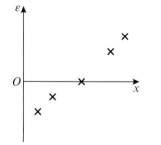 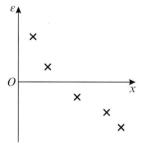 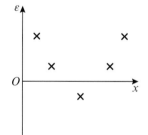

Example 5

The table shows the relationship between the temperature, $t\,°C$, and the sales of ice cream, s, on five days in June:

Temp, $t\,(°C)$	15	16	18	19	21
Sales, s (100s)	12.0	15.0	17.5	p	24.0

Online Explore residuals of data points and reasonableness of fit using GeoGebra.

The equation of the regression line of s on t is given as $s = -17.154 + 1.9693t$.

a Calculate the residuals for the given regression line and hence find the value of p.

b By considering the residuals, comment on whether a linear regression model is suitable for these data.

a

t	s	$s = -17.154 + 1.9693t$	ε
15	12.0	12.3855	−0.3855
16	15.0	14.3548	0.6452
18	17.5	18.2934	−0.7934
19	p	20.2627	$p - 20.2627$
21	24.0	24.2013	−0.2013

−0.3855 + 0.6452 − 0.7934 + (p − 20.2627) − 0.2013 = 0

⇒ p = 21.0 (3 s.f.)

The residual for t = 19 is 0.7373.

b The residuals appear to be randomly distributed around zero therefore it is likely that the linear regression model is suitable.

> Calculate the predicted values using the regression line equation.

> Write the residual for $t = 19$ in terms of p.

> Use the fact that the sum of residuals adds up to zero.

> Look at the distribution of the residuals around zero. They alternate signs, and don't follow an obvious pattern.

You can use residuals to identify possible outliers.

Example 6

The table shows the time taken, t minutes, to produce y litres of paint in a factory.

t	2.1	3.7	4.8	6.1	7.2
y	19.2	27.3	26.9	38.5	40.9

The regression line of y on t is given as $y = 9.7603 + 4.3514t$.

One of the y-values was incorrectly recorded.

a Calculate the residuals and write down the outlier.

b Comment on the validity of ignoring this outlier in your analysis.

c Ignoring the outlier, produce a new model.

d Use the new model to estimate the amount of paint that is produced in 4.8 minutes.

a

t	y	$y = 9.7603 + 4.3514t$	ε
2.1	19.2	18.8982	0.3018
3.7	27.3	25.8605	1.4395
4.8	26.9	30.6470	−3.7470
6.1	38.5	36.3038	2.1962
7.2	40.9	41.0904	−0.1904

The incorrect value is 26.9.

b The residuals suggest that this data point does not follow the pattern of the rest of the data, so it is valid to remove it.

c New model: y = 10.669 + 4.3573t

d y = 10.669 + 4.3573 × 4.8 = 31.6 litres (3 s.f.)

> Look for a data point with a residual that is far larger than the other residuals.

Problem-solving

You could also say that the data point **is** a valid piece of data so it should be used, or that there are only five data points so you should retain them all. You can make any reasonable conclusion as long as you give a reason.

> Use your calculator to find the new values of a and b.

> Substitute 4.8 into your new equation.

It is often useful to have a numerical value to indicate how closely a given set of data fits a linear regression model. Because the sum of the residuals is 0, you find the **square** of each residual and work out the sum of these values. This is called the **residual sum of squares (RSS)**.

- **You can calculate the residual sum of squares (RSS) for a linear regression model using the formula**

$$\text{RSS} = S_{yy} - \frac{(S_{xy})^2}{S_{xx}}$$

Note The formula is given in the formulae booklet. You are not expected to be able to derive it.

The linear regression model is the linear model which minimises the RSS for a given set of data. Unlike the product moment correlation coefficient, which takes values between -1 and 1, the units of the RSS are same as the units of the response variable squared. For this reason, you should only use the RSS to compare goodness of fit for data recorded in the same units.

Links In your A-level course, you used the product moment correlation coefficient, r, to measure the strength and type of linear correlation. ← **SM2, Section 1.2**

The RSS is also linked to the product moment correlation coefficient, r, by the equation
$$\text{RSS} = S_{yy} (1 - r^2)$$
→ **Section 2.1**

Example 7

The data shows the sales, in 100s, y, of *Slush* at a riverside café and the number of hours of sunshine, x, on five random days during August.

x	8	10	11.5	12	12.2
y	7.1	8.2	8.9	9.2	9.5

Given that $\sum x = 53.7$ $\sum y = 42.9$ $\sum x^2 = 589.09$ $\sum y^2 = 371.75$ $\sum xy = 467.45$

a calculate the residual sum of squares (RSS).

The RSS for five random days in December is 0.0562.

b State, with a reason, which month is more likely to have a linear fit between the number of hours of sunshine and the sales of *Slush*.

a $S_{yy} = \sum y^2 - \frac{(\sum y)^2}{n} = 371.75 - \frac{42.9^2}{5}$

  $= 3.668$

  $S_{xx} = \sum x^2 - \frac{(\sum x)^2}{n} = 589.09 - \frac{53.7^2}{5}$ ⎫

  $= 12.352$ ⎬ Calculate the summary statistics.

  $S_{xy} = \sum xy - \frac{\sum x \sum y}{n}$ ⎪

  $= 467.45 - \frac{53.7 \times 42.9}{5} = 6.704$ ⎭

  $\text{RSS} = S_{yy} - \frac{(S_{xy})^2}{S_{xx}} = 3.668 - \frac{6.704^2}{12.352}$ — Calculate the RSS using the given formula.

  $= 0.0294 \ (3 \text{ s.f.})$

b $0.0294 < 0.0562$ therefore August is more likely to have a linear fit. — The smaller the value of the RSS, the more likely a linear fit.

Exercise 1C

1 The table shows the relationship between two variables, x and y:

x	1.1	1.3	1.4	1.7	1.9
y	12.2	14.5	16.9	p	23.5

The equation of the regression line of y on x is given as $y = -3.633 + 14.33x$.

Calculate the residuals for the given regression line and hence find the value of p.

E/P 2 The table shows the masses of six baby elephants, m kg, against the number of days premature they were born, x.

x	2	5	8	9	11	15
m	110	105	103	101	96	88

The equation of the regression line of m on x is given as $m = 114.3 - 1.655x$.

a Calculate the residual values.   **(2 marks)**

b Draw a residual plot for this data.   **(2 marks)**

c With reference to your residual plot, comment on the suitability of a linear model for this data.   **(1 mark)**

> **Hint** There is an example of a residual plot on page 11.

E/P 3 Sarah completes a crossword each day. She measures both the time taken, t minutes, and the accuracy of her answers, given as a percentage, p. She records this data for 10 days and the results are shown in the table below:

t	5.1	5.7	6.3	6.4	7.1	7.2	8.0	8.3	8.7	9.1
p	79	81	85	86	89	84	95	96	98	99

The regression line of p on t is given as $p = 51.04 + 5.308t$.

a Calculate the residuals and use your results to identify an outlier.   **(3 marks)**

b State, with a reason, whether this outlier should be included in the data.   **(1 mark)**

c Ignoring the outlier, produce another model.   **(2 marks)**

d Use this model to predict the percentage of correct answers if the crossword takes Sarah 7.8 minutes to complete.   **(1 mark)**

E/P 4 The table shows the age, x years, of a particular model of car and the value, y, in £1000s.

x	1.2	1.7	2.4	3.1	3.8	4.2	5.1
y	13.1	12.5	10.9	9.4	7.9	a	5.8

The regression line of y on x is given as $y = 15.7 - 2.02x$.

a Calculate the residuals and hence find the value of a, correct to three significant figures.   **(3 marks)**

b By considering the signs of the residuals, explain whether or not the linear regression model is suitable for this data.   **(1 mark)**

E/P **5** The table shows the ages of runners, a, against the times taken to complete an obstacle course, t minutes.

a	17	19	22	25	30	38	41	44
t	18	19	21	22	25	28	29	31

$\sum a = 236$ $\sum t = 193$ $\sum a^2 = 7720$ $\sum t^2 = 4821$ $\sum at = 6046$

a State what is measured by the residual sum of squares. **(1 mark)**

b Calculate the residual sum of squares (RSS). **(5 marks)**

The runners then complete a cross-country course. The RSS for the new set of data is 1.154.

c State, with a reason, which data is more likely to have a linear fit. **(1 mark)**

E/P **6** The table shows the amount of rainfall, d mm, against the relative humidity, h%, for Stratford-upon-Avon on 7 random days during September.

h	67	69	74	77	79	81	87
d	1.3	1.7	1.9	2.0	2.2	2.4	3.1

Given that $S_{hh} = 289.4$ $S_{dd} = 1.949$ $S_{hd} = 23.13$

a calculate the residual sum of squares (RSS). **(2 marks)**

The RSS for a random sample of 7 days in October is 0.0965.

b State, with a reason, which sample is more likely to have a linear fit. **(1 mark)**

E/P **7** A particular model of car depreciates in value as it gets older. The table below shows the ages, x years, and the values, y £1000s of a random sample of these cars.

x	0.7	1.3	1.8	2.3	2.9	3.8
y	15.4	13.5	12.1	10.1	8.5	5.8

$\sum x = 12.8$ $\sum y = 65.4$ $\sum x^2 = 33.56$ $\sum y^2 = 773.72$ $\sum xy = 120.03$

a Calculate the equation of the regression line of y on x, giving your answer in the form $y = a + bx$. Give the values of a and b correct to 4 significant figures. **(3 marks)**

b Give an interpretation of the value of a. **(1 mark)**

c Use your regression line to estimate the value of a car that is 2 years old. **(1 mark)**

d Calculate the values of the residuals. **(2 marks)**

e Use your answer to part **d** to explain whether a linear model is suitable for these data. **(1 mark)**

f Calculate the residual sum of squares (RSS). **(2 marks)**

A sample for a second model of car has an RSS of 0.2548.

g State, with a reason, which sample is more likely to have a linear fit. **(1 mark)**

Challenge

The table shows the relationship between two variables, x and y.

x	1	5	7
y	9	p	q

Given that the equation of the regression line of y on x is $y = 2 + 4x$,
Find the values of p and q.

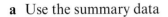
E **1** Two variables s and t are thought to be connected by an equation of the form $t = a + bs$, where a and b are constants.

    a Use the summary data

      $\sum s = 553$ $\sum t = 549$ $\sum st = 31\,185$ $n = 12$ $\bar{s} = 46.0833$

      $\bar{t} = 45.75$ $S_{ss} = 6193$

      to work out the regression line of t on s.   **(3 marks)**

    b Find the value of t when s is 50.   **(1 mark)**

E **2** A biologist recorded the breadth (x cm) and the length (y cm) of 12 beech leaves. The data collected can be summarised as follows.

      $\sum x^2 = 97.73$ $\sum x = 33.1$ $\sum y = 66.8$ $\sum xy = 195.94$

    a Calculate S_{xx} and S_{xy}.   **(2 marks)**

    b Find the equation of the regression line of y on x in the form $y = a + bx$.   **(3 marks)**

    c Predict the length of a beech leaf that has a breadth of 3.0 cm.   **(1 mark)**

E/P **3** Energy consumption is claimed to be a good predictor of Gross National Product. An economist recorded the energy consumption (x) and the Gross National Product (y) for eight countries. The data are shown in the table.

Energy consumption, x	3.4	7.7	12.0	75	58	67	113	131
Gross National Product, y	55	240	390	1100	1390	1330	1400	1900

    a Calculate S_{xy} and S_{xx}.   **(2 marks)**

    b Find the equation of the regression line of y on x in the form $y = a + bx$.   **(3 marks)**

    c Estimate the Gross National Product of a country that has an energy consumption of 100.   **(1 mark)**

    d Estimate the energy consumption of a country that has a Gross National Product of 3500.   **(1 mark)**

    e Comment on the reliability of your answer to **d**.   **(1 mark)**

E **4** In an environmental survey on the survival of mammals, the tail length t (cm) and body length m (cm) of a random sample of six small mammals of the same species were measured.

    These data are coded such that $x = \dfrac{m}{2}$ and $y = t - 2$.

    The data from the coded records are summarised below.

      $\sum y = 13.5$ $\sum x = 25.5$ $\sum xy = 84.25$ $S_{xx} = 59.88$

    a Find the equation of the regression line of y on x in the form $y = ax + b$.   **(3 marks)**

    b Hence find the equation of the regression line of t on m.   **(2 marks)**

    c Predict the tail length of a mammal that has a body length of 10 cm.   **(2 marks)**

E/P 5 A sports scientist recorded the number of breaths per minute (r) and the pulse rate per minute (p) for 10 athletes at different levels of physical exertion. The data are shown in the table.

The data are coded such that $x = \dfrac{r - 10}{2}$ and $y = \dfrac{p - 50}{2}$

x	3	5	5	7	8	9	9	10	12	13
y	4	9	10	11	17	15	17	19	22	27

(You may use $\sum x = 81$ $\quad \sum x^2 = 747$ $\quad \sum y = 151$ $\quad \sum y^2 = 2695$ $\quad \sum xy = 1413$)

a Calculate S_{xy} and S_{xx}. **(2 marks)**

b Find the equation of the regression line of y on x in the form $y = a + bx$. **(3 marks)**

c Find the equation of the regression line for p on r. **(2 marks)**

d Estimate the number of pulse beats per minute for someone who is taking 22 breaths per minute. **(2 marks)**

e Comment on the reliability of your answer to **d**. **(1 mark)**

E/P 6 A farm food supplier monitors the number of hens kept (x) against the weekly consumption of hen food (y kg) for a sample of 10 small holders. He records the data and works out the regression line for y on x to be $y = 0.16 + 0.79x$.

a Write down a practical interpretation of the figure 0.79. **(1 mark)**

b Estimate the amount of food that is likely to be needed by a small holder who has 30 hens. **(2 marks)**

c If food costs £12 for a 10 kg bag, estimate the weekly cost of feeding 50 hens. **(2 marks)**

E/P 7 Water voles are becoming very rare. A naturalist society decided to record details of the water voles in their area. The members measured the mass (y) to the nearest 10 grams, and the body length (x) to the nearest millimetre, of eight active healthy water voles. The data they collected are in the table.

Body length, x (mm)	140	150	170	180	180	200	220	220
Mass, y (grams)	150	180	190	220	240	290	300	310

a Draw a scatter diagram of these data. **(2 marks)**

b Give a reason to support the calculation of a regression line for these data. **(1 mark)**

c Use the coding $l = \dfrac{x}{10}$ and $w = \dfrac{y}{10}$ to work out the regression line of w on l. **(3 marks)**

d Find the equation of the regression line for y on x. **(2 marks)**

e Draw the regression line on the scatter diagram. **(1 mark)**

f Use your regression line to calculate an estimate for the mass of a water vole that has a body length of 210 mm. Write down, with a reason, whether or not this is a reliable estimate. **(2 marks)**

The members of the society remove any water voles that seem unhealthy from the river and take them into care until they are fit to be returned.

They find three water voles on one stretch of river which have the following measurements.

    A: Mass 235 g and body length 180 mm
    B: Mass 180 g and body length 200 mm
    C: Mass 195 g and body length 220 mm

g Write down, with a reason, which of these water voles were removed from the river. **(1 mark)**

(E/P) **8** A mail order company pays for postage of its goods partly by destination and partly by total weight sent out on a particular day. The number of items sent out and the total weights were recorded over a seven-day period. The data are shown in the table.

Number of items, n	10	13	22	15	24	16	19
Weight, w (kg)	2800	3600	6000	3600	5200	4400	5200

   a Use the coding $x = n - 10$ and $y = \dfrac{w}{400}$ to work out S_{xy} and S_{xx}. **(4 marks)**

   b Work out the equation of the regression line for y on x. **(3 marks)**

   c Work out the equation of the regression line for w on n. **(2 marks)**

   d Use your regression equation to estimate the weight of 20 items. **(2 marks)**

   e State why it would be unwise to use the regression equation to estimate the weight of 100 items. **(1 mark)**

   f Use your equation of the regression line found in part **b** to work out the residuals of the coded data points (x, y). **(2 marks)**

   g Use your equation of the regression line found in part **c** to work out the residuals of the original data points (n, w). **(2 marks)**

   h Explain how your answers to parts **f** and **g** are related to the coding used. **(1 mark)**

(E/P) **9** The table shows the time, t hours, against the temperature, $T°C$, of a chemical reaction.

t	2	3	5	6	7	9	10
T	72	68	59	54	50	42	38

Given that the equation of the regression line of T on t is $T = 80.445 - 4.289t$,

   a calculate the residual values. **(2 marks)**

   b State, with a reason, whether a linear model is suitable in this case. **(1 mark)**

Given that $S_{tt} = 52$, $S_{TT} = 957.43$ and $S_{tT} = -223$,

   c calculate the residual sum of squares (RSS). **(2 marks)**

A second chemical reaction has a RSS of 0.8754.

   d State, with a reason, which reaction is most likely to have a linear fit. **(1 mark)**

(E/P) **10** A meteorologist is developing a model to describe the relationship between the number of hours of sunshine, s, and the daily rainfall, f mm, in summer.

A random sample of the number of hours sunshine and the daily rainfall is taken from 8 days and are summarised below:

$$\sum s = 53.4 \qquad \sum s^2 = 395.76 \qquad \sum f = 29.9 \qquad \sum f^2 = 131.93 \qquad \sum sf = 171.66$$

   a Calculate S_{ss} and S_{sf}. **(2 marks)**

   b Find the equation of the regression line of f on s. **(3 marks)**

   c Use your equation to estimate the daily rainfall when there is 7.5 hours of sunshine. **(1 mark)**

   d Calculate the residual sum of squares (RSS). **(3 marks)**

The table shows the residual for each value of s.

s	3.1	4.2	5.4	6.2	7.1	8.8	9.1	9.5
Residual	−0.177	−0.196	0.256	0.124	x	−0.129	−0.216	−0.032

e Find the value of x. **(2 marks)**

f By considering the signs of the residuals, explain whether or not the linear regression model is suitable for these data. **(1 mark)**

(E/P) 11 A random sample of 9 baby southern hairy-nosed wombats was taken. The age, x, in days, and the mass, y grams, was recorded. The results were as follows:

x	2	3	4	5	6	7	8	9	10
y	4	5	7	8	9	11	12	11	15

(You may use $S_{xx} = 60$ $S_{yy} = 98.89$ $S_{xy} = 75$)

a Find the equation of the regression line of y on x in the form $y = a + bx$ as a model for these results. Give the values of a and b correct to three significant figures. **(2 marks)**

b Show that the residual sum of squares is 5.14 to three significant figures. **(2 marks)**

c Calculate the residual values. **(2 marks)**

d Write down the outlier. **(1 mark)**

e i Comment on the validity of ignoring this outlier.

   ii Ignoring the outlier, produce another model.

   iii Use this model to estimate the mass of a baby wombat after 20 days.

   iv Comment, giving a reason, on the reliability of your estimate. **(5 marks)**

(E/P) 12 The annual turnover, £t million of eight randomly selected UK companies, and the number of staff employed in 100s, s, is recorded and the data shown in the table below:

t, £million	1.2	1.5	1.8	2.1	2.5	2.7	2.8	3.1
s, 00s	1.1	1.4	1.7	2.2	2.4	2.6	2.9	3.2

($\sum t = 17.7$ $\sum s = 17.5$ $\sum t^2 = 42.33$ $\sum s^2 = 42.07$ $\sum ts = 42.16$)

a Calculate the equation of the regression line of s on t, giving your answer in the form $s = a + bt$. Give the values of a and b correct to three significant figures. **(3 marks)**

b Use your regression line to predict the number of employees in a UK company with an annual turnover of £2 300 000. **(2 marks)**

The table shows the residuals for each value of t:

t	1.2	1.5	1.8	2.1	2.5	2.7	2.8	3.1
Residual	0.0121	−0.0137	−0.0395	0.1347	−0.0997	p	0.0745	0.0487

c Find the value of p. **(2 marks)**

d By considering the signs of the residuals, or otherwise, comment on the suitability of the linear regression model for these data. **(1 mark)**

e Calculate the residual sum of squares (RSS). **(2 marks)**

A random sample of equivalent companies in France is taken and the residual sum of squares is found to be 0.421.

f State, with a reason, which sample is likely to have the better linear fit. **(1 mark)**

Challenge

A set of bivariate data (x_1, y_1), (x_2, y_2), (x_3, y_3), ... (x_n, y_n) is modelled by the linear regression equation $y = a + bx$, where $a = \overline{y} - b\overline{x}$.

a Prove that the sum of the residuals of the data points is 0.

b By means of an example, or otherwise, explain why this condition does not guarantee that the model closely fits the data.

Summary of key points

1 The **residual** of a given data point is the difference between the observed value of the dependent variable and the predicted value of the dependent variable.

2 The equation of the regression line of y on x is:

$$y = a + bx$$

where $b = \dfrac{S_{xy}}{S_{xx}}$ and $a = \overline{y} - b\overline{x}$

3 $S_{xy} = \sum xy - \dfrac{\sum x \sum y}{n}$

$S_{xx} = \sum x^2 - \dfrac{(\sum x)^2}{n}$

$S_{yy} = \sum y^2 - \dfrac{(\sum y)^2}{n}$

4 If a set of bivariate data has regression equation $y = a + bx$, then the residual of the data point (x_i, y_i) is given by $y_i - (a + bx_i)$. The sum of the residuals of all data points is 0.

5 You can calculate the residual sum of squares (RSS) for a linear regression model using the formula

$$\text{RSS} = S_{yy} - \dfrac{(S_{xy})^2}{S_{xx}}$$

Correlation

Objectives

After completing this chapter, you should be able to:

● Calculate the value of the product moment correlation coefficient, understand the effect of coding on it and understand the conditions for its use → **pages 22–26**

● Calculate and interpret Spearman's rank correlation coefficient → **pages 26–32**

● Carry out hypothesis tests for zero correlation using either Spearman's rank correlation coefficient or the product moment correlation coefficient → **pages 33–38**

Spearman's rank correlation coefficient can be used to determine the degree to which two ice skating judges agree or disagree about the relative performance of the skaters. It considers rankings, rather than specific values. → **Exercise 2B Q8**

Prior knowledge check

1 The scatter diagram shows data on the age and value of a particular model of car.

a Describe the type of correlation shown.

b Interpret the correlation in context.
← **SM1, Chapter 4**

2 A set of 8 bivariate data points (x, y) is summarised using $\sum x = 48$, $\sum y = 92$, $\sum x^2 = 340$, $\sum y^2 = 1142$ and $\sum xy = 616$.
Calculate:

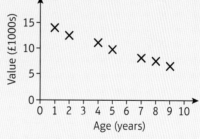

a S_{xx} b S_{yy} c S_{xy} ← **Chapter 1**

2.1 The product moment correlation coefficient

The **product moment correlation coefficient (PMCC)** measures the **linear correlation** between two variables. The PMCC can take values between 1 and −1, where 1 is perfect positive linear correlation and −1 is perfect negative linear correlation.

Watch out The PMCC was designed to analyse **continuous data** that comes from a population having a **bivariate normal distribution**. (That is, when considered separately, both the x and y data sets are normally distributed.) For data sets which do not satisfy these conditions, other correlation coefficients may be more valid. → **Section 2.2**

In Chapter 1 you used the summary statistics S_{xx}, S_{yy} and S_{xy} to calculate the coefficients of a regression line equation and to calculate the residual sum of squares (RSS). You can also use these summary statistics to calculate the product moment correlation coefficient.

- **The product moment correlation coefficient, r, is given by**

$$r = \frac{S_{xy}}{\sqrt{S_{xx} S_{yy}}}$$

This formula is given in the formulae booklet.

Online Explore linear correlation between two variables, measured by the PMCC, using GeoGebra.

Links You can calculate the value of the PMCC from raw data using your calculator. ← **SM2, Section 1.2**

Sometimes the original data is coded to make it easier to manage. Coding affects different statistics in different ways. As long as the coding is **linear**, the product moment correlation coefficient will be unaffected by the coding. Examples of linear coding of a data set x_i are $p_i = ax_i + b$ and $p_i = \dfrac{x_i - a}{b}$

You can think of linear coding as a change in scale on the axes of a scatter graph.

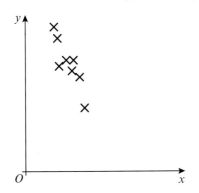

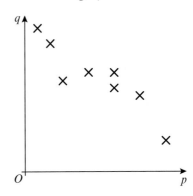

Raw data is often tightly grouped or contains very small or very large values. Changing the scale (which is equivalent to linear coding) can make the scatter graph easier to read.

The degree of linear correlation is unaffected by the change of scale. The value of r for the uncoded and coded values will be the same.

Example 1

The number of vehicles, x millions, and the number of accidents, y thousands, in 15 different countries were recorded. The following summary statistics were calculated and a scatter graph of the data is given to the right:

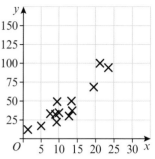

$\sum x = 176.9$ $\sum y = 679$ $\sum x^2 = 2576.47$ $\sum y^2 = 39\,771$ $\sum xy = 9915.3$

a Calculate the product moment correlation coefficient between x and y.

b With reference to your answer to part **a** and the scatter graph, comment on the suitability of a linear regression model for these data.

a $S_{xx} = \sum x^2 - \dfrac{(\sum x)^2}{n}$

Use the standard formula given in the formulae booklet.

$= 2576.47 - \dfrac{176.9^2}{15} = 490.23$

$S_{yy} = \sum y^2 - \dfrac{(\sum y)^2}{n}$

$= 39\,771 - \dfrac{679^2}{15} = 9034.93$

$S_{xy} = \sum xy - \dfrac{\sum x \sum y}{n}$

$= 9915.3 - \dfrac{176.9 \times 679}{15} = 1907.63$

$r = \dfrac{S_{xy}}{\sqrt{S_{xx}S_{yy}}}$

$= \dfrac{1907.63}{\sqrt{490.23 \times 9034.93}} = 0.906 \ (3 \text{ s.f.})$

The value of the correlation coefficient is 0.906. This is a positive correlation.

The greater the number of vehicles the higher the number of accidents.

b From the scatter graph, the data appear to be linearly distributed and the correlation coefficient calculated in part **a** is close to 1, so a linear regression model appears suitable for these data.

Example 2

Data are collected on the amount of dietary supplement, d grams, given to a sample of 8 cows and their milk yield, m litres. The data were coded using $x = \dfrac{d}{2} - 6$ and $y = \dfrac{m}{20}$. The following summary statistics were obtained:

$\sum d^2 = 4592$ $S_{dm} = 90.6$ $\sum x = 44$ $S_{yy} = 0.05915$

a Use the formula for S_{yy} to show that $S_{mm} = 23.66$.

b Find the value of the product moment correlation coefficient between d and m.

a $S_{yy} = \sum\left(\dfrac{m}{20}\right)^2 - \dfrac{\left(\sum \dfrac{m}{20}\right)^2}{8}$ ────── Substitute the code for y into the formula for S_{yy}.

$0.05915 = \dfrac{1}{400}\sum m^2 - \dfrac{\dfrac{1}{400}\left(\sum m\right)^2}{8}$

$= \dfrac{1}{400}\left(\sum m^2 - \dfrac{\left(\sum m\right)^2}{8}\right)$

$= \dfrac{1}{400}S_{mm}$

Hence $S_{mm} = 0.05915 \times 400 = 23.66$ ────── Substitute and simplify to find the value of S_{mm}.

Problem-solving

If you take a factor of $\dfrac{1}{20^2} = \dfrac{1}{400}$ out of each term on the right-hand side you are left with the formula for S_{mm}.

b $\sum x = \sum\left(\dfrac{d}{2} - 6\right) \Rightarrow 44 = \dfrac{1}{2}\sum d - 8 \times 6$

Hence $\sum d = 184$ ────── Substitute the code into $\sum x$ and rearrange to find $\sum d$.

$S_{dd} = 4592 - \dfrac{184^2}{8} = 360$ ────── Find S_{dd} using the standard formula.

$r = \dfrac{90.6}{\sqrt{360 \times 23.66}} = 0.982$ (3 s.f.)

Exercise 2A

1 Given that $S_{xx} = 92$, $S_{yy} = 112$ and $S_{xy} = 100$ find the value of the product moment correlation coefficient between x and y.

2 Given the following summary data,

$\sum x = 367 \qquad \sum y = 270 \qquad \sum x^2 = 33\,845 \qquad \sum y^2 = 12\,976 \qquad \sum xy = 17\,135 \qquad n = 6$

calculate the product moment correlation coefficient, r, using the formula

$r = \dfrac{S_{xy}}{\sqrt{S_{xx}S_{yy}}}$

(E) 3 The ages, a years, and heights, h cm, of seven members of a team were recorded. The data were summarised as follows:

$\sum a = 115 \qquad \sum a^2 = 1899 \qquad S_{hh} = 571.4 \qquad S_{ah} = 72.1$

  a Find S_{aa}.   **(1 mark)**

  b Find the value of the product moment correlation coefficient between a and h.   **(1 mark)**

  c Describe and interpret the correlation between the age and height of these seven people based on these data.   **(2 marks)**

(E) 4 In research on the quality of bacon produced by different breeds of pig, data were obtained about the leanness, l, and taste, t, of the bacon. The data are shown in the table.

Leanness, l	1.5	2.6	3.4	5.0	6.1	8.2
Taste, t	5.5	5.0	7.7	9.0	10.0	10.2

  a Find S_{ll}, S_{tt} and S_{lt}.   **(3 marks)**

  b Calculate the product moment correlation coefficient between l and t using the values found in part **a**.   **(2 marks)**

A scatter graph is drawn of the data.

c With reference to your answer to part **b** and the scatter graph, comment on the suitability of a linear regression model for these data. **(2 marks)**

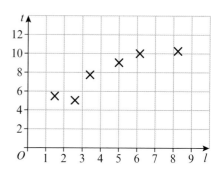

E **5** Eight children had their IQ measured and then took a general knowledge test. Their IQ, x, and their marks, y, for the test were summarised as follows:

$$\sum x = 973 \qquad \sum x^2 = 120\,123 \qquad \sum y = 490 \qquad \sum y^2 = 33\,000 \qquad \sum xy = 61\,595.$$

a Calculate the product moment correlation coefficient. **(3 marks)**

b Describe and interpret the correlation coefficient between IQ and general knowledge. **(2 marks)**

6 Two variables, x and y, were coded using $A = x - 7$ and $B = y - 100$.
The product moment correlation coefficient between A and B is found to be 0.973.
Find the product moment correlation coefficient between x and y.

7 The following data are to be coded using the coding $p = x$ and $q = y - 100$.

x	0	5	3	2	1
y	100	117	112	110	106

a Complete a table showing the values of p and q.

b Use your values of p and q to find the product moment correlation coefficient between p and q.

c Hence write down the product moment correlation coefficient between x and y.

8 The product moment correlation is to be worked out for the following data set using coding.

x	50	40	55	45	60
y	4	3	5	4	6

a Using the coding $p = \dfrac{x}{5}$ and $t = y$ find the values of S_{pp}, S_{tt} and S_{pt}.

b Calculate the product moment correlation coefficient between p and t.

c Write down the product moment correlation coefficient between x and y.

E **9** A shopkeeper thinks that the more newspapers he sells in a week the more sweets he sells. He records the amount of money (m pounds) that he takes in newspaper sales and also the amount of money he takes in sweet sales (s pounds) each week for seven weeks. The data are shown in the following table.

Newspaper sales, m pounds	380	402	370	365	410	392	385
Sweet sales, s pounds	560	543	564	573	550	544	530

a Use the coding $x = m - 365$ and $y = s - 530$ to find S_{xx}, S_{yy} and S_{xy}. **(4 marks)**

b Calculate the product moment correlation coefficient for m and s. **(1 mark)**

c State, with a reason, whether or not what the shopkeeper thinks is correct. **(1 mark)**

E/P **10** A student vet collected 8 blood samples from a horse with an infection. For each sample, she recorded the amount of drug, f, given to the horse and the amount of antibodies present in the blood, g. She coded the data using $f = 10x$ and $g = 5(y + 10)$ and drew a scatter diagram of x against y.

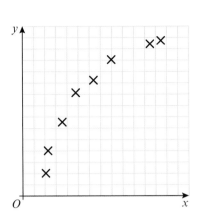

$\sum g^2 = 74\,458.75$

$S_{fg} = 5667.5$

$\sum y = 70.9$

$S_{xx} = 111.48$

Unfortunately, she forgot to label the axes on her scatter diagram and left the summary data calculations incomplete.

A second student was asked to complete the analysis of the data.

a Show that $S_{ff} = 11\,148$. **(3 marks)**

> **Problem-solving**
>
> Substitute the code into the formula for S_{xx}.

b Find the value of the product moment correlation coefficient between f and g. **(4 marks)**

c With reference to the scatter diagram, comment on the result in part **b**. **(1 mark)**

E/P **11** Alice, a market gardener, measures the amount of fertiliser, x litres, that she adds to the compost for a random sample of 7 chilli plant beds. She also measures the yield of chillies, y kg. The data are shown in the table below:

x, litres	1.1	1.3	1.4	1.7	1.9	2.1	2.5
y, kg	6.2	10.5	12	15	17	18	19

$\left(\sum x = 12 \quad \sum x^2 = 22.02 \quad \sum y = 97.7 \quad \sum y^2 = 1491.69 \quad \sum xy = 180.37 \right)$

a Show that the product moment correlation coefficient for these data is 0.946, correct to three significant figures. **(4 marks)**

The equation of the regression line of y on x is given as $y = -1.2905 + 8.8945x$.

b Calculate the residuals. **(3 marks)**

Alice thinks that because the PMCC is close to 1, a linear relationship is a good model for these data.

c With reference to the residuals, evaluate Alice's conclusion. **(2 marks)**

2.2 Spearman's rank correlation coefficient

In cases where the correlation is not linear, or where the data are not measurable on a continuous scale, the PMCC may not be a good measure of the correlation between two variables.

For example, suppose a manufacturer of tea produced a number of different blends; you could taste each blend and place the blends in order of preference. You do not, however, have a continuous numerical scale for measuring your preference. Similarly, it may be quicker to arrange a group of individuals in order of height than to measure each one. Under these circumstances, Spearman's rank correlation coefficient is used.

Spearman's rank correlation coefficient is denoted by r_s. In principle, r_s is simply a special case of the product moment correlation coefficient in which the data are converted to **rankings** before calculating the coefficient.

- **To rank two sets of data, X and Y, you give the rank 1 to the highest of the x_i values, 2 to the next highest, 3 to the next highest, and so on. You do the same for the y_i values.**

Note Spearman's rank correlation coefficient is sometimes used as an approximation for the product moment correlation coefficient as it is easier to calculate.

Note It makes no difference if you rank the smallest as 1, the next smallest as 2, etc., provided you do the same to both X and Y.

In general, you might choose to use Spearman's rank correlation coefficient rather than the product moment correlation coefficient if one of the following conditions is true:

- one or both data sets are not from a normally distributed population
- there is a non-linear relationship between the two data sets
- one or both data sets already represent a ranking (as in Example 3 below).

Example 3

Two tea tasters were asked to rank nine blends of tea in their order of preference. The tea they liked best was ranked 1. Their orders of preference are shown in the table:

Blend	A	B	C	D	E	F	G	H	I
Taster 1 (x)	3	6	2	8	5	9	7	1	4
Taster 2 (y)	5	6	4	2	7	8	9	1	3

Calculate Spearman's rank correlation coefficient for these data.

Online Explore Spearman's rank correlation coefficient using GeoGebra.

x_i	y_i	x_i^2	y_i^2	$x_i y_i$
3	5	9	25	15
6	6	36	36	36
2	4	4	16	8
8	2	64	4	16
5	7	25	49	35
9	8	81	64	72
7	9	49	81	63
1	1	1	1	1
4	3	16	9	12
$\sum x_i = 45$	$\sum y_i = 45$	$\sum x_i^2 = 285$	$\sum y_i^2 = 285$	$\sum x_i y_i = 258$

Find x_i^2, y_i^2 and $x_i y_i$

Find $\sum x_i$, $\sum y_i$, $\sum x_i^2$, $\sum y_i^2$ and $\sum x_i y_i$

$$r_s = \frac{S_{xy}}{\sqrt{S_{xx}S_{yy}}} = \frac{\sum x_i y_i - \frac{\sum x_i \sum y_i}{n}}{\sqrt{\left(\sum x_i^2 - \frac{(\sum x_i)^2}{n}\right)\left(\sum y_i^2 - \frac{(\sum y_i)^2}{n}\right)}}$$

Use the standard formula to calculate r_s

$$= \frac{258 - \frac{45 \times 45}{9}}{\sqrt{\left(285 - \frac{45 \times 45}{9}\right)\left(285 - \frac{45 \times 45}{9}\right)}} = \frac{33}{\sqrt{60 \times 60}} = 0.55$$

Since Spearman's rank correlation coefficient is derived from the PMCC:

- **A Spearman's rank correlation coefficient of:**
    - **+1** **means that rankings are in perfect agreement**
    - **−1** **means that the rankings are in exact reverse order**
    - **0** **means that there is no correlation between the rankings.**

You can calculate Spearman's rank correlation coefficient more quickly by looking at the **differences** between the ranks of each observation.

- **If there are no tied ranks, Spearman's rank correlation coefficient, r_s, is calculated using**

$$r_s = 1 - \frac{6\sum d^2}{n(n^2 - 1)}$$

  where d is the difference between the ranks of each observation, and n is the number of pairs of observations.

> **Watch out** **Tied ranks** occur when two or more data values in one of the data sets are the same. If there are only one or two tied ranks, this formula gives a reasonable estimate for r_s but if there are many tied ranks then you should use the PMCC formula with the ranked data.

Example 4

During a cattle show, two judges ranked ten cattle for quality according to the following table.

Cattle	A	B	C	D	E	F	G	H	I	J
Judge A	1	5	2	6	4	8	3	7	10	9
Judge B	3	6	2	7	5	8	1	4	9	10

Find Spearman's rank correlation coefficient between the two judges and comment on the result.

A	B	d	d^2
1	3	−2	4
5	6	−1	1
2	2	0	0
6	7	−1	1
4	5	−1	1
8	8	0	0
3	1	2	4
7	4	3	9
10	9	1	1
9	10	−1	1
		Total	22

In this case the data are already ranked, and there are no tied ranks.

Find d and d^2 for each pair of ranks.

Find $\sum d^2$.

$$r_s = 1 - \frac{6 \times 22}{10(100 - 1)} = 0.867$$

Calculate r_s.

There is a reasonable degree of agreement between the two judges.

Draw a conclusion.

If you are ranking data, and two or more data values are equal, then these data values will have a **tied rank**.

- **Equal data values should be assigned a rank equal to the mean of the tied ranks.**

For example:

Data value	200	350	350	400	700	800	800	800	1200
Rank	1	2.5	2.5	4	5	7	7	7	9

The 2nd and 3rd rank are tied, so assign a rank of 2.5 to each of these data values.

The 6th, 7th and 8th ranks are tied, so assign a rank of $\dfrac{6+7+8}{3} = 7$ to each of these data values.

When ranks are tied, the formula for the Spearman's rank correlation coefficient only gives an approximate value of r_s. This approximation is sufficient when there are only a small number of tied ranks. If there are many tied ranks, then you should use the PMCC formula with the ranked data.

Example 5

The marks of eight pupils in French and German tests were as follows:

	A	B	C	D	E	F	G	H
French, $f\%$	52	25	86	33	55	55	54	46
German, $g\%$	40	48	65	57	40	39	63	34

a Use the formula $r_s = 1 - \dfrac{6\sum d^2}{n(n^2-1)}$ to find an estimate for Spearman's rank correlation coefficient, showing clearly how you deal with tied ranks. Give your answer to 2 decimal places.

b Without recalculating the correlation coefficient, state how your answer to part **a** would change if:

  i pupil H's mark for German was changed to 38%
  ii a ninth pupil was included who scored 95% in French and 89% in German.

The teacher collects extra data from other students in the class and finds that there are now many tied ranks.

c Describe how she would now find a measure of the correlation.

a

f	g	Rank, f	Rank, g	d	d^2
52	40	4	3.5	0.5	0.25
25	48	1	5	−4	16
86	65	8	8	0	0
33	57	2	6	−4	16
55	40	6.5	3.5	3	9
55	39	6.5	2	4.5	20.25
54	63	5	7	−2	4
46	34	3	1	2	4

$\sum d^2 = 69.5$

$r_s = 1 - \dfrac{6 \times 69.5}{8(8^2-1)} = 0.17$ (2 d.p.)

Both sets of data should be ranked in the same order, in this case smallest to largest.

Since the two 40s are tied in rank, both data points are allocated the mean of ranks 3 and 4.

Since the two 55s are tied in rank, both data points are allocated the mean of ranks 6 and 7.

Calculate d^2 and use the formula for r_s.

b **i** Pupil rank does not change so no effect on r_s.

  ii Pupil will have the same rank for both tests (9) and hence $d = 0$.
  This means $\sum d^2$ will not change but since n has increased by 1, the denominator is bigger so r_s will increase.

c Give the mean of the tied ranks to each pupil and calculate the PMCC directly from the ranked data, rather than using the Spearman's rank correlation coefficient formula.

Exercise 2B

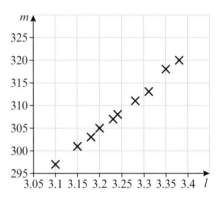

P **1** The scatter graph shows the length, l m, and mass, m kg, of 10 randomly selected male Siberian tigers.

A student wishes to analyse the correlation between l and m. Give one reason why the student might choose to use:

a the product moment correlation coefficient

b Spearman's rank correlation coefficient.

P **2** A college is trying to determine whether a published placement test (PPT) gives a good indicator of the likely student performance in a final exam. Data on past performances are shown in the scatter graph:

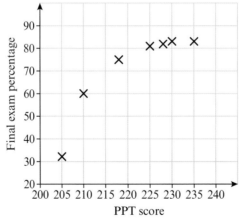

Give a reason why the college should not use the product moment correlation coefficient to measure the strength of the correlation of the two variables.

P **3** A sports science researcher is investigating whether there is a correlation between the height of a basketball player and the number of attempts it takes them to score a free throw. The researcher proposes to collect a random sample of data and then calculate the product moment correlation coefficient between the two variables.

Give a reason why the PMCC would not be appropriate in this situation and state an alternative method that the researcher can use.

4 For each of the data sets of ranks given below, calculate the Spearman's rank correlation coefficient and interpret the result.

a

r_x	1	2	3	4	5	6
r_y	3	2	1	5	4	6

b

r_x	1	2	3	4	5	6	7	8	9	10
r_y	2	1	4	3	5	8	7	9	6	10

c

r_x	5	2	6	1	4	3	7	8
r_y	5	6	3	8	7	4	2	1

5 Match the scatter graphs with the given values of Spearman's rank correlation coefficient.

a

b

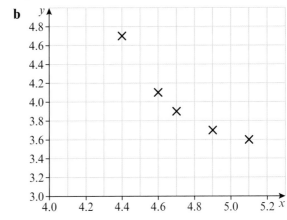

c

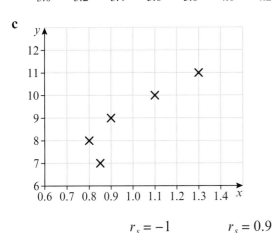

d

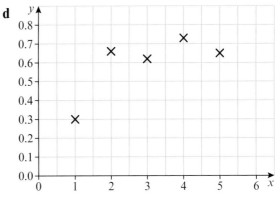

$$r_s = -1 \qquad r_s = 0.9 \qquad r_s = 1 \qquad r_s = 0.5$$

(E) **6** The number of goals scored by football teams and their positions in the league were recorded as follows for the top 12 teams.

Team	A	B	C	D	E	F	G	H	I	J	K	L
Goals	49	44	43	36	40	39	29	21	28	30	33	26
League position	1	2	3	4	5	6	7	8	9	10	11	12

a Find $\sum d^2$, where d is the difference between the ranks of each observation. **(3 marks)**

b Calculate Spearman's rank correlation coefficient for these data.
What conclusions can be drawn from this result? **(3 marks)**

7 A veterinary surgeon and a trainee veterinary surgeon both rank a small herd of cows for quality. Their rankings are shown below.

Cow	A	D	F	E	B	C	H	J
Qualified vet	1	2	3	4	5	6	7	8
Trainee vet	1	2	5	6	4	3	8	7

Find Spearman's rank correlation coefficient for these data, and comment on the experience of the trainee vet.

(P) 8 Two adjudicators at an ice dance skating competition award marks as follows.

Competitor	A	B	C	D	E	F	G	H	I	J
Judge 1	7.8	6.6	7.3	7.4	8.4	6.5	8.9	8.5	6.7	7.7
Judge 2	8.1	6.8	8.2	7.5	8.0	6.7	8.5	8.3	6.6	7.8

a Explain why you would use Spearman's rank correlation coefficient in this case.

b Calculate Spearman's rank correlation coefficient r_s, and comment on how well the judges agree.

It turns out that Judge 1 incorrectly recorded their score for competitor A and it should have been 7.7.

c Explain how you would now deal with equal data values if you had to recalculate the Spearman's rank correlation coefficient.

(E) 9 In a diving competition, two judges scored each of 7 divers on a forward somersault with twist.

Diver	A	B	C	D	E	F	G
Judge 1 score	4.5	5.1	5.2	5.2	5.4	5.7	5.8
Judge 2 score	5.2	4.8	4.9	5.1	5.0	5.3	5.4

a Give one reason to support the use of Spearman's rank correlation coefficient in this case. **(1 mark)**

b Calculate Spearman's rank correlation coefficient for these data. **(4 marks)**

The judges also scored the divers on a back somersault with two twists. Spearman's rank correlation coefficient for their ranks in this case was 0.676.

c Compare the judges' ranks for the two dives. **(1 mark)**

(E/P) 10 Two tea tasters sample 6 different teas and give each a score out of 10.

Tea	A	B	C	D	E	F
Taster 1 score	7	8	6	7	9	10
Taster 2 score	8	9	5	7	10	10

a Calculate Spearman's rank correlation coefficient for these data. **(4 marks)**

b Without recalculating the correlation coefficient, explain how your answer to part a would change if:
  i taster 2 changed his score for tea C to 6
  ii a seventh tea was tasted and both tasters scored it as 4. **(3 marks)**

The tasters tried lots of other types of tea and found that there were now many tied ranks.

c Describe how you would now find the correlation. **(1 mark)**

2.3 Hypothesis testing for zero correlation

You might need to carry out a **hypothesis test** to determine whether the correlation for a particular sample indicates that there is likely to be a non-zero correlation within the whole population. Hypothesis tests for zero correlation can use either the product moment correlation coefficient or the Spearman's rank correlation coefficient.

To test whether the population correlation, ρ, is greater than zero or to test whether the population correlation, ρ, is less than zero, use a **one-tailed test:**

- **For a one-tailed test, use either:**
  - **$H_0 : \rho = 0$, $H_1 : \rho > 0$ or**
  - **$H_0 : \rho = 0$, $H_1 : \rho < 0$**

To test whether the population correlation is **not equal** to zero, use a two-tailed test:

- **For a two-tailed test, use:**
  - **$H_0 : \rho = 0$, $H_1 : \rho \neq 0$**

Determine the critical region for r or r_s using the tables of critical values given in the formulae booklet and on page 216, or using your calculator. This is the table of critical values for the **product moment correlation coefficient**.

Notation r is usually used to denote the correlation for a **sample**. The Greek letter rho (ρ) is used to denote the correlation for the **whole population**. If you need to distinguish between the PMCC and the Spearman's rank correlation coefficient, use r and ρ for the PMCC and use r_s and ρ_s for Spearman's.

Links You carried out hypothesis tests for zero correlation with the product moment correlation coefficient in your A level course.

← SM2, Section 1.3

Watch out The table for the Spearman's rank correlation coefficient is different. Read the question carefully and choose the correct table.

Product moment coefficient					
Level					Sample
0.10	0.05	0.025	0.01	0.005	size
0.8000	0.9000	0.9500	0.9800	0.9900	4
0.6870	0.8054	0.8783	0.9343	0.9587	5
0.6084	0.7293	0.8114	0.8822	0.9172	6
0.5509	0.6694	0.7545	0.8329	0.8745	7
0.5067	0.6215	0.7067	0.7887	0.8343	8
0.4716	0.5822	0.6664	0.7498	0.7977	9

For a sample size of 8 you see from the table that the critical value of r to be significant at the 5% level on a one-tailed test is 0.6215. An observed value of r greater than 0.6215 from a sample of size 8 would provide sufficient evidence to reject the null hypothesis and conclude that $\rho > 0$. Similarly, an observed value of r less than -0.6215 would provide sufficient evidence to conclude that $\rho < 0$.

Example 6

A chemist observed 20 reactions, and recorded the mass of the reactant, x grams, and the duration of a reaction, y minutes.

She summarised her findings as follows:

$$\sum x = 20 \quad \sum y = 35 \quad \sum xy = 65 \quad \sum x^2 = 35 \quad \sum y^2 = 130$$

Test, at the 5% significance level, whether these results show evidence of any correlation between the mass of the reactant and the duration of the reaction.

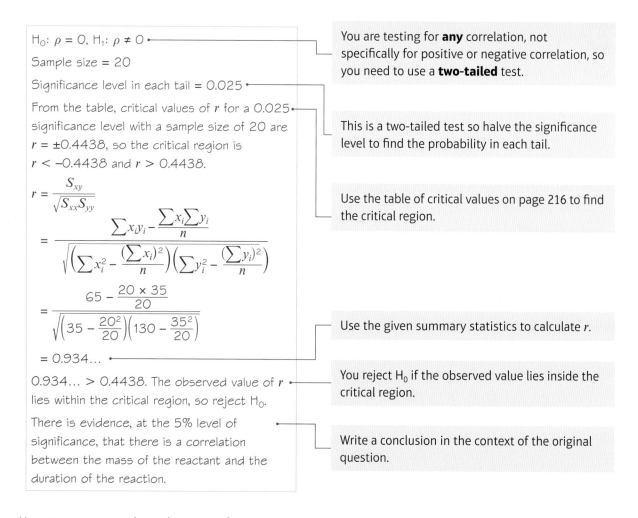

$H_0: \rho = 0$, $H_1: \rho \neq 0$

Sample size = 20

Significance level in each tail = 0.025

From the table, critical values of r for a 0.025 significance level with a sample size of 20 are $r = \pm 0.4438$, so the critical region is $r < -0.4438$ and $r > 0.4438$.

$$r = \frac{S_{xy}}{\sqrt{S_{xx}S_{yy}}}$$

$$= \frac{\sum x_i y_i - \frac{\sum x_i \sum y_i}{n}}{\sqrt{\left(\sum x_i^2 - \frac{(\sum x_i)^2}{n}\right)\left(\sum y_i^2 - \frac{(\sum y_i)^2}{n}\right)}}$$

$$= \frac{65 - \frac{20 \times 35}{20}}{\sqrt{\left(35 - \frac{20^2}{20}\right)\left(130 - \frac{35^2}{20}\right)}}$$

$$= 0.934\ldots$$

$0.934\ldots > 0.4438$. The observed value of r lies within the critical region, so reject H_0.

There is evidence, at the 5% level of significance, that there is a correlation between the mass of the reactant and the duration of the reaction.

You are testing for **any** correlation, not specifically for positive or negative correlation, so you need to use a **two-tailed** test.

This is a two-tailed test so halve the significance level to find the probability in each tail.

Use the table of critical values on page 216 to find the critical region.

Use the given summary statistics to calculate r.

You reject H_0 if the observed value lies inside the critical region.

Write a conclusion in the context of the original question.

You can carry out a hypothesis test for zero correlation using Spearman's rank correlation coefficient in the same way. The table of critical values for Spearman's coefficient is also given in the formulae booklet and on page 216.

Sample size	Spearman's coefficient		
	Level		
	0.05	0.025	0.01
4	1.0000	–	–
5	0.9000	1.0000	1.0000
6	0.8286	0.8857	0.9429
7	0.7143	0.7857	0.8929
8	0.6429	0.7381	0.8333
9	0.6000	0.7000	0.7833

For a sample size of 8 you see from the table that the critical value of r_s to be significant at the 0.025 level on a one-tailed test is ± 0.7381.

Example 7

The popularity of 16 subjects at a comprehensive school was found by counting the number of boys and the number of girls who chose each subject and then ranking the subjects. The results are shown in the table below.

Subject	*A*	*B*	*C*	*D*	*E*	*F*	*G*	*H*	*I*	*J*	*K*	*L*	*M*	*N*	*O*	*P*
Boys' ranks, *b*	2	5	9	8	1	3	15	16	6	10	12	14	4	7	11	13
Girls' ranks, *g*	4	7	11	3	6	9	12	16	5	13	10	8	2	1	15	14

a Calculate Spearman's rank correlation coefficient.

b Using a suitable test, at the 1% level of significance, test the assertion that boys' and girls' choices are positively correlated.

a

Subject	*A*	*B*	*C*	*D*	*E*	*F*	*G*	*H*	*I*	*J*	*K*	*L*	*M*	*N*	*O*	*P*
Boys' ranks, *b*	2	5	9	8	1	3	15	16	6	10	12	14	4	7	11	13
Girls' ranks, *g*	4	7	11	3	6	9	12	16	5	13	10	8	2	1	15	14
d	-2	-2	-2	5	-5	-6	3	0	1	-3	2	6	2	6	-4	-1
d²	4	4	4	25	25	36	9	0	1	9	4	36	4	36	16	1

Extend the table to include rows for *d* and d^2.

$\sum d^2 = 214$

$$r_s = 1 - \frac{6\sum d^2}{n(n^2 - 1)}$$

$$= 1 - \frac{6 \times 214}{16(16^2 - 1)}$$

$$= 0.685\ldots$$

State your hypotheses. You are testing for positive correlation so this is a one-tailed test.

b $H_0: \rho = 0$

$H_1: \rho > 0$

From the tables for a sample size of 16 the critical value is 0.5824.

Since 0.685... > 0.5824, the result is significant at the 1% level.

Find the critical value.

See if your value of r_s is significant.

You reject H_0 and accept H_1: there is evidence that boys' and girls' choices are positively correlated.

Draw a conclusion.

Exercise 2C

1 A sample of 7 observations (*x*, *y*) was taken, and the following values were calculated:

$\sum x = 29$ $\sum x^2 = 131$ $\sum y = 28$ $\sum y^2 = 140$ $\sum xy = 99$

a Calculate the product moment correlation coefficient for this sample.

b Test $H_0: \rho = 0$ against $H_1: \rho \neq 0$. Use a 1% significance level and state any assumptions you have made.

E 2 The ages, X years, and heights, Y cm, of 11 members of an athletics club were recorded and the following statistics were used to summarise the results.

$$\sum X = 168 \quad \sum Y = 1275 \quad \sum XY = 20\,704 \quad \sum X^2 = 2585 \quad \sum Y^2 = 320\,019$$

a Calculate the product moment correlation coefficient for these data. **(3 marks)**

b Test the assertion that the ages and heights of the club members are positively correlated. State your conclusion in words and any assumptions you have made. Use a 5% level of significance. **(5 marks)**

3 A sample of 30 compact cars was taken, and the fuel consumption and engine sizes of the cars were ranked.

A consumer group wants to test whether fuel consumption and engine size are related.

a Find the critical region for a hypothesis test based on Spearman's rank correlation coefficient. Use a 5% level of significance.

A Spearman's rank correlation coefficient of $r_s = 0.5321$ was calculated for the sample.

b Comment on this value in light of your answer to part **a**.

E/P 4 For one of the activities at a gymnastics competition, 8 gymnasts were awarded marks out of 10 for artistic performance and for technical ability. The results were as follows.

Gymnast	A	B	C	D	E	F	G	H
Technical ability	8.5	8.6	9.5	7.5	6.8	9.1	9.4	9.2
Artistic performance	6.2	7.5	8.2	6.7	6.0	7.2	8.0	9.1

The value of the product moment correlation coefficient for these data is 0.774.

a Stating your hypotheses clearly, and using a 1% level of significance, test for evidence of a positive association between technical ability and artistic performance. Interpret this value. **(4 marks)**

b Calculate the value of Spearman's rank correlation coefficient for these data. **(3 marks)**

c Give one reason why a hypothesis test based on Spearman's rank correlation coefficient might be more suitable for this data set. **(1 mark)**

d Use your answer to part **b** to carry out a second hypothesis test for evidence of a positive correlation between technical ability and artistic performance. Use a 1% significance level. **(4 marks)**

E/P 5 Two judges ranked 8 ice skaters in a competition according to the table below.

Skater / Judge	i	ii	iii	iv	v	vi	vii	viii
A	2	5	3	7	8	1	4	6
B	3	2	6	5	7	4	1	8

A test is to be carried out to see if there is a positive association between the rankings of the judges.

a Give a reason to support the use of Spearman's rank correlation coefficient in this case. **(1 mark)**

b Evaluate Spearman's rank correlation coefficient. **(3 marks)**

c Carry out the test at the 5% level of significance, stating your hypotheses clearly. **(4 marks)**

(P) **6** Each of the teams in a school hockey league had the total number of goals scored by them and against them recorded, with the following results.

Team	A	B	C	D	E	F	G
Goals for	39	40	28	27	26	30	42
Goals against	22	28	27	42	24	38	23

Investigate whether there is any association between the goals for and those against by using Spearman's rank correlation coefficient. Use a suitable test at the 1% level to investigate the statement, 'A team that scores a lot of goals concedes very few goals'.

(E) **7** The weekly takings and weekly profits for six different branches of a kebab restaurant are shown in the table below.

Shop	1	2	3	4	5	6
Takings (£)	400	6200	3600	5100	5000	3800
Profits (£)	400	1100	450	750	800	500

a Calculate Spearman's rank correlation coefficient, r_s, between the takings and profit. **(3 marks)**

b Test, at the 5% significance level, the assertion that profits and takings are positively correlated. **(4 marks)**

(E) **8** The rankings of 12 students in mathematics and music were as follows.

Mathematics	1	2	3	4	5	6	7	8	9	10	11	12
Music	6	4	2	3	1	7	5	9	10	8	11	12

a Calculate Spearman's rank correlation coefficient, r_s, showing your value of $\sum d^2$. **(3 marks)**

b Test the assertion that there is no correlation between these subjects. State the null and alternative hypotheses used. Use a 5% significance level. **(4 marks)**

(E/P) **9** A child is asked to place 10 objects in order and gives the ordering

    A   C   H   F   B   D   G   E   J   I

The correct ordering is

    A   B   C   D   E   F   G   H   I   J

Conduct, at the 5% level of significance, a suitable hypothesis test to determine whether there is a positive association between the child's order and the correct ordering. You must state clearly which correlation coefficient you are using and justify your selection. **(8 marks)**

(E/P) **10** The crop of a root vegetable was measured over six consecutive years, the years being ranked for wetness. The results are given in the table below.

Year	1	2	3	4	5	6
Crop (10 000 tons)	62	73	52	77	63	61
Rank of wetness	5	4	1	6	3	2

A seed producer claims that crop yield and wetness are not correlated. Test this assertion using a 5% significance level. You must state which correlation coefficient you are using and justify your selection. **(8 marks)**

(P) **11** A researcher collects data on the heights and masses of a random sample of gorillas. She finds that the correlation coefficient between the data is 0.546.

   a Explain which measure of correlation the researcher is likely to have used.

Given that the value of the correlation coefficient provided sufficient evidence to accept the alternative hypothesis that there is positive correlation between the variables,

   b find the smallest possible significance level given that she collected data from 14 gorillas

   c find the smallest possible sample size given that she carried out the test at the 5% level of significance.

Mixed exercise 2

(E) **1** Wai wants to know whether the 10 people in her group are as good at science as they are at art. She collected the end of term test marks for science (s), and art (a), and coded them using $x = \dfrac{s}{10}$ and $y = \dfrac{a}{10}$

The data she collected can be summarised as follows,

$$\sum x = 67 \quad \sum x^2 = 465 \quad \sum y = 65 \quad \sum y^2 = 429 \quad \sum xy = 434.$$

   a Work out the product moment correlation coefficient for x and y. **(3 marks)**

   b Write down the product moment correlation coefficient for s and a. **(1 mark)**

   c Write down whether or not it is it true to say that the people in Wai's group who are good at science are also good at art. Give a reason for your answer. **(1 mark)**

(E) **2** Nimer thinks that oranges that are very juicy cost more than those that are not very juicy. He buys 20 oranges from different places, and measures the amount of juice (j ml), that each orange produces. He also notes the price (p) of each orange.

The data can be summarised as follows,

$$\sum j = 979 \quad \sum p = 735 \quad \sum j^2 = 52\,335 \quad \sum p^2 = 32\,156 \quad \sum jp = 39\,950.$$

   a Find S_{jj}, S_{pp} and S_{jp}. **(3 marks)**

   b Using your answers to part **a**, calculate the product moment correlation coefficient. **(1 mark)**

   c Describe the type of correlation between the amount of juice and the cost and state, with a reason, whether or not Nimer is correct. **(2 marks)**

(E/P) **3** A geography student collected data on GDP per capita, x (in \$1000s), and infant mortality rates, y (deaths per 1000), from a sample of 8 countries. She coded the data using $p = x - 10$ and $q = \dfrac{y}{20}$ and drew a scatter diagram of p against q.

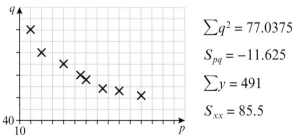

$$\sum q^2 = 77.0375$$
$$S_{pq} = -11.625$$
$$\sum y = 491$$
$$S_{xx} = 85.5$$

Unfortunately, she spilt coffee on her work and the only things still legible were her unfinished scatter diagram and a few summary data calculations.

   a Show that $S_{pp} = S_{xx}$. **(3 marks)**

   b Find the value of the product moment correlation coefficient between p and q. **(4 marks)**

   c Write down the value of the correlation coefficient between x and y. **(1 mark)**

   d With reference to the scatter diagram, comment on the result in part **b**. **(1 mark)**

(E) 4 Two judges at a cat show place the 10 entries in the following rank orders.

Cat	A	B	C	D	E	F	G	H	I	J
First judge	4	6	1	2	5	3	10	9	8	7
Second judge	2	9	3	1	7	4	6	8	5	10

   a Explain why Spearman's rank correlation coefficient is appropriate for these data. **(1 mark)**

   b Find the value of Spearman's rank correlation coefficient. **(3 marks)**

   c Explain briefly the role of the null and alternative hypotheses in a test of significance. **(1 mark)**

   d Stating your hypotheses clearly, carry out a test at the 5% level of significance and use your result to comment on the extent of the agreement between the two judges. **(4 marks)**

(E/P) 5 a Explain briefly the conditions under which you would measure association using Spearman's rank correlation coefficient. **(1 mark)**

   b Nine applicants for places at a college were interviewed by two tutors. Each tutor ranked the applicants in order of merit. The rankings are shown below.

Applicant	A	B	C	D	E	F	G	H	I
Tutor 1	1	2	3	4	5	6	7	8	9
Tutor 2	1	3	5	4	2	7	9	8	6

     By carrying out a suitable hypothesis test, investigate the extent of the agreement between the two tutors. **(7 marks)**

(E) 6 In a ski jumping contest each competitor made two jumps. The order of merit for the 10 competitors who completed both jumps are shown.

Ski jumper	A	B	C	D	E	F	G	H	I	J
First jump	2	9	7	4	10	8	6	5	1	3
Second jump	4	10	5	1	8	9	2	7	3	6

   a Calculate, to 2 decimal places, Spearman's rank correlation coefficient for the performance of the ski jumpers in the two jumps. **(3 marks)**

   b Using a 5% significance, and quoting from the table of critical values, investigate whether there is a positive association between performance on the first and second jumps. State your null and alternative hypotheses clearly. **(4 marks)**

(E/P) 7 An expert on porcelain is asked to place seven china bowls in date order of manufacture, assigning the rank 1 to the oldest bowl. The actual dates of manufacture and the order given by the expert are shown below.

Bowl	A	B	C	D	E	F	G
Date of manufacture	1920	1857	1710	1896	1810	1690	1780
Order given by expert	7	3	4	6	2	1	5

Carry out a hypothesis test to determine whether the expert is able to judge relative age accurately. You must state:
- the significance level of your test
- your null and alternative hypotheses
- which correlation coefficient, with justification. **(8 marks)**

(E) **8** A small bus company provides a service for a small town and some neighbouring villages. In a study of their service a random sample of 20 journeys was taken and the distances x, in kilometres, and journey times t, in minutes, were recorded. The average distance was 4.535 km and the average journey time was 15.15 minutes.

   a Using $\sum x^2 = 493.77$, $\sum t^2 = 4897$, $\sum xt = 1433.8$, calculate the product moment correlation coefficient for these data. **(3 marks)**

   b Stating your hypotheses clearly test, at the 5% level, whether or not there is evidence of a positive correlation between journey time and distance. **(4 marks)**

   c State any assumptions that have to be made to justify the test in part **b**. **(1 mark)**

(E) **9** A group of students scored the following marks in their statistics and geography exams.

Student	A	B	C	D	E	F	G	H
Statistics	64	71	49	38	72	55	54	68
Geography	55	50	51	47	65	45	39	82

   a Find the value of Spearman's rank correlation coefficient between the marks of these students. **(3 marks)**

   b Stating your hypotheses and using a 5% level of significance, test whether marks in statistics and marks in geography are associated. **(4 marks)**

(E/P) **10** An international study of female literacy investigated whether there was any correlation between the life expectancy of females and the percentage of adult females who were literate. A random sample of 8 countries was taken and the following data were collected.

Life expectancy (years)	49	76	69	71	50	64	78	74
Literacy (%)	25	88	80	62	37	86	89	67

   a Find Spearman's rank correlation coefficient for these data. **(3 marks)**

   b Stating your hypotheses clearly test, at the 5% level of significance, whether or not there is evidence of a correlation between the rankings of literacy and life expectancy for females. **(4 marks)**

   c Give one reason why Spearman's rank correlation coefficient and not the product moment correlation coefficient has been used in this case. **(1 mark)**

   d Without recalculating the correlation coefficient, explain how your answer to part **a** would change if:

   i the literacy percentage for the eighth country was actually 77

   ii a ninth country was added to the sample with life expectancy 79 years and literacy percentage 92%. **(3 marks)**

   A much larger sample is taken and it is found that there are many tied ranks.

   e Describe how you would find the correlation with many tied ranks. **(2 marks)**

(E) **11** Six Friesian cows were ranked in order of merit at an agricultural show by the official judge and by a student vet. The ranks were as follows:

Official judge	1	2	3	4	5	6
Student vet	1	5	4	2	6	3

    a Calculate Spearman's rank correlation coefficient between these rankings. **(3 marks)**

    b Investigate whether or not there was agreement between the rankings of the judge and the student.

    State clearly your hypotheses, and carry out an appropriate one-tailed significance test at the 5% level. **(4 marks)**

(E) **12** As part of a survey in a particular profession, age, x years, and salary, £y thousands, were recorded. The values of x and y for a randomly selected sample of ten members of the profession are as follows:

x	30	52	38	48	56	44	41	25	32	27
y	22	38	40	34	35	32	28	27	29	41

$(\sum x = 393 \quad \sum x^2 = 16483 \quad \sum y = 326 \quad \sum y^2 = 10968 \quad \sum xy = 13014)$

    a Calculate, to 3 decimal places, the product moment correlation coefficient between age and salary. **(3 marks)**

    b State two conditions under which it might be appropriate to use Spearman's rank correlation coefficient. **(1 mark)**

    c Calculate, to 3 decimal places, Spearman's rank correlation coefficient between age and salary. **(3 marks)**

It is suggested that there is no correlation between age and salary.

    d Set up appropriate null and alternative hypotheses and carry out an appropriate test to evaluate this suggestion. Use a 5% significance level. **(4 marks)**

(E) **13** A machine hire company kept records of the age, X months, and the maintenance costs, £Y, of one type of machine. The following table summarises the data for a random sample of 10 machines.

Machine	A	B	C	D	E	F	G	H	I	J
Age, x	63	12	34	81	51	14	45	74	24	89
Maintenance costs, y	111	25	41	181	64	21	51	145	43	241

    a Calculate, to 3 decimal places, the product moment correlation coefficient.
    (You may use $\sum x^2 = 30625 \quad \sum y^2 = 135481 \quad \sum xy = 62412$.) **(3 marks)**

    b Calculate, to 3 decimal places, Spearman's rank correlation coefficient. **(3 marks)**

For a different type of machine similar data were collected. From a large population of such machines a random sample of 10 was taken and Spearman's rank correlation coefficient, based on $\sum d^2 = 36$, was 0.782.

    c Using a 5% level of significance and quoting from the tables of critical values, interpret this rank correlation coefficient. Use a two-tailed test and state clearly your null and alternative hypotheses. **(4 marks)**

(E) **14** The data below show the height above sea level, x metres, and the temperature, $y\,°C$, at 7.00 a.m., on the same day in summer at nine places in Europe.

Height, x (m)	1400	400	280	790	390	590	540	1250	680
Temperature, y (°C)	6	15	18	10	16	14	13	7	13

    a Use your calculator to find the product moment correlation coefficient for this sample. **(1 mark)**

    b Test, at the 5% significance level, whether height above sea level and temperature are negatively correlated. **(4 marks)**

On the same day the number of hours of sunshine was recorded and Spearman's rank correlation coefficient between hours of sunshine and temperature, based on $\sum d^2 = 28$, was 0.767.

    c Stating your hypotheses and using a 5% two-tailed test, interpret this rank correlation coefficient. **(4 marks)**

(E) **15** **a** Explain briefly the conditions under which you would measure association using Spearman's rank correlation coefficient rather than the product moment correlation coefficient. **(1 mark)**

At an agricultural show 10 Shetland sheep were ranked by a qualified judge and by a trainee judge. Their rankings are shown in the table.

Qualified judge	1	2	3	4	5	6	7	8	9	10
Trainee judge	1	2	5	6	7	8	10	4	3	9

    b Calculate Spearman's rank correlation coefficient for these data. **(3 marks)**

    c Using a suitable table and a 5% significance level, state your conclusions as to whether there is some degree of agreement between the two sets of ranks. **(4 marks)**

(E) **16** The positions in a league table of 8 rugby clubs at the end of a season are shown, together with the average attendance (in hundreds) at home matches during the season.

Club	A	B	C	D	E	F	G	H
Position	1	2	3	4	5	6	7	8
Average attendance	30	32	12	19	27	18	15	25

Calculate Spearman's rank correlation coefficient between position in the league and home attendance. Comment on your results. **(4 marks)**

(E/P) **17** The ages, in months, and the weights, in kg, of a random sample of nine babies are shown in the table below.

Baby	A	B	C	D	E	F	G	H	I
Age (x)	1	2	2	3	3	3	4	4	5
Weight (y)	4.4	5.2	5.8	6.4	6.7	7.2	7.6	7.9	8.4

    a The product moment correlation coefficient between weight and age for these babies was found to be 0.972. By testing for positive correlation at the 5% significance level interpret this value. **(4 marks)**

A boy who does not know the weights or ages of these babies is asked to list them, by guesswork, in order of increasing weight. He puts them in the order

  A C E B G D I F H

b Obtain, to 3 decimal places, a rank correlation coefficient between the boy's order and the true weight order. **(3 marks)**

c By carrying out a suitable hypothesis test at the 5% significance level, assess the boy's ability to correctly rank the babies by weight. **(4 marks)**

Challenge

x_i and y_i are ranked variables with no ties, so that each takes the values 1, 2, 3, … n exactly once. The difference for each pair of data values is defined as $d_i = y_i - x_i$.

a Explain why $\sum x_i^2 = \sum y_i^2 = \sum_{r=1}^{n} r^2$, and hence express the quantity in terms of n.

b Explain why $\sqrt{\sum(x_i - \bar{x})^2 \sum(y_i - \bar{y})^2} = \sum(x_i - \bar{x})^2$, and hence show that $\sqrt{\sum(x_i - \bar{x})^2 \sum(y_i - \bar{y})^2} = \dfrac{n(n^2 - 1)}{12}$

c By expanding $\sum(y_i - x_i)^2$, show that $\sum x_i y_i = \sum x_i^2 - \dfrac{\sum d_i^2}{2}$.

d Hence show that $\sum(x_i - \bar{x})(y_i - \bar{y}) = \dfrac{n(n^2 - 1)}{12} - \dfrac{\sum d_i^2}{2}$

e Hence, or otherwise, prove that: $\dfrac{\sum(x_i - \bar{x})(y_i - \bar{y})}{\sqrt{\sum(x_i - \bar{x})^2 \sum(y_i - \bar{y})^2}} = 1 - \dfrac{6\sum d_i^2}{n(n^2 - 1)}$

Summary of key points

1 The **product moment correlation coefficient**, r, is given by $r = \dfrac{S_{xy}}{\sqrt{S_{xx} S_{yy}}}$

2 **Spearman's rank correlation coefficient** is a special case of the product moment correlation coefficient in which the data are converted to rankings before calculating the coefficient. To rank two sets of data, X and Y, you give rank 1 to the highest of the x_i values, 2 to the next highest and so on. You do the same for the y_i values.

3 A **Spearman's rank correlation coefficient** of:
+1 means that rankings are in perfect agreement
−1 means that the rankings are in exact reverse order
 0 means that there is no correlation between the rankings.

4 If there are no tied ranks, Spearman's rank correlation coefficient, r_s, is calculated using:

$r_s = 1 - \dfrac{6\sum d^2}{n(n^2 - 1)}$ where d is the difference between the ranks of each observation,

and n is the number of pairs of observations.

5 Equal data values should be assigned a rank equal to the mean of the tied ranks.

6 For a one-tailed test, use either: For a two-tailed test, use:
- $H_0 : \rho = 0$, $H_1 : \rho > 0$ or   - $H_0 : \rho = 0$, $H_1 : \rho \neq 0$
- $H_0 : \rho = 0$, $H_1 : \rho < 0$

3

Continuous distributions

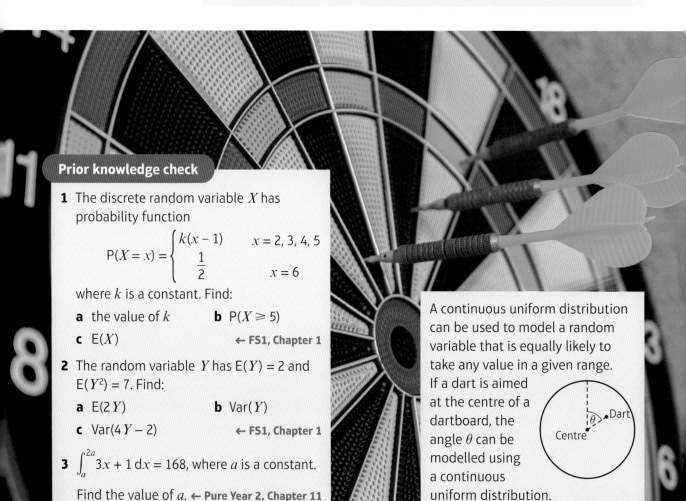

Prior knowledge check

1 The discrete random variable X has probability function

$$P(X = x) = \begin{cases} k(x - 1) & x = 2, 3, 4, 5 \\ \dfrac{1}{2} & x = 6 \end{cases}$$

where k is a constant. Find:

  a the value of k **b** $P(X \geqslant 5)$

  c $E(X)$ ← **FS1, Chapter 1**

2 The random variable Y has $E(Y) = 2$ and $E(Y^2) = 7$. Find:

  a $E(2Y)$ **b** $Var(Y)$

  c $Var(4Y - 2)$ ← **FS1, Chapter 1**

3 $\displaystyle\int_{a}^{2a} 3x + 1 \, dx = 168$, where a is a constant.

  Find the value of a. ← **Pure Year 2, Chapter 11**

A continuous uniform distribution can be used to model a random variable that is equally likely to take any value in a given range. If a dart is aimed at the centre of a dartboard, the angle θ can be modelled using a continuous uniform distribution.

3.1 Continuous random variables

A **continuous random variable** can take any one of infinitely many values. The probability that a continuous random variable takes any one specific value is 0, but you can write the probability that it takes values within a given range.

If ten coins are flipped:

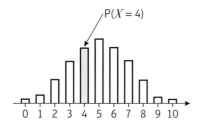

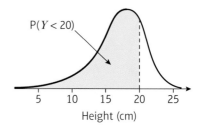

X = number of heads

Probability of getting 4 heads is written as P(X = 4)

X is a **discrete** random variable

Y = maximum height achieved by a flipped coin

Probability that the maximum height is less than 20 cm is written as P($Y < 20$)

Y is a **continuous** random variable

To describe the probability distribution of a discrete random variable you usually give a table of values, or a **probability mass function**, to define the probability of the random variable taking each value in its sample space.

Because the probability of a continuous random variable taking a specific value is 0, you cannot define its distribution in this way. Instead, you can use a **probability density function (p.d.f.)** to define the probability of the random variable taking values within a given range.

> **Links** A normally distributed random variable is an example of a continuous random variable. This curve shows the probability density function for a normal distribution:
>
>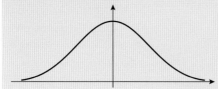
>
> ← SM2, Section 3.1

- **If X is a continuous random variable with probability density function f(x), then**

  - **f(x) ⩾ 0 for all $x \in \mathbb{R}$** ———— Probabilities must always be non-negative.

  - **P($a < X < b$) = $\int_a^b$ f(x) dx** ———— This is the area under the p.d.f. between the limits a and b.

    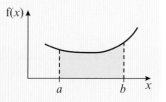

  - **$\int_{-\infty}^{\infty}$ f(x) dx = 1** ———— The total area under the p.d.f. must equal 1.

In practice, probability density functons are often non-zero on a limited subset of the real numbers. For the final condition, you only have to integrate the probability density function across all values of x for which f(x) is non-zero.

Example 1

In each case, state whether the function could be a probability density function, where k is a positive constant.

a $f(x) = \begin{cases} 2x + 1 & 0 \leqslant x \leqslant 3 \\ 0 & \text{otherwise} \end{cases}$

b $f(x) = \begin{cases} k(2 - x) & -2 \leqslant x \leqslant 2 \\ 0 & \text{otherwise} \end{cases}$

c $f(x) = \begin{cases} kx(5 - x) & 0 \leqslant x \leqslant 6 \\ 0 & \text{otherwise} \end{cases}$

Online Explore probability density functions using GeoGebra.

a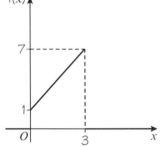

Area under $f(x) = 12 \neq 1$, so $f(x)$ cannot be a p.d.f.

Sketch the function. The total area under the function between $x = 0$ and $x = 3$ must equal 1 if it is to be a p.d.f.

b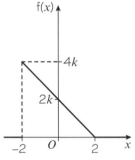

Area under $f(x) = 8k$, so $f(x)$ could be a p.d.f. if $k = \frac{1}{8}$

Watch out x can take positive or negative values, but $f(x)$ must be non-negative for all possible values of x.

c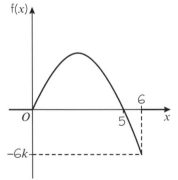

For all positive values of k, there are values in $0 \leqslant x \leqslant 6$ for which $f(x) < 0$, so $f(x)$ cannot be a p.d.f.

Example 2

The random variable X has probability density function

$$f(x) = \begin{cases} kx(4-x) & 2 \leqslant x \leqslant 4 \\ 0 & \text{otherwise} \end{cases}$$

Find:

a the value of k and sketch the p.d.f.

b $P(2.5 < X < 3)$, giving your answer to 3 decimal places.

a
$$\int_2^4 k(4x - x^2)\, dx = 1$$

$$k\left[2x^2 - \frac{x^3}{3}\right]_2^4 = 1$$

$$k\left(\left(32 - \frac{64}{3}\right) - \left(8 - \frac{8}{3}\right)\right) = 1$$

$$k\left(\frac{16}{3}\right) = 1$$

$$k = \frac{3}{16}$$

Area under the curve must equal 1.

$$\int_2^4 f(x)\, dx = 1$$

The p.d.f. is equal to zero everywhere other than [2, 4], so you only need to integrate between these limits.

$$\int_{-\infty}^{\infty} f(x)\, dx = \int_2^4 f(x)\, dx$$

Sketching the graph:

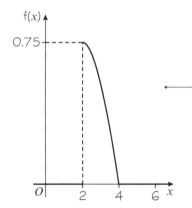

When sketching the graph, remember to label the axes and label the important values. These are the boundaries of the given range of x values and their corresponding y values. Here they are the 2, 4 and 0.75.

Watch out $f(x) = 0$ for $x < 2$ and $x > 4$. Make sure you only draw your curve between $x = 2$ and $x = 4$.

b $P(2.5 < X < 3) = \displaystyle\int_{2.5}^3 \frac{3x(4-x)}{16}\, dx$

$$= \frac{3}{16} \int_{2.5}^3 (4x - x^2)\, dx$$

$$= \frac{3}{16}\left[2x^2 - \frac{1}{3}x^3\right]_{2.5}^3$$

$$= \frac{3}{16}\left(9 - \frac{175}{24}\right) = \frac{41}{128}$$

$$= 0.320 \ (3 \text{ d.p.})$$

The probability that X takes a value between a and b is given by $\int_a^b f(x)\, dx$. Because the random variable is continuous, it doesn't matter whether you use strict or non-strict inequalities.

Watch out Your answer is a probability. Make sure it is between 0 and 1.

Example 3

The random variable X has probability density function

$$f(x) = \begin{cases} k & 1 < x < 2 \\ k(x-1) & 2 \leqslant x \leqslant 4 \\ 0 & \text{otherwise} \end{cases}$$

a Sketch f(x).　　　　**b** Find the value of k.　　　　**c** Find $P(X > 3)$.

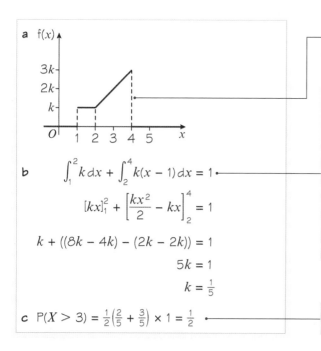

a

Use dotted lines to show where the function changes value.

b $\displaystyle\int_1^2 k\,dx + \int_2^4 k(x-1)\,dx = 1$

$$[kx]_1^2 + \left[\frac{kx^2}{2} - kx\right]_2^4 = 1$$

$$k + ((8k - 4k) - (2k - 2k)) = 1$$

$$5k = 1$$

$$k = \tfrac{1}{5}$$

The total area is 1. Here the area has been found by integrating but it is sometimes easier to find the value of k by looking at the sketch.

Area $= k + \frac{1}{2} \times 2 \times (k + 3k) = 5k$

(using the area of a rectangle and trapezium)

Area $= 1$

$5k = 1$

$k = \tfrac{1}{5}$

c $P(X > 3) = \tfrac{1}{2}\left(\tfrac{2}{5} + \tfrac{3}{5}\right) \times 1 = \tfrac{1}{2}$

Use the formula for the area of a trapezium.

Exercise 3A

1 Give reasons why the following are not valid probability density functions.

a $f(x) = \begin{cases} \frac{1}{4}x & -1 \leqslant x \leqslant 2 \\ 0 & \text{otherwise} \end{cases}$　　**b** $f(x) = \begin{cases} x^2 & 1 \leqslant x \leqslant 3 \\ 0 & \text{otherwise} \end{cases}$　　**c** $f(x) = \begin{cases} x^3 - 2 & -1 \leqslant x \leqslant 3 \\ 0 & \text{otherwise} \end{cases}$

2 For what value of k is the following a valid probability density function?

$$f(x) = \begin{cases} k(x^2 - 1) & -4 \leqslant x \leqslant -2 \\ 0 & \text{otherwise} \end{cases}$$

3 Sketch the following probability density functions.

a $f(x) = \begin{cases} \frac{1}{8}(x-2) & 2 \leqslant x \leqslant 6 \\ 0 & \text{otherwise} \end{cases}$　　　**b** $f(x) = \begin{cases} \frac{2}{15}(5-x) & 1 \leqslant x \leqslant 4 \\ 0 & \text{otherwise} \end{cases}$

4 Find the value of k so that each of the following is a valid probability density function.

a $f(x) = \begin{cases} kx & 1 \leqslant x \leqslant 3 \\ 0 & \text{otherwise} \end{cases}$　　**b** $f(x) = \begin{cases} kx^2 & 0 \leqslant x \leqslant 3 \\ 0 & \text{otherwise} \end{cases}$　　**c** $f(x) = \begin{cases} k(1 + x^2) & -1 \leqslant x \leqslant 2 \\ 0 & \text{otherwise} \end{cases}$

5 The continuous random variable X has probability density function given by

$$f(x) = \begin{cases} k(4 - x) & 0 \leqslant x \leqslant 2 \\ 0 & \text{otherwise} \end{cases}$$

a Find the value of k.

b Sketch the probability density function for all values of x.

c Find $P(X > 1)$.

(E) **6** The continuous random variable X has probability density function given by

$$f(x) = \begin{cases} kx^2(2 - x) & 0 \leqslant x \leqslant 2 \\ 0 & \text{otherwise} \end{cases}$$

a Find the value of k. **(3 marks)**

b Find $P(0 < X < 1)$. **(3 marks)**

(E) **7** The continuous random variable X has probability density function given by

$$f(x) = \begin{cases} kx^3 & 1 \leqslant x \leqslant 4 \\ 0 & \text{otherwise} \end{cases}$$

a Find the value of k. **(3 marks)**

b Find $P(1 < X < 2)$ **(3 marks)**

(E/P) **8** The continuous random variable X has probability density function given by

$$f(x) = \begin{cases} k & 0 < x < 2 \\ k(2x - 3) & 2 \leqslant x \leqslant 3 \\ 0 & \text{otherwise} \end{cases}$$

a Find the value of k. **(2 marks)**

b Sketch the probability density function for all values of x. **(2 marks)**

A different continuous random variable Y has probability density function given by

$$f(y) = \begin{cases} \frac{3}{16} y^2 & -2 \leqslant x \leqslant 2 \\ 0 & \text{otherwise} \end{cases}$$

c Given that X and Y are independent, find the probability that X and Y are both less than 1. **(4 marks)**

(E/P) **9** The length of time visitors spent on a news website, X minutes, is modelled using the probability density function

$$f(x) = \begin{cases} \frac{1}{60}(x + 1) & 0 \leqslant x \leqslant 10 \\ 0 & \text{otherwise} \end{cases}$$

a Use this model to find the probability that a randomly chosen visitor spends less than 30 seconds on the website. **(3 marks)**

b Sketch the probability density function. **(2 marks)**

In reality, a small number of customers are found to spend more than 10 minutes on the website.

c Sketch a probability density function that might provide a better model for X. **(1 mark)**

A **E** **10** The continuous random variable X has probability density function

$$f(x) = \begin{cases} \dfrac{k}{x} & 1 \leqslant x \leqslant 5 \\ 0 & \text{otherwise} \end{cases}$$

a Find the value of k. **(3 marks)**

b Find $P(2 < X < 4)$, giving your answer in the form $\dfrac{\ln a}{\ln b}$ where a and b are integers to be determined. **(3 marks)**

E **11** The continuous random variable X has probability density function

$$f(x) = \begin{cases} \dfrac{k}{x + 2} & -1 \leqslant x \leqslant 4 \\ 0 & \text{otherwise} \end{cases}$$

a Find the value of k. **(3 marks)**

b Find $P(1 < X < 3)$, giving your answer to 3 decimal places. **(3 marks)**

E **12** The continuous random variable X has probability density function

$$f(x) = \begin{cases} k \sin(\pi x) & 0 \leqslant x \leqslant 1 \\ 0 & \text{otherwise} \end{cases}$$

a Find the value of k. **(3 marks)**

b Sketch the probability density function for all values of x. **(2 marks)**

c Find $P(0 < X < \frac{1}{3})$. **(3 marks)**

Challenge

The length, T minutes, of a telephone call to a customer service department has probability density function,

$$f(t) = \begin{cases} \dfrac{k}{t^3} & t \geqslant 1 \\ 0 & \text{otherwise} \end{cases}$$

a Find the value of k.

b Find the probability that a call will be:

i less than 3 minutes **ii** more than 20 minutes.

c Given that $P(p < T < 2p) = 0.12$, find the value of p.

Problem-solving

There is no upper limit on the sample space of T. You will need to use an improper integral to find the area under the p.d.f.

3.2 The cumulative distribution function

Calculating probabilities from the p.d.f. can be time consuming, and requires integration. You can save time by finding the **cumulative distribution function (c.d.f.)** of a random variable.

- **For a random variable X, the cumulative distribution function $F(x) = P(X \leqslant x)$.**

This is equivalent to the area under the p.d.f. to the left of x.

$$F(x) = P(X \leqslant x) = \int_{-\infty}^{x} f(t)\, dt$$

Notation The p.d.f. is written with a **lower case f**, and the c.d.f. is written with a **capital F**.

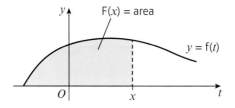

You use t rather than x as the variable in the integration to avoid confusion with x as the limit of the integration.

- **If X is a continuous random variable with c.d.f. $F(x)$ and p.d.f. $f(x)$:**

$$f(x) = \frac{d}{dx}F(x) \text{ and } F(x) = \int_{-\infty}^{x} f(t)\, dt$$

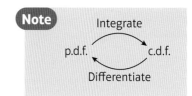

Note Integrate p.d.f. → c.d.f. Differentiate

Example 4

The random variable X has probability density function

$$f(x) = \begin{cases} \frac{1}{4}x & 1 \leqslant x \leqslant 3 \\ 0 & \text{otherwise} \end{cases}$$

Find $F(x)$.

Online Explore cumulative distribution functions using GeoGebra.

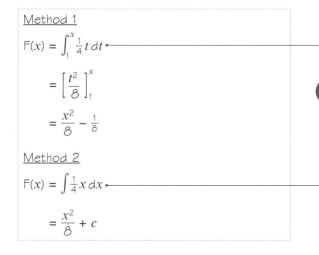

Method 1

$$F(x) = \int_{1}^{x} \frac{1}{4}t\, dt$$

$$= \left[\frac{t^2}{8}\right]_{1}^{x}$$

$$= \frac{x^2}{8} - \frac{1}{8}$$

Method 2

$$F(x) = \int \frac{1}{4}x\, dx$$

$$= \frac{x^2}{8} + c$$

Use $F(x) = \int_{-\infty}^{x} f(t)\, dt$. This definition is given in the formulae booklet.

Problem-solving

The p.d.f. is 0 for all values of $x < 1$, so you can use 1 as the lower limit for the integration.

An alternative method is to use an indefinite integral and put $+ c$. You can use $f(x)$ here since there are no limits in the integration.

$$\frac{3^2}{8} + c = 1$$

$$c = -\frac{1}{8}$$

$$F(x) = \begin{cases} 0 & x < 1 \\ \dfrac{x^2}{8} - \dfrac{1}{8} & 1 \leqslant x \leqslant 3 \\ 1 & x > 3 \end{cases}$$

c can be found using F(3) = 1 or F(1) = 0 as 3 and 1 are the upper and lower limits of the given range.

You must define F(x) over all of ℝ. F(x) = 0 for all values less than 1 and F(x) = 1 for all values greater than 3.

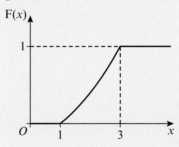

Example 5

The random variable X has probability density function

$$f(x) = \begin{cases} \frac{1}{5} & 1 < x < 2 \\ \frac{1}{5}(x - 1) & 2 \leqslant x \leqslant 4 \\ 0 & \text{otherwise} \end{cases}$$

Specify fully the cumulative distribution function of X.

Method 1

If $x \leqslant 1$

F(x) = 0   so F(1) = 0

If $1 < x < 2$

$$F(x) = F(1) + \int_1^x \frac{1}{5} dt$$

$$= \left[\frac{1}{5} t \right]_1^x$$

$$= \frac{1}{5}x - \frac{1}{5}$$

so $F(2) = \frac{1}{5}$

If $2 \leqslant x \leqslant 4$

$$F(x) = F(2) + \int_2^x \frac{1}{5}(t - 1) dt$$

$$= \frac{1}{5} + \left[\frac{t^2}{10} - \frac{t}{5} \right]_2^x$$

$$= \left(\frac{1}{5} \right) + \left(\left(\frac{x^2}{10} - \frac{x}{5} \right) - \left(\frac{4}{10} - \frac{2}{5} \right) \right)$$

$$= \frac{x^2}{10} - \frac{x}{5} + \frac{1}{5}$$

Watch out Integrate each section of the p.d.f. separately. The c.d.f. is **cumulative** so for each section, you need to add on the value of the c.d.f. at the upper limit of the previous section.

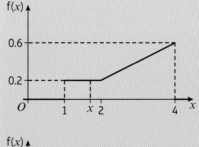

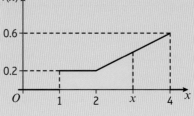

Method 2

If $1 < x < 2$

$$F(x) = \int \frac{1}{5} dx = \frac{1}{5}x + c$$

$\frac{1}{5} + c = 0$ so $c = -\frac{1}{5}$ •————— For the first part use F(1) = 0.

If $2 \leq x \leq 4$

$$F(x) = \int \frac{1}{5}(x - 1)\,dx = \frac{x^2}{10} - \frac{x}{5} + d$$

$1 = \frac{4^2}{10} - \frac{4}{5} + d$ so $d = \frac{1}{5}$ •————— For the second part use F(4) = 1.

$$F(x) = \begin{cases} 0 & x \leq 1 \\ \frac{1}{5}x - \frac{1}{5} & 1 < x < 2 \\ \frac{x^2}{10} - \frac{x}{5} + \frac{1}{5} & 2 \leq x \leq 4 \\ 1 & x > 4 \end{cases}$$

Watch out Remember to write the cumulative distribution in full. This means you need to define $F(x)$ for all values of $x \in \mathbb{R}$.

Example 6

The random variable X has cumulative distribution function

$$F(x) = \begin{cases} 0 & x < 0 \\ \frac{1}{5}x + \frac{3}{20}x^2 & 0 \leq x \leq 2 \\ 1 & x > 2 \end{cases}$$

Find:

a $P(X \leq 1.5)$

b $P(0.5 \leq X \leq 1.5)$

c $P(X = 1)$

d the probability density function, f(x).

a $P(X \leq 1.5) = F(1.5)$ •————— Using $F(x) = P(X \leq x)$

$\qquad = \frac{1}{5} \times 1.5 + \frac{3}{20} \times 1.5^2$

$\qquad = 0.6375$

b $P(0.5 \leq X \leq 1.5) = F(1.5) - F(0.5)$ •————— $P(0.5 \leq 1.5) = P(X \leq 1.5) - P(X \leq 0.5)$

$\qquad = 0.6375 - 0.1375$

$\qquad = 0.5$

c $P(X = 1) = 0$ •————— In a continuous distribution $P(X = x) = 0$.

d $\frac{d}{dx}\left(\frac{1}{5}x + \frac{3}{20}x^2\right) = \frac{1}{5} + \frac{3}{10}x$ •————— Use $\frac{d}{dx}F(x) = f(x)$

$$f(x) = \begin{cases} \frac{1}{5} + \frac{3}{10}x & 0 \leq x \leq 2 \\ 0 & \text{otherwise} \end{cases}$$

Watch out If $F(x)$ is constant on a given interval, then $f(x) = 0$ on that interval. For $x > 2$, $F(x) = 1$ and $\frac{d}{dx}(1) = 0$.

Exercise 3B

1 The continuous random variable X has probability density function given by

$$f(x) = \begin{cases} \frac{3}{8}x^2 & 0 \leqslant x \leqslant 2 \\ 0 & \text{otherwise} \end{cases}$$

Find $F(x)$.

2 The continuous random variable X has probability density function given by

$$f(x) = \begin{cases} \frac{1}{4}(4 - x) & 1 \leqslant x \leqslant 3 \\ 0 & \text{otherwise} \end{cases}$$

Find $F(x)$.

(E) 3 The continuous random variable X has probability density function given by

$$f(x) = \begin{cases} \frac{1}{9}x & 0 < x < 3 \\ \frac{1}{9}(6 - x) & 3 \leqslant x \leqslant 6 \\ 0 & \text{otherwise} \end{cases}$$

Define fully the cumulative distribution function of X. **(6 marks)**

4 The continuous random variable X has probability density function given by

$$f(x) = \begin{cases} k & 0 \leqslant x < 3 \\ k(2x - 5) & 3 \leqslant x \leqslant 5 \\ 0 & \text{otherwise} \end{cases}$$

a Sketch $f(x)$.   b Find the value of k.   c Find $F(x)$.

(E) 5 The continuous random variable X has cumulative distribution function given by

$$F(x) = \begin{cases} 0 & x < 2 \\ \frac{1}{5}(x^2 - 4) & 2 \leqslant x \leqslant 3 \\ 1 & x > 3 \end{cases}$$

Find the probability density function, $f(x)$. **(3 marks)**

6 The continuous random variable X has cumulative distribution function given by

$$F(x) = \begin{cases} 0 & x < 1 \\ \frac{1}{2}(x - 1) & 1 \leqslant x \leqslant 3 \\ 1 & x > 3 \end{cases}$$

Find:

a $P(X \leqslant 2.5)$ b $P(X > 1.5)$ c $P(1.5 \leqslant X \leqslant 2.5)$

E/P **7** The continuous random variable X has cumulative distribution function

$$F(x) = \begin{cases} 0 & x < 2 \\ \frac{1}{6}x^p + q & 2 \leqslant x \leqslant 4 \\ 1 & x > 4 \end{cases}$$

Find the exact values of p and q, showing your working clearly. **(7 marks)**

E **8** The continuous random variable X has cumulative distribution function given by

$$F(x) = \begin{cases} 0 & x < 1 \\ \frac{1}{2}(x^3 - 2x^2 + x) & 1 \leqslant x \leqslant 2 \\ 1 & x > 2 \end{cases}$$

  a Find the probability density function f(x). **(3 marks)**
  b Sketch the probability density function. **(2 marks)**
  c Find $P(X < 1.5)$. **(1 mark)**

E **9** The continuous random variable X has probability density function given by

$$f(x) = \begin{cases} k(4 - x^2) & 0 \leqslant x \leqslant 2 \\ 0 & \text{otherwise} \end{cases}$$

  a Show that $k = \frac{3}{16}$ **(3 marks)**
  b Find the cumulative distribution function of X. **(3 marks)**
  c Find $P(0.69 < X < 0.70)$. Give your answer correct to one significant figure. **(2 marks)**

E/P **10** A student attempts to define a random variable X by means of the following cumulative distribution function:

$$F(x) = \begin{cases} 0 & x < 1 \\ \ln x & x \geqslant 1 \end{cases}$$

Without using calculus, explain why this is not a valid cumulative distribution function. **(2 marks)**

E/P **11** The continuous random variable X has cumulative distribution function

$$F(x) = \begin{cases} 0 & x < 0 \\ \frac{1}{120}(kx - x^3) & 0 \leqslant x \leqslant 3 \\ 1 & x > 3 \end{cases}$$

where k is a constant.
Find:
  a the value of k **(2 marks)**
  b $P(X > 2)$ **(2 marks)**

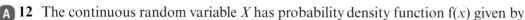

A **12** The continuous random variable X has probability density function $f(x)$ given by

E

$$f(x) = \begin{cases} \dfrac{1}{x \ln 7} & 1 \leqslant x < 7 \\ 0 & \text{otherwise} \end{cases}$$

Find $F(x)$. **(3 marks)**

E/P **13** The continuous random variable X has probability density function given by

$$f(x) = \begin{cases} \pi \cos(\pi x) & 0 \leqslant x < \frac{1}{2} \\ 0 & \text{otherwise} \end{cases}$$

Find $F(x)$. **(3 marks)**

E/P **14** The continuous random variable X has cumulative distribution function $F(x)$ given by

$$F(x) = \begin{cases} 0 & x < 1 \\ k(x - 1 + \ln x) & 1 \leqslant x \leqslant 3 \\ 1 & x > 3 \end{cases}$$

Find:

a the exact value of k **(3 marks)**

b the probability density function $f(x)$. **(3 marks)**

> **Challenge**
>
> The lifetime, in years, of a light bulb is modelled by the random variable T with probability density function
>
> $$f(t) = \begin{cases} 1.25e^{-1.25t} & t \geqslant 0 \\ 0 & t < 0 \end{cases}$$
>
> Find:
>
> **a** an expression for $F(t)$
>
> **b** the probability that a light bulb lasts for between 1 and 2 years
>
> **c** the probability that a light bulb lasts for more than 3 years.

3.3 Mean and variance of a continuous distribution

You can extend the ideas of mean and variance of a random variable to continuous random variables.

■ **If X is a continuous random variable with probability density function $f(x)$:**

• **the mean or expected value of X is given by**

$$E(X) = \mu = \int_{-\infty}^{\infty} x f(x)\, dx$$

> **Links** These definitions correspond to the mean and variance of a discrete random variable, with $\sum$ replaced with $\int_{-\infty}^{\infty}$ and the probability mass function replaced with the p.d.f.
>
> ← **FS1, Chapter 1**

• **the variance of X is given by**

$$\text{Var}(X) = \sigma^2 = \int_{-\infty}^{\infty} (x - \mu)^2 f(x)\, dx$$

$$= \int_{-\infty}^{\infty} x^2 f(x)\, dx - \mu^2$$

These definitions will be given in the formulae booklet in your exam.

You can also find the mean of a function of a continuous random variable in a similar way as with discrete random variables:

- **If X is a continuous random variable, then $E(g(X)) = \int_{-\infty}^{\infty} g(x)\, f(x)\, dx$**

This gives the following convenient way to calculate $\text{Var}(X)$:

- **$\text{Var}(X) = E(X^2) - (E(X))^2$**

In the case where $g(X)$ is a linear function of the form $aX + b$, it is useful to learn the following results:

- **$E(aX + b) = aE(X) + b$**
- **$\text{Var}(aX + b) = a^2\text{Var}(X)$**

Links These results are the same as those for discrete random variables.
← **FS1, Section 1.3**

Example 7

A random variable Y has probability density function

$$f(y) = \begin{cases} \frac{1}{4}y & 1 \leqslant y \leqslant 3 \\ 0 & \text{otherwise} \end{cases}$$

Find:

a $E(Y)$ **b** $\text{Var}(Y)$ **c** $E(2Y - 3)$ **d** $\text{Var}(2Y - 3)$

a $E(Y) = \int_1^3 \frac{1}{4}y^2 dy$ ———— $yf(y) = y \times \frac{1}{4}y.$

$= \left[\frac{1}{12}y^3 \right]_1^3$

$= \frac{27}{12} - \frac{1}{12} = \frac{26}{12} = \frac{13}{6}$ ————

If an exact answer is required you must leave your answer as a fraction. Otherwise you may write the answer as a fraction or as a decimal to 3 significant figures.

b $\text{Var}(Y) = \int_1^3 \frac{1}{4}y^3 dy - \left(\frac{13}{6} \right)^2$ ————

$= \left[\frac{1}{16}y^4 \right]_1^3 - \left(\frac{13}{6} \right)^2$

———— $y^2 f(y) = y^2 \times \frac{1}{4}y$

$= \frac{81}{16} - \frac{1}{16} - \frac{169}{36} = \frac{11}{36}$

c $E(2Y - 3) = 2E(Y) - 3$ ———— $E(aX + b) = aE(X) + b.$

$= 2 \times \frac{13}{6} - 3$

$= \frac{4}{3}$

d $\text{Var}(2Y - 3) = 4\text{Var}(Y)$ ———— $\text{Var}(aX + b) = a^2\text{Var}(X).$

$= \frac{44}{36} = \frac{11}{9}$

Example 8

A random variable X has probability density function

$$f(x) = \begin{cases} \frac{3}{32}(3 + 2x - x^2) & -1 \leqslant x \leqslant 3 \\ 0 & \text{otherwise} \end{cases}$$

a Sketch the probability density function.

b Find $E(X)$.

a The sketch of the p.d.f. is

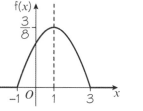

b By symmetry $E(X) = 1$ •

Problem-solving

If the p.d.f. of a continuous random variable, X, is **symmetric** about some point $x = a$, then $E(X) = a$.

You could also find $E(X)$ by calculating
$$\frac{3}{32}\int_{-1}^{3} x(3 + 2x - x^2)\,dx,$$ but symmetry saves you time in this question.

Example 9

The random variable X has probability density function

$$f(x) = \begin{cases} \frac{2}{15}x & 0 \leqslant x < 3 \\ \frac{1}{5}(5 - x) & 3 \leqslant x \leqslant 5 \\ 0 & \text{otherwise} \end{cases}$$

Find:

a $E(X)$ **b** $Var(X)$

a $E(X) = \int_0^3 \frac{2}{15}x^2\,dx + \int_3^5 \frac{1}{5}(5x - x^2)\,dx$

$= \left[\frac{2}{45}x^3\right]_0^3 + \left[\frac{1}{5}\left(\frac{5}{2}x^2 - \frac{x^3}{3}\right)\right]_3^5$

$= \left(\frac{6}{5} - 0\right) + \left(\left(\frac{25}{2} - \frac{25}{3}\right) - \left(\frac{9}{2} - \frac{9}{5}\right)\right)$

$= \frac{8}{3}$

b $Var(X) = \int_0^3 \frac{2}{15}x^3\,dx + \int_3^5 \frac{1}{5}(5x^2 - x^3)\,dx - \left(\frac{8}{3}\right)^2$

$= \left[\frac{2}{60}x^4\right]_0^3 + \left[\frac{1}{5}\left(\frac{5}{3}x^3 - \frac{x^4}{4}\right)\right]_3^5 - \frac{64}{9}$

$= \left(\frac{27}{10} - 0\right) + \left(\left(\frac{125}{3} - \frac{125}{4}\right) - \left(9 - \frac{81}{20}\right)\right) - \frac{64}{9}$

$= \frac{19}{18}$

To find the expectation or variance of a piecewise function, you need to integrate each part separately, then add.

Sketching the p.d.f. shows that it is not symmetrical so you need to integrate to find $E(X)$.

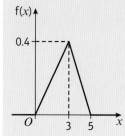

Example 10

In a block-building video game, a cube of side length X pixels is randomly generated. X is a continuous random variable with probability density function

$$f(x) = \begin{cases} \frac{1}{800}(x+10) & 20 \leqslant x \leqslant 40 \\ 0 & \text{otherwise} \end{cases}$$

a Find: **i** $E(X)$ **ii** $E(X^2)$ **iii** $\text{Var}(X)$

b Find the expected volume of the cube.

a **i** $E(X) = \frac{1}{800}\int_{20}^{40} x(x+10)\,dx$

$\qquad = \frac{1}{800}\left[\frac{1}{3}x^3 + 5x^2\right]_{20}^{40}$

$\qquad = \frac{1}{800}\left(\frac{88\,000}{3} - \frac{14\,000}{3}\right)$

$\qquad = \frac{185}{6}$

ii $E(X^2) = \frac{1}{800}\int_{20}^{40} x^2(x+10)\,dx$ — $\qquad E(X^2) = \int_{-\infty}^{\infty} x^2\,f(x)\,dx$

$\qquad = \frac{1}{800}\left[\frac{1}{4}x^4 + \frac{10}{3}x^3\right]_{20}^{40}$

$\qquad = \frac{1}{800}\left(\frac{2\,560\,000}{3} - \frac{200\,000}{3}\right)$

$\qquad = \frac{2950}{3}$

iii $\text{Var}(X) = \frac{2950}{3} - \left(\frac{185}{6}\right)^2 = \frac{1175}{36}$

> You already know $E(X)$ and $E(X^2)$, so you can work out the variance using the formula
> $$\text{Var}(X) = E(X^2) - (E(X))^2$$

b $E(X^3) = \frac{1}{800}\int_{20}^{40} x^3(x+10)\,dx$

$\qquad = \frac{1}{800}\left[\frac{1}{5}x^5 + \frac{5}{2}x^4\right]_{20}^{40}$

$\qquad = \frac{1}{800}(26\,880\,000 - 1\,040\,000)$

$\qquad = 32\,300$

> **Watch out** The expected volume is $E(X^3)$. This is not the same as $(E(X))^3$. You need to use the rule $E(g(X)) = \int_{-\infty}^{\infty} g(x)\,f(x)\,dx$

Exercise 3C

1 The continuous random variable X has a probability density function given by

$$f(x) = \begin{cases} kx^2 & 0 \leqslant x \leqslant 2 \\ 0 & \text{otherwise} \end{cases}$$

Find:

a k **b** $E(X)$ **c** $\text{Var}(X)$

2 The continuous random variable Y has a probability density function given by

$$f(y) = \begin{cases} \dfrac{y^2}{9} & 0 \leqslant y \leqslant 3 \\ 0 & \text{otherwise} \end{cases}$$

Find:

a $E(Y)$ **b** $\text{Var}(Y)$ **c** the standard deviation of Y.

3 The continuous random variable Y has a probability density function given by

$$f(y) = \begin{cases} \dfrac{y}{8} & 0 \leqslant y \leqslant 4 \\ 0 & \text{otherwise} \end{cases}$$

Find:

a $E(Y)$

b $\text{Var}(Y)$

c the standard deviation of Y

d $P(Y > \mu)$

e $\text{Var}(3Y + 2)$

f $E(Y + 2)$

> **Hint** $\mu = E(Y)$. In part **d**, use your answer from part **a**.

(E) **4** The continuous random variable X has a probability density function given by

$$f(x) = \begin{cases} k(1 - x) & 0 \leqslant x \leqslant 1 \\ 0 & \text{otherwise} \end{cases}$$

a Find k.   **(3 marks)**

b Find $E(X)$.   **(3 marks)**

c Show that $\text{Var}(X) = \dfrac{1}{18}$ **(2 marks)**

d Find $P(X > \mu)$.   **(3 marks)**

5 The continuous random variable X has a probability density function given by

$$f(x) = \begin{cases} 12x^2(1 - x) & 0 \leqslant x \leqslant 1 \\ 0 & \text{otherwise} \end{cases}$$

Find:

a $P(X < 0.5)$ **b** $E(X)$

(E/P) **6** The continuous random variable X has a probability density function given by

$$f(x) = \begin{cases} \frac{3}{8}(1 + x^2) & -1 \leqslant x \leqslant 1 \\ 0 & \text{otherwise} \end{cases}$$

a Sketch the probability density function of X. **(2 marks)**

b Write down $E(X)$.   **(1 mark)**

c Show that $\sigma^2 = 0.4$.   **(3 marks)**

d Find $P(-\sigma < X < \sigma)$.   **(3 marks)**

> **Problem-solving**
>
> Use symmetry to answer part **b**.

7 The continuous random variable T has a probability density function given by

$$f(t) = \begin{cases} kt^3 & 0 \leqslant t \leqslant 2 \\ 0 & \text{otherwise} \end{cases}$$

where k is a positive constant.

a Find k.   **b** Show that $E(T)$ is 1.6.

Find:

c $E(2T + 3)$ **d** $\text{Var}(T)$ **e** $\text{Var}(2T + 3)$ **f** $P(T < 1)$

8 The continuous random variable X has a probability density function given by

$$f(x) = \begin{cases} \dfrac{x^2}{27} & 0 \leqslant x < 3 \\ \dfrac{1}{3} & 3 \leqslant x \leqslant 5 \\ 0 & \text{otherwise} \end{cases}$$

a Draw a sketch of $f(x)$.

Find:

b $E(X)$ **c** $\text{Var}(X)$ **d** the standard deviation, σ, of X.

(E/P) **9** The continuous random variable X has a probability density function given by

$$f(x) = \begin{cases} \frac{1}{2}(x - 1) & 1 \leqslant x < 2 \\ \frac{1}{6}(5 - x) & 2 \leqslant x \leqslant 5 \\ 0 & \text{otherwise} \end{cases}$$

a Sketch $f(x)$.   **(2 marks)**

Find:

b $E(X)$ **(5 marks)**

c $\text{Var}(X)$ **(4 marks)**

(E) **10** Telephone calls arriving at a company are referred immediately by the telephonist to other people working in the company. The time a call takes, in minutes, is modelled by a continuous random variable T, having a probability density function given by

$$f(t) = \begin{cases} kt^2 & 0 \leqslant t \leqslant 10 \\ 0 & \text{otherwise} \end{cases}$$

a Show that $k = 0.003$.   **(3 marks)**

Find:

b $E(T)$ **(3 marks)**

c $\text{Var}(T)$ **(2 marks)**

d the probability of a call lasting between 7 and 9 minutes.   **(3 marks)**

e Sketch the probability density function.   **(2 marks)**

E/P **11** A continuous random variable X has probability density function

$$f(x) = \begin{cases} \frac{3}{4} - \frac{3}{16}x^2 & 0 \leqslant x \leqslant 2 \\ 0 & \text{otherwise} \end{cases}$$

Find:

a $E(X)$ **(3 marks)**

b $E(X^2)$ **(3 marks)**

c $\text{Var}(X)$ **(2 marks)**

Problem-solving

$$E(X^2) = \int_{-\infty}^{\infty} x^2 f(x)\, dx$$

E/P **12** The random variable X has cumulative distribution function

$$F(x) = \begin{cases} 0 & x < 0 \\ \dfrac{x^2}{100} & 0 \leqslant x \leqslant 10 \\ 1 & x > 10 \end{cases}$$

a Find $\text{Var}(X)$. **(4 marks)**

b Show that $E(X^3) = 400$ **(3 marks)**

A **13** A continuous random variable X has a probability density function given by

E/P

$$f(x) = \begin{cases} \dfrac{k}{x} & 1 \leqslant x \leqslant 3 \\ 0 & \text{otherwise} \end{cases}$$

Find:

a the value of k **(3 marks)**

b $E(X)$ **(3 marks)**

c $\text{Var}(X)$ **(3 marks)**

E/P **14** A continuous random variable X has a probability density function given by

$$f(x) = \begin{cases} \dfrac{c}{x(3-x)} & 1 \leqslant x \leqslant 2 \\ 0 & \text{otherwise} \end{cases}$$

a Show that $c = \dfrac{3}{\ln 4}$ **(3 marks)**

b Calculate the mean and variance of X. **(6 marks)**

E/P **15** A continuous random variable X has probability density function

$$f(x) = \begin{cases} 2x & 0 \leqslant x \leqslant 1 \\ 0 & \text{otherwise} \end{cases}$$

Find $E(\ln X)$. **(5 marks)**

Challenge

Given that $f(x)$ is the probability density function of a continuous random variable X, prove, from the following definitions, that $\text{Var}(X) = E(X^2) - (E(X))^2$

$$E(X) = \mu = \int_{-\infty}^{\infty} x f(x)\, dx$$

$$\text{Var}(X) = \sigma^2 = \int_{-\infty}^{\infty} (x - \mu)^2 f(x)\, dx$$

Watch out Only use the definitions given in the question in your proof.

3.4 Mode, median, percentiles and skewness

You need to be able to find the mode of a continuous random variable.

■ **The mode of a continuous random variable is the value of x for which the p.d.f. is a maximum.**

This is the value of x for which the probability distribution is 'most dense'. A random variable can have more than one modal value, though you will usually only be asked to find the mode in cases where the probability density function has a unique maximum value.

Example 11

The random variables X and Y have probability density functions $f(x)$ and $g(y)$ respectively.

$$f(x) = \begin{cases} 12x^2(1-x) & 0 \leqslant x \leqslant 1 \\ 0 & \text{otherwise} \end{cases} \qquad g(y) = \begin{cases} 2y & 0 \leqslant y \leqslant 1 \\ 0 & \text{otherwise} \end{cases}$$

Find the mode of:

a X **b** Y

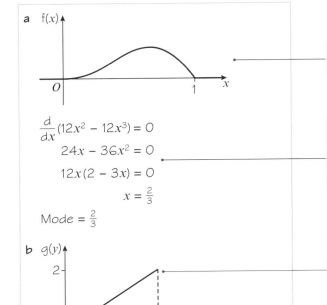

a

$$\frac{d}{dx}(12x^2 - 12x^3) = 0$$
$$24x - 36x^2 = 0$$
$$12x(2 - 3x) = 0$$
$$x = \tfrac{2}{3}$$
$$\text{Mode} = \tfrac{2}{3}$$

b

$$\text{Mode} = 1$$

Always sketch the graph when finding the mode.

From the sketch the mode occurs at the maximum point.

To find the maximum solve $\dfrac{d}{dx}f(x) = 0$.

You need to justify your answer. You can do this with a sketch, or by observing that $g'(y) = 2 > 0$ so $g(y)$ is strictly increasing on the interval $[0, 1]$, and 0 elsewhere, so the mode must be 1.

Watch out The mode does not need to occur at or even near the 'middle' of a probability distribution.

You can use the cumulative distribution function to define measures of location for a continuous random variable.

■ **If X is a continuous random variable with c.d.f. F(x):**
  • **the median of X is the value m such that F(m) = 0.5**
  • **the lower quartile of X is the value Q_1 such that F(Q_1) = 0.25**
  • **the upper quartile of X is the value Q_3 such that F(Q_3) = 0.75**
  • **the nth percentile of X is the value P_n such that F(P_n) = $\dfrac{n}{100}$**

> **Note** The median is also sometimes written as Q_2. For a symmetrical distribution, the median is equal to the mean.

Example 12

A continuous random variable X has probability density function

$$f(x) = \begin{cases} 4x - 4x^3 & 0 \leqslant x \leqslant 1 \\ 0 & \text{otherwise} \end{cases}$$

Find:

a the c.d.f. of X **b** the median value of X.

a Method 1

$$F(x) = \int_0^x (4t - 4t^3)\, dt$$

$$= [2t^2 - t^4]_0^x$$

$$= 2x^2 - x^4$$

     $F(x) = \int_0^x f(t)\, dt$

Method 2

$$F(x) = \int (4x - 4x^3)\, dx$$

$$= 2x^2 - x^4 + c$$

$$F(0) = 0$$

$$c = 0$$

$$F(x) = \begin{cases} 0 & x < 0 \\ 2x^2 - x^4 & 0 \leqslant x \leqslant 1 \\ 1 & x > 1 \end{cases}$$

b $2m^2 - m^4 = 0.5$

$2m^4 - 4m^2 + 1 = 0$

$$m^2 = \frac{4 \pm \sqrt{16 - 8}}{4}$$

$$m^2 = 1 \pm \frac{\sqrt{2}}{2}$$

$$m = \sqrt{1 \pm \frac{\sqrt{2}}{2}}$$

$$= 1.31 \text{ or } 0.541 \text{ (3 s.f.)}$$

median = 0.541 (3 s.f.)

     $F(x) = 0.5$

     This is a quadratic equation in m^2.

     Use the quadratic formula
$$\frac{-b \pm \sqrt{b^2 - 4ac}}{2a}$$

     Select the value that is in the range $0 \leqslant x \leqslant 1$.

Example 13

A continuous random variable X has the cumulative distribution function

$$F(x) = \begin{cases} 0 & x < 1 \\ \frac{1}{5}x - \frac{1}{5} & 1 \leqslant x \leqslant 2 \\ \frac{x^2}{10} - \frac{x}{5} + \frac{1}{5} & 2 < x < 4 \\ 1 & x \geqslant 4 \end{cases}$$

Find:

a the interquartile range

b the 10th percentile.

a $F(2) = \frac{1}{5}(2) - \frac{1}{5} = 0.2$

Lower quartile

$\frac{Q_1^2}{10} - \frac{Q_1}{5} + \frac{1}{5} = 0.25$

$Q_1^2 - 2Q_1 + 2 = 2.5$

$Q_1^2 - 2Q_1 - 0.5 = 0$

$Q_1 = \frac{2 \pm \sqrt{4 + 2}}{2}$

$Q_1 = 2.22$ or -0.225 (3 s.f.)

$Q_1 = 2.22$ (3 s.f.)

Upper quartile

$\frac{Q_3^2}{10} - \frac{Q_3}{5} + \frac{1}{5} = 0.75$

$Q_3^2 - 2Q_3 + 2 = 7.5$

$Q_3^2 - 2Q_3 - 5.5 = 0$

$Q_3 = \frac{2 \pm \sqrt{4 + 22}}{2}$

$Q_3 = 3.55$ or -1.55 (3 s.f.)

$Q_3 = 3.55$ (3 s.f.)

Interquartile range $= 3.55 - 2.22$

$\qquad\qquad\qquad\quad = 1.33$ (3 s.f.)

b $\frac{1}{5}(P_{10}) - \frac{1}{5} = 0.1$

$P_{10} = 1.5$

Problem-solving

If a c.d.f. is defined piecewise it is a good idea to find the boundary values for each section of $F(x)$. This will tell you which section of the function to use when calculating the median, quartiles or percentiles.

F(2) = 0.2, so the lower quartile lies in the second section of the function. Set $F(Q_1) = 0.25$ using this section of the c.d.f.

Multiply through by 10.

Select the value that is in the range $2 < x < 4$.

Select the value that is in the range $2 < x < 4$.

Interquartile range $= Q_3 - Q_1$

Look at the boundary value of the c.d.f.
F(2) = 0.2 so the 10th percentile, P_{10}, lies in the **first** section of the function.

Check that $P_{10} < Q_1$

You can use the concept of **skewness** to describe the symmetry of a distribution, or lack thereof. These probability density functions show examples of different types of skewness.

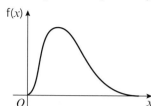

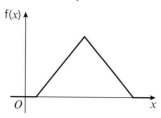

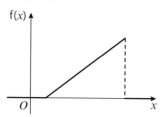

This distribution has a longer 'tail' at its right-hand end, and the mass of the distribution is concentrated at the left-hand end. It is **positively skewed**.

If a distribution is symmetrical it can be described as being **unskewed** or having **no skew**.

This distribution has a longer 'tail' at its left-hand end, and the mass of the distribution is concentrated at the right-hand end. It is **negatively skewed**.

In many cases, you can use the following rules to determine whether a distribution is positively or negatively skewed.

- **Positive skew: mode < median < mean**
- **Negative skew: mean < median < mode**

Note You can also refer to a sketch of the p.d.f. to describe skewness. Comparing **one pair** of measures of central tendency, or using a sketch, will be sufficient to justify skewness in your exam.

Example 14

A continuous random variable X has probability density function

$$f(x) = \begin{cases} \frac{1}{10}x & 0 \leqslant x < 2 \\ \frac{1}{4} - \frac{1}{40}x & 2 \leqslant x < 10 \\ 0 & \text{otherwise} \end{cases}$$

a Find:

i the mean of X

ii the mode of X.

b Comment on the skewness of the distribution.

a i $\int_0^2 \frac{1}{10}x^2 \, dx + \int_2^{10} \left(\frac{1}{4}x - \frac{1}{40}x^2\right) dx$

$= \left[\frac{1}{30}x^3\right]_0^2 + \left[\frac{1}{8}x^2 - \frac{1}{120}x^3\right]_2^{10}$

$= \frac{8}{3} + \left(\frac{25}{6} - \frac{13}{30}\right)$

$= \frac{32}{5}$

Mean of $X = \frac{32}{5}$ or 6.4

When the p.d.f. is defined as a piecewise function, you have to integrate each section of the function separately between the appropriate limits.

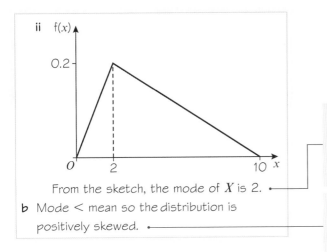

ii

From the sketch, the mode of X is 2.

When the sections of a probability density function are linear, it is usually easiest to find the mode by drawing a sketch of the p.d.f. The mode is the value of x at the point where f(x) is a maximum.

b Mode < mean so the distribution is positively skewed.

You could also say that from the sketch, the distribution is more dense for smaller values of x, so it is positively skewed.

Exercise 3D

1 The continuous random variable X has probability density function given by

$$f(x) = \begin{cases} \frac{3}{80}(8 + 2x - x^2) & 0 \leqslant x \leqslant 4 \\ 0 & \text{otherwise} \end{cases}$$

a Sketch the probability density function of X.

b Find the mode of X.

2 The continuous random variable X has probability density function given by

$$f(x) = \begin{cases} \frac{1}{8}x & 0 \leqslant x \leqslant 4 \\ 0 & \text{otherwise} \end{cases}$$

a Find the cumulative distribution function of X.

b Find the following, giving your answers to 3 significant figures:
  i the median of X
  ii the 10th percentile of X
  iii the 80th percentile of X.

3 The continuous random variable X has cumulative distribution function given by

$$F(x) = \begin{cases} 0 & x < 0 \\ \dfrac{x^2}{6} & 0 \leqslant x < 2 \\ -\dfrac{x^2}{3} + 2x - 2 & 2 \leqslant x \leqslant 3 \\ 1 & x > 3 \end{cases}$$

Find the following, giving your answers to 3 decimal places:

a the median value of X

b the quartiles and the interquartile range of X.

4 The continuous random variable X has probability density function given by

$$f(x) = \begin{cases} 1 - \frac{1}{2}x & 0 \leqslant x \leqslant 2 \\ 0 & \text{otherwise} \end{cases}$$

a Sketch the probability density function of X.

b Write down the mode of X.

Find:

c the cumulative distribution function of X

d the median value of X

e the upper quartile

f the 5th percentile, giving your answer correct to 3 significant figures.

5 The continuous random variable Y has probability density function given by

$$f(y) = \begin{cases} \frac{1}{2} - \frac{1}{9}y & 0 \leqslant y \leqslant 3 \\ 0 & \text{otherwise} \end{cases}$$

a Sketch the probability density function of Y.

b Use your sketch to describe the skewness of Y.

c Write down the mode of Y.

Find:

d the cumulative distribution function of Y

e the median value of Y correct to 3 significant figures

f the 10th to 90th percentile range, correct to 3 significant figures.

6 The continuous random variable X has probability density function given by

$$f(x) = \begin{cases} \frac{1}{4}x^3 & 0 \leqslant x \leqslant 2 \\ 0 & \text{otherwise} \end{cases}$$

a Sketch the probability density function of X.

b Write down the mode of X.

Find:

c the cumulative distribution function of X

d the median value of X.

7 The continuous random variable X has probability density function given by

$$f(x) = \begin{cases} \frac{3}{8}(x^2 + 1) & -1 \leqslant x \leqslant 1 \\ 0 & \text{otherwise} \end{cases}$$

a Sketch the probability density function of X.

b What can you say about the mode of X?

c Write down the median value of X.

d Find the cumulative distribution function of X.

8 The continuous random variable X has probability density function given by

$$f(x) = \begin{cases} \frac{3}{10}(3x - x^2) & 0 \leqslant x \leqslant 2 \\ 0 & \text{otherwise} \end{cases}$$

a Sketch the probability density function of X.

b Use your sketch to describe the skewness of X.

Find:

c the mode of X

d the cumulative distribution function of X.

e Show that the median value of X lies between 1.23 and 1.24.

(E/P) **9** The continuous random variable X has cumulative distribution function given by

$$F(x) = \begin{cases} 0 & x < 1 \\ \frac{1}{8}(x^2 - 1) & 1 \leqslant x \leqslant 3 \\ 1 & x > 3 \end{cases}$$

Find:

a the probability density function of the random variable X **(2 marks)**

b the mode of X **(2 marks)**

c the median of X.   **(2 marks)**

d Describe the skewness of X, giving a reason for your answer.   **(2 marks)**

e Find the value k such that $P(k < x < k + 1) = 0.6$ **(3 marks)**

(E) **10** The continuous random variable X has cumulative distribution function given by

$$F(x) = \begin{cases} 0 & x < 0 \\ 4x^3 - 3x^4 & 0 \leqslant x \leqslant 1 \\ 1 & x > 1 \end{cases}$$

Find:

a the probability density function of the random variable X **(3 marks)**

b the mode of X **(2 marks)**

c $P(0.2 < X < 0.5)$ **(3 marks)**

E/P **11** The amount of vegetables eaten by a family in a week is a continuous random variable W kg. The continuous random variable W has probability density function given by

$$f(w) = \begin{cases} \dfrac{20}{5^5} w^3(5 - w) & 0 \leqslant w \leqslant 5 \\ 0 & \text{otherwise} \end{cases}$$

    a Find the cumulative distribution function of the random variable W. **(3 marks)**

    b Show that the median of W lies between 3.4 kg and 3.5 kg. **(3 marks)**

    c Find the mode of W, fully justifying your answer. **(4 marks)**

    d Hence describe the skewness of the distribution, giving a reason for your answer. **(1 mark)**

E/P **12** The continuous random variable X has a probability density function given by

$$f(x) = \begin{cases} \dfrac{1}{4} & 0 \leqslant x < 1 \\ \dfrac{x^3}{5} & 1 \leqslant x \leqslant 2 \\ 0 & \text{otherwise} \end{cases}$$

    Find:

    a $E(X)$ **(5 marks)**

    b the cumulative distribution function **(4 marks)**

    c to 3 decimal places, the median and the interquartile range of the distribution. **(5 marks)**

    d Describe the skewness of this distribution, giving a reason for your answer. **(1 mark)**

P **13** For each of the following sets of conditions, sketch the probability density function of a distribution which satisfies all the conditions:

    a the distribution is symmetrical but mode $\neq$ median

    b there is a unique mode which lies outside the interquartile range.

E/P **14** By fully specifying a suitable p.d.f. give an example of a non-symmetrical distribution in which the median and the mode are equal. **(3 marks)**

A **15** The continuous random variable X has probability density function given by

E

$$f(x) = \begin{cases} \dfrac{1}{x \ln 5} & 2 \leqslant x \leqslant 10 \\ 0 & \text{otherwise} \end{cases}$$

    a Find the mode of X, fully justifying your answer. **(2 marks)**

    b Specify the cumulative distribution function of X. **(3 marks)**

    Find:

    c the exact value of the median of X **(2 marks)**

    d the quartiles and the interquartile range of X. **(3 marks)**

16 The life, X, of the *Nitelite* light bulb is modelled by the probability density function

$$f(x) = \begin{cases} 2.5e^{-2.5x} & x \geqslant 0 \\ 0 & \text{otherwise} \end{cases}$$

where X is measured in thousands of hours.

Find:

a the median of X **(5 marks)**

b the quartiles and the interquartile range of X. **(3 marks)**

 17 The continuous random variable X has probability density function given by

$$f(x) = \begin{cases} k\sec^2(\pi x) & 0 \leqslant x \leqslant 0.25 \\ 0 & \text{otherwise} \end{cases}$$

Find:

a the value of k **(3 marks)**

b the cumulative distribution function of X **(2 marks)**

c the median of X. **(2 marks)**

18 The continuous random variable X has probability density function given by

$$f(x) = \begin{cases} \dfrac{k}{x(5-x)} & 2 \leqslant x \leqslant 4 \\ 0 & \text{otherwise} \end{cases}$$

Find:

a the exact value of k **(4 marks)**

b $E(X)$ **(3 marks)**

c $\text{Var}(X)$ **(4 marks)**

d the c.d.f. of X **(4 marks)**

e the median of X. **(2 marks)**

f Write down the mode of X. **(1 mark)**

g Comment on the skewness of this distribution. **(2 marks)**

3.5 The continuous uniform distribution

- **A random variable having a continuous uniform distribution over the interval [a, b] has p.d.f.**

$$f(x) = \begin{cases} \dfrac{1}{b-a} & a \leqslant x \leqslant b \\ 0 & \text{otherwise} \end{cases}$$

A sketch of the p.d.f. is shown.

Notation If X has the continuous uniform distribution over the interval [a, b] you write $X \sim U[a, b]$.

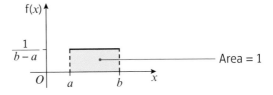

Example 15

The continuous random variable X is uniformly distributed on [3, 5]. Find:

a $P(3.2 < X < 4.3)$ **b** k such that $P(2X < k - X) = 0.2$.

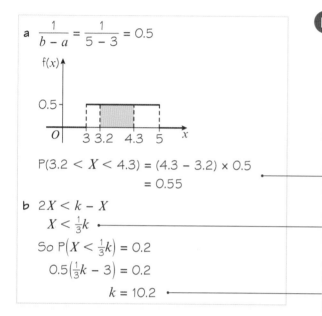

a $\dfrac{1}{b-a} = \dfrac{1}{5-3} = 0.5$

$P(3.2 < X < 4.3) = (4.3 - 3.2) \times 0.5$
$= 0.55$

b $2X < k - X$
$X < \tfrac{1}{3}k$
So $P\left(X < \tfrac{1}{3}k\right) = 0.2$
$0.5\left(\tfrac{1}{3}k - 3\right) = 0.2$
$k = 10.2$

Note When dealing with a uniform distribution it is easier to sketch the p.d.f. and work out the area of the rectangle. You can also use proportion: you are interested in a range of values of width 1.1 out of an interval of width 2, so the probability is $\dfrac{1.1}{2} = 0.55$.

$P(3.2 < X < 4.3)$ is the area of the shaded section on the sketch.

Solve the inequality.

Write an expression for the area under the rectangle to the left of $\tfrac{1}{3}k$. Set this equal to 0.2 then solve to find k.

Example 16

The continuous random variable X has p.d.f. as shown in the diagram. Find:

a the value of k **b** $P(3 < X < 3.5)$
c $P(X > 3 \mid X > 2)$

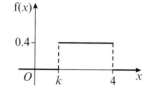

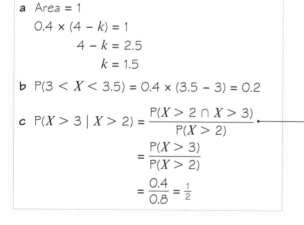

a Area = 1
$0.4 \times (4 - k) = 1$
$4 - k = 2.5$
$k = 1.5$

b $P(3 < X < 3.5) = 0.4 \times (3.5 - 3) = 0.2$

c $P(X > 3 \mid X > 2) = \dfrac{P(X > 2 \cap X > 3)}{P(X > 2)}$
$= \dfrac{P(X > 3)}{P(X > 2)}$
$= \dfrac{0.4}{0.8} = \tfrac{1}{2}$

Use $P(A|B) = \dfrac{P(A \cap B)}{P(B)}$. The probability that X is greater than 2 **and** greater than 3 is just the probability that X is greater than 3.
← SM2, Section 2.4

Problem-solving

To solve **conditional probability** problems with a continuous uniform distribution, you can use a continuous uniform distribution on a restricted sample space. Given that $X > 2$, the value of X is uniformly distributed on [2, 4].

Example 17

The continuous random variable X has probability density function

$$f(x) = \begin{cases} \frac{1}{5} & 3 \leqslant x \leqslant 8 \\ 0 & \text{otherwise} \end{cases}$$

a Write down the name of this distribution.

The continuous random variable $Y = 12 - 3X$.

Find:

b $E(Y)$ **c** $P(Y > 0)$ **d** Find $P(X < 7 \mid Y < 0)$

a Continuous uniform distribution

b $E(Y) = E(12 - 3X)$ Use $E(aX + b) = aE(X) + b$

    $= 12 - 3E(X)$

    $= 12 - 3 \times 5.5$

    $= -4.5$

c $P(Y > 0) = P(12 - 3X > 0)$ Convert the probability in terms of Y into a probability in terms of X.

    $= P(12 > 3X)$

    $= P(4 > X)$ Use proportion. $X < 4$ is an interval of width 1 out of a total interval width of 5.

    $= \frac{1}{5}$

d $P(X < 7 \mid Y < 0) = P(X < 7 \mid X > 4)$ Write both probabilities in terms of the same random variable.

    $= \dfrac{P(4 < X < 7)}{P(X > 4)}$ The probability that $X < 7$ **and** $X > 4$ is $P(4 < X < 7)$

    $= \dfrac{0.6}{0.8}$

    $= \dfrac{3}{4}$

Problem-solving

You could also tackle this problem by finding the distribution of Y. A linear transformation of a uniform distribution will be uniform, so $Y \sim U[-12, 3]$.

Example 18

The continuous random variable X has probability density function

$$f(x) = \begin{cases} \dfrac{1}{b - a} & a \leqslant x \leqslant b \\ 0 & \text{otherwise} \end{cases}$$

Find:

a $E(X)$ **b** $Var(X)$ **c** $F(x)$

a

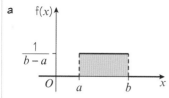

By symmetry $E(X) = \dfrac{a + b}{2}$

The formula for $E(X)$ for the continuous uniform distribution is given in the formulae booklet. However, you should be able to derive it from first principles.

b $Var(X) = \displaystyle\int_a^b \dfrac{(x - \mu)^2}{b - a}\,dx$

This is $\displaystyle\int_{-\infty}^{\infty} (x - \mu)^2\, f(x)\, dx$. You could use $E(X^2) - (E(X))^2$ but it is more difficult in this case.

$$= \int_a^b \left(x - \left(\dfrac{a + b}{2}\right)\right)^2 \times \dfrac{1}{(b - a)}\,dx$$

$$= \left[\dfrac{\left(x - \left(\dfrac{a + b}{2}\right)\right)^3}{3(b - a)}\right]_a^b$$

Using $\displaystyle\int (x + a)^n\,dx = \dfrac{(x + a)^{n+1}}{n + 1}$

$$= \dfrac{\left(b - \left(\dfrac{a + b}{2}\right)\right)^3}{3(b - a)} - \dfrac{\left(a - \left(\dfrac{a + b}{2}\right)\right)^3}{3(b - a)}$$

$$= \dfrac{\left(\dfrac{b - a}{2}\right)^3 - \left(\dfrac{a - b}{2}\right)^3}{3(b - a)}$$

$$= \dfrac{\left(\dfrac{b - a}{2}\right)^3 + \left(\dfrac{b - a}{2}\right)^3}{3(b - a)}$$

Using $-(a - b)^3 = (b - a)^3$

$$= \dfrac{\dfrac{(b - a)^3}{8} + \dfrac{(b - a)^3}{8}}{3(b - a)}$$

$$= \dfrac{\dfrac{(b - a)^3}{4}}{3(b - a)}$$

$$= \dfrac{(b - a)^3}{12(b - a)}$$

$$= \dfrac{(b - a)^2}{12}$$

The formula for $Var(X)$ for the continuous uniform distribution is given in the formulae booklet. However, you should be able to derive it from first principles.

c If $a \leqslant x \leqslant b$, $F(x) = \displaystyle\int_a^x \dfrac{1}{b - a}\,dt$

$$= \left[\dfrac{t}{b - a}\right]_a^x$$

$$= \dfrac{x - a}{b - a}$$

$$F(x) = \begin{cases} 0 & x < a \\[2mm] \dfrac{x - a}{b - a} & a \leqslant x \leqslant b \\[2mm] 1 & x > b \end{cases}$$

Watch out The cumulative distribution function for a uniform continuous distribution is not given in the formulae booklet. It can be useful to remember it, but make sure you know how to derive it from first principles as shown here.

- **For a continuous uniform distribution U[a, b]**

  - **E(X)** $= \dfrac{a + b}{2}$

  - **Var(X)** $= \dfrac{(b - a)^2}{12}$

  - **F(X)** $= \begin{cases} 0 & x < a \\ \dfrac{x - a}{b - a} & a \leqslant x \leqslant b \\ 1 & x > b \end{cases}$

Example 19

The continuous random variable X is uniformly distributed over the interval [4, 7].
Find:

a E(X)   **b** Var(X)

c the cumulative distribution function of X, for all x.

a E(X) $= \dfrac{4 + 7}{2} = 5.5$

b Var(X) $= \dfrac{(7 - 4)^2}{12} = \dfrac{3}{4}$

c F(x) $= \displaystyle\int_4^x \dfrac{1}{7 - 4} dt = \left[\dfrac{t}{3}\right]_4^x = \dfrac{x - 4}{3}$

F(x) $= \begin{cases} 0 & x < 4 \\ \dfrac{x - 4}{3} & 4 \leqslant x \leqslant 7 \\ 1 & x > 7 \end{cases}$

> You can write this down straight away if you learn the formula for F(X) of a continuous uniform distribution.

Example 20

The continuous random variable Y is uniformly distributed over the interval [a, b].
Given that E(Y) = 1 and Var(Y) = $\frac{16}{3}$, find the value of a and the value of b.

E(Y) $= \dfrac{a + b}{2} = 1$

$a + b = 2$ (1)

Var(Y) $= \dfrac{(b - a)^2}{12} = \dfrac{16}{3}$

$(b - a)^2 = 64$ (2)

> **Problem-solving**
>
> Use the formulae for the mean and variance of a continuous uniform distribution to form simultaneous equations in a and b.

Solving equations (1) and (2) simultaneously

$b = 2 - a$

$(2 - a - a)^2 = 64$

$(2 - 2a) = \pm 8$

$2 - 2a = 8$	$2 - 2a = -8$
$a = -3$	$a = 5$
$b = 2 - (-3)$	$b = 2 - 5$
$= 5$	$= -3$

Since $a < b$, $a = -3$ and $b = 5$.

Example 21

The continuous variable X is uniformly distributed over the interval $[-3, 5]$.

a Write down $E(X)$.

b Use integration to find the variance of X.

a $E(X) = \dfrac{-3 + 5}{2} = 1$

b $Var(X) = E(X^2) - (E(X))^2$

$= \displaystyle\int_{-3}^{5} \dfrac{x^2}{8}\,dx - 1^2$

$= \left[\dfrac{x^3}{24}\right]_{-3}^{5} - 1$

$= \dfrac{125}{24} + \dfrac{27}{24} - 1$

$= \dfrac{16}{3}$

You could have used

$\displaystyle\int_{-3}^{5} \dfrac{(x-1)^2}{8}\,dx = \left[\dfrac{(x-1)^3}{24}\right]_{-3}^{5}$

$= \dfrac{4^3}{24} - \dfrac{(-4)^3}{24}$

$= \dfrac{16}{3}$

Exercise 3E

1 The continuous random variable $X \sim U[2, 7]$.

Find:

a $P(3 < X < 5)$ **b** $P(X > 4)$

2 The continuous random variable X has a probability density function as shown in the diagram.

Find:

a the value of k **b** $P(4 < X < 7.9)$

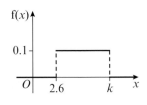

(P) 3 The continuous random variable X has probability density function

$$f(x) = \begin{cases} k & -2 \leqslant x \leqslant 6 \\ 0 & \text{otherwise} \end{cases}$$

Find:

a the value of k b $P(-1.3 < X < 4.2)$ c p such that $P(3X < X + p) = 0.5$

d $P(X > 5 \mid X > 0)$ e $P(X > 0 \mid X < 3)$ f $P(X < 1 \mid 0 < X < 2)$

(P) 4 The continuous random variable $Y \sim U[a, b]$. Given that $P(Y < 5) = \frac{1}{4}$ and $P(Y > 7) = \frac{1}{2}$, find the value of a and the value of b.

(P) 5 The continuous random variable $X \sim U[2, 8]$.

a Write down the distribution of $Y = 2X + 5$.

b Find $P(12 < Y < 20)$.

Hint If X has the continuous uniform distribution then $aX + b$ where a and b are constants will also have the continuous uniform distribution.

(E/P) 6 The continuous random variable X has probability density function

$$f(x) = \begin{cases} \frac{1}{10} & 2 \leqslant x \leqslant 12 \\ 0 & \text{otherwise} \end{cases}$$

a Write down the name of this distribution.   **(1 mark)**

The continuous random variable $Y = 20 - 2X$.
Find:

b $E(Y)$ **(2 marks)**

c $P(Y < 4)$ **(2 marks)**

d $P(Y > 4 \mid X < 10)$ **(3 marks)**

7 The continuous variable Y is uniformly distributed over the interval $[-3, 5]$.
Find:

a $E(X)$ **(1 mark)**

b $\text{Var}(X)$ **(1 mark)**

c $E(X^2)$ **(2 marks)**

d the cumulative distribution function of X, for all x.   **(3 marks)**

8 Find $E(X)$ and $\text{Var}(X)$ for the following probability density functions.

a
$$f(x) = \begin{cases} \frac{1}{4} & 1 \leqslant x \leqslant 5 \\ 0 & \text{otherwise} \end{cases}$$

b
$$f(x) = \begin{cases} \frac{1}{8} & -2 \leqslant x \leqslant 6 \\ 0 & \text{otherwise} \end{cases}$$

9 The continuous random variable X has probability density function as shown in the diagram.
Find:

a $E(X)$ b $\text{Var}(X)$

c $E(X^2)$ d the cumulative distribution function of X, for all x.

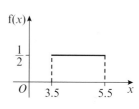

E/P 10 The continuous random variable $Y \sim U[a, b]$. Given that $E(Y) = 1$ and $Var(Y) = \frac{4}{3}$, find the value of a and the value of b. **(3 marks)**

E/P 11 The continuous random variable X has probability density function

$$f(x) = \begin{cases} \frac{1}{6} & -1 \leqslant x \leqslant 5 \\ 0 & \text{otherwise} \end{cases}$$

Given that $Y = 4X - 6$, find $E(Y)$ and $Var(Y)$. **(3 marks)**

12 The continuous random variable X is uniformly distributed over the interval $[\alpha, \beta]$.
Find:
a $P(X < \frac{3}{7}\alpha + \frac{4}{7}\beta)$ **b** $P(X > \frac{2}{5}\alpha + \frac{3}{5}\beta)$

E/P 13 The continuous random variable R is uniformly distributed on the interval $\alpha \leqslant R \leqslant \beta$.
Given that $E(R) = 5$ and $Var(R) = \frac{4}{3}$, find:
a the value of α and the value of β **(3 marks)**
b $P(R < 5.2)$ **(1 mark)**

E/P 14 The continuous random variable X is uniformly distributed over the interval $\alpha \leqslant R \leqslant \beta$.
a Write down the probability density function of X, for all x. **(1 mark)**
b Given that $E(X) = 2.5$ and $P(X < 1) = \frac{4}{11}$, find the value of α and the value of β. **(3 marks)**

E 15 The continuous random variable X is uniformly distributed over the interval $[-5, 4]$.
a Write down fully the probability density function $f(x)$ of X. **(2 marks)**
b Sketch the probability density function $f(x)$ of X. **(2 marks)**
Find:
c $E(X^2)$ **(2 marks)**
d $P(-0.2 < X < 0.6)$ **(2 marks)**

E 16 A continuous random variable X has cumulative distribution function

$$F(x) = \begin{cases} 0 & x < -3 \\ \dfrac{x + 3}{7} & -3 \leqslant x \leqslant 4 \\ 1 & x > 4 \end{cases}$$

a Find $P(X < 0)$. **(1 mark)**
b Find the probability density function $f(x)$ of X. **(2 marks)**
c Write down the name of the distribution of X. **(1 mark)**
d Find the mean and the variance of X. **(3 marks)**

E/P 17 The continuous random variable X is uniformly distributed over the interval $[-1, 4]$.
Find:

   a $E(X)$ **(2 marks)**

   b $\text{Var}(X)$ **(2 marks)**

   c $E(X^2)$ **(2 marks)**

   d $P(X < 1.4)$ **(1 mark)**

   A total of 6 observations of X are made.

   e Find the probability that exactly 4 of these observations are less than 1.4.   **(2 marks)**

E/P 18 The continuous random variable X is uniformly distributed over the interval $[\alpha, \beta]$.
Given that $E(X) = 7.5$ and $P(X > 10.5) = 0.25$

   a find the value of α and the value of β.   **(3 marks)**

   b Given that $P(X < c) = \frac{1}{3}$, find:

     i the value of c

     ii $P(c < X < 9)$ **(3 marks)**

3.6 Modelling with the continuous uniform distribution

The continuous uniform distribution is frequently used to model real-life situations. For example, if you know that trains leave from a station hourly, but you arrive not knowing when the next train will leave, then the length of time you have to wait after arriving at the station, X minutes, could be modelled as $X \sim U[0, 60]$.

Example 22

The trunk of a small tree varies in diameter from 10 cm at the bottom to 2 cm at the top. The tree is cut horizontally at a randomly chosen point, and the radius R cm of the cross-section is modelled as $R \sim U[1, 5]$.

Find the expected value of the area, A, of the cross-section of the tree.

$A = \pi R^2$

$E(A) = E(\pi R^2)$

$\quad = \pi E(R^2)$

$\text{Var}(R) = E(R^2) - (E(R))^2$

Rearranging gives

$E(R^2) = \text{Var}(R) + (E(R))^2$

$\text{Var}(R) = \dfrac{(5-1)^2}{12} = \dfrac{4}{3}$

$E(R) = \dfrac{5+1}{2} = 3$

$E(R^2) = \dfrac{4}{3} + 9 = \dfrac{31}{3}$

$E(A) = \dfrac{31\pi}{3}$

Watch out $E(R^2)$ is not the same as $(E(R))^2$

To find $E(R^2)$ you could have used

$E(X^2) = \int x^2 f(x)\, dx$

$E(R^2) = \int_1^5 \frac{1}{4} r^2 dr = \left[\frac{1}{12} r^3\right]_1^5$ since $\dfrac{1}{b-a} = \dfrac{1}{5-1} = \dfrac{1}{4}$

$\qquad = \dfrac{125}{12} - \dfrac{1}{12}$

$\qquad = \dfrac{31}{3}$

Example 23

The length of a pencil is measured to the nearest cm. Write down the distribution of the rounding errors R.

The error is the difference between the true length and the recorded length.

If a pencil is recorded as 20 cm long then its length is anywhere in the interval

$19.5 \text{ cm} \leqslant \text{length} < 20.5 \text{ cm}$

The error is therefore in the interval

$-0.5 \leqslant \text{error} < 0.5$

As it is reasonable to assume that the error is equally likely to take any of the values in this range

$R \sim U[-0.5, 0.5]$

Note The uniform distribution is often used as a model for errors made by rounding up or down when recording measurements.

Example 24

Write down the name of the distribution you would recommend as a suitable model for each of the following situations.

a The masses of 200 g tins of tomatoes produced on a production line.

b The difference between the true length and the length of metal rods measured to the nearest centimetre.

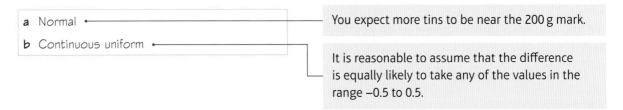

a Normal — You expect more tins to be near the 200 g mark.

b Continuous uniform — It is reasonable to assume that the difference is equally likely to take any of the values in the range −0.5 to 0.5.

Exercise 3F

1 The random variable X is the length of a side of a square and is modelled as $X \sim U[4.5, 5.5]$. The random variable Y is the area of the square.
Find $E(Y)$.

2 The random variable R has a continuous uniform distribution over the interval $[5, 11]$.

a Specify fully the probability density function of R.

b Find $P(7 < R < 10)$.

The random variable A is the area of a circle radius R cm.

c Find $E(A)$.

E **3** In a computer game, an alien appears every 2 seconds. The player stops the alien by pressing a key. The object of the game is to stop the alien as soon as it appears. Given that the player actually presses the key T seconds after the alien first appears, a simple model of the game assumes that T is a continuous uniform random variable defined over the interval [0, 1].

   a Write down P($T < 0.2$). **(1 mark)**

   b Write down E(T). **(2 marks)**

   c Use integration to find Var (T). **(3 marks)**

E/P **4** The time in minutes that Priya takes to check out at her local supermarket follows a continuous uniform distribution defined over the interval [2, 10].

   Find:

   a the probability that Priya will take more than 7 minutes to check out on one visit to the supermarket **(2 marks)**

   b the probability that Priya will take less than 5 minutes to check out on each of three successive visits to the supermarket **(3 marks)**

   Given that Priya has already spent 5 minutes at the checkout,

   c find the probability that she will take a total of less than 8 minutes to check out. **(2 marks)**

E/P **5** A drinks machine dispenses coffee into cups. It is electronically controlled to cut off the flow of coffee randomly between 175 ml and 215 ml. The random variable X is the volume of coffee dispensed into a cup, and is uniformly distributed.

   a Specify the probability density function of X and sketch its graph. **(3 marks)**

   b Find the probability that the machine dispenses:
     i less than 187 ml
     ii exactly 187 ml. **(3 marks)**

   c Calculate the interquartile range of X. **(3 marks)**

   d Determine the value of x such that P($X \geqslant x$) = 0.65 **(2 marks)**

   Brenda buys five cups of coffee from the drinks machine for people in her office. **Hint** Use a binomial distribution.

   e Find the probability that exactly three of the cups contain less than 187 ml. **(3 marks)**

E/P **6** The continuous random variable X represents the error, in mm, made when a machine cuts iron rods to a target length. X has a continuous uniform distribution over the interval [−3.0, 3.0].

   Find:

   a P($X < -2.3$) **(2 marks)**

   b P($|X| > 2.0$) **(2 marks)**

   Ten rods are cut.

   c Calculate the probability that exactly six are cut within 2 mm of the target length. **(3 marks)**

E/P **7** A manufacturer produces sweets of length Y mm where Y has a continuous uniform distribution with range [20, 28].

    a Find the probability that a randomly selected sweet has length greater than 26 mm. **(2 marks)**

    These sweets are randomly packed in bags of 20 sweets.

    b Find the probability that a randomly selected bag will contain at least 7 sweets with length greater than 26 mm. **(3 marks)**

E/P **8** The waiting times, in minutes, between flight take-offs at an airport are modelled by the continuous random variable X with probability density function

$$f(x) = \begin{cases} \frac{1}{5} & 2 \leqslant x \leqslant 7 \\ 0 & \text{otherwise} \end{cases}$$

    A randomly selected flight takes off at 10 am.

    a Find the probability that the next flight takes off before 10:05 am **(2 marks)**

    b Find the probability that at least 3 of the next 10 flights have a waiting time of more than 6 minutes. **(3 marks)**

E/P **9** A wooden dowelling rod of length 20 cm is cut into two pieces at a randomly chosen point. The length of the longer piece, X cm, is modelled as having a continuous uniform distribution over the interval [10, 20].

    The two pieces of the dowelling rod are used to from the base and height of a rectangle, as shown below.

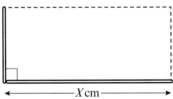

    Find the expected area of the rectangle. **(6 marks)**

Mixed exercise 3

1 The random variable X has probability density function $f(x)$ given by

$$f(x) = \begin{cases} \frac{1}{3}\left(1 + \frac{x}{2}\right) & 0 \leqslant x \leqslant 2 \\ 0 & \text{otherwise} \end{cases}$$

    Find:

    a $E(X)$ and $E(3X + 2)$ **b** $\text{Var}(X)$ and $\text{Var}(3X + 2)$

    c $P(X < 1)$ **d** $P(X > E(X))$

    e $P(0.5 < X < 1.5)$

2 The random variable X has probability density function f(x) given by

$$f(x) = \begin{cases} 2 - 2x & 0 \leqslant x \leqslant 1 \\ 0 & \text{otherwise} \end{cases}$$

  a Evaluate E(X).

  b Evaluate Var(X).

  c Write down the values of E($2X + 1$) and Var($2X + 1$).

  d Specify fully the cumulative distribution function of X.

  e Work out the median value of X.

3 The continuous random variable Y has cumulative distribution function given by

$$F(y) = \begin{cases} 0 & y < 1 \\ k(y^2 - y) & 1 \leqslant y \leqslant 2 \\ 1 & y > 2 \end{cases}$$

  where k is a positive constant.

  a Show that $k = \frac{1}{2}$ **b** Find P($Y < 1.5$).

  c Find the value of the median.   **d** Specify fully the probability density function f(y).

4 The continuous random variable X has cumulative distribution function

$$F(x) = \begin{cases} 0 & x < 2 \\ \frac{1}{5}(x^2 - 4) & 2 \leqslant x \leqslant 3 \\ 1 & x > 3 \end{cases}$$

  Find:

  a P($X > 2.4$)

  b the median

  c the probability density function, f(x).

  d Evaluate E(X).

  e Find the mode of X.

(E) **5** The random variable X has probability density function f(x) given by

$$f(x) = \begin{cases} kx^2 & 0 \leqslant x \leqslant 2 \\ 0 & \text{otherwise} \end{cases}$$

  where k is a positive constant.

  a Show that $k = \frac{3}{8}$ **(1 mark)**

  b Calculate E(X).   **(3 marks)**

  c Specify fully the cumulative distribution function of X.   **(4 marks)**

  d Find the value of the median.   **(2 marks)**

  e Find the value of the mode.   **(1 mark)**

E **6** The random variable Y has probability density function f(y) given by

$$f(y) = \begin{cases} k(y^2 + 2y + 2) & 1 \leqslant y \leqslant 3 \\ 0 & \text{otherwise} \end{cases}$$

where k is a positive constant.

a Show that $k = \frac{3}{62}$ **(2 marks)**

b Specify fully the cumulative distribution function of Y.   **(4 marks)**

c Evaluate $P(Y \leqslant 2)$.   **(3 marks)**

E **7** A random variable X has probability density function f(x) given by

$$f(x) = \begin{cases} \frac{3}{32}(4 - x^2) & -2 \leqslant x \leqslant 2 \\ 0 & \text{otherwise} \end{cases}$$

a Sketch the probability density function of X.   **(2 marks)**

b Write down the mode of X.   **(1 mark)**

c Specify fully the cumulative distribution function of X.   **(4 marks)**

d Find $P(0.5 < X < 1.5)$.   **(3 marks)**

E **8** A random variable X has probability density function f(x) given by

$$f(x) = \begin{cases} \frac{1}{3} & 0 \leqslant x < 1 \\ \frac{2}{7}x^2 & 1 \leqslant x \leqslant 2 \\ 0 & \text{otherwise} \end{cases}$$

a Find $E(X)$.   **(3 marks)**

b Specify fully the cumulative distribution function of X.   **(4 marks)**

c Find:

   i the median of X

   ii the 15th percentile of X.   **(3 marks)**

d Describe the skewness of the distribution, giving a reason for your answer.   **(1 mark)**

E/P **9** The continuous random variable X has cumulative distribution function

$$F(x) = \begin{cases} 0 & x < 1 \\ 0.05a^x - b & 1 \leqslant x \leqslant 2 \\ 1 & x > 2 \end{cases}$$

where a and b are positive constants.

Find a and b, showing your working clearly.   **(7 marks)**

E/P **10** A student writes the following cumulative distribution function for a continuous random variable X.

$$F(x) = \begin{cases} 0 & x < 5 \\ \frac{1}{5}(16x - x^2 - 55) & 5 \leqslant x \leqslant 10 \\ 1 & x > 10 \end{cases}$$

Explain why this cannot be a cumulative distribution function. **(2 marks)**

E/P **11** A continuous random variable X has probability density function f(x) given by

$$f(x) = \begin{cases} kx - k & 1 \leqslant x \leqslant 3 \\ 0 & \text{otherwise} \end{cases}$$

where k is a positive constant.

a Show that $k = \frac{1}{2}$ **(2 marks)**

b Find E(X). **(3 marks)**

c Work out the cumulative distribution function, F(x). **(4 marks)**

d Show that the median value lies between 2.4 and 2.5. **(3 marks)**

e Hence comment on the skewness of the distribution. **(1 mark)**

f Find the 10th to 90th percentile range, giving your answer correct to 3 significant figures. **(2 marks)**

E/P **12** The continuous random variable X has probability density function given by

$$f(x) = \begin{cases} x & 0 \leqslant x \leqslant 1 \\ \dfrac{3x^2}{14} & 1 < x \leqslant 2 \\ 0 & \text{otherwise} \end{cases}$$

a Sketch the probability density function of X. **(2 marks)**

b Find the mode of X. **(1 mark)**

c Find E$(2X)$. **(3 marks)**

d Find Var$(2X + 1)$. **(3 marks)**

e Specify fully the cumulative distribution function of X. **(4 marks)**

f Using your answer to part **e**, find the median of X. **(2 marks)**

E/P **13** The continuous random variable X has probability density function $f(x)$ given by

$$f(x) = \begin{cases} \dfrac{x^3}{16} & 0 \leqslant x < 2 \\ \dfrac{5-x}{6} & 2 \leqslant x \leqslant 5 \\ 0 & \text{otherwise} \end{cases}$$

    a Sketch the graph of $f(x)$ for all values of x.   **(2 marks)**

    b Write down the mode of X.   **(1 mark)**

    c Show that $P(X > 2) = 0.75$ **(2 marks)**

    d Define fully the cumulative distribution function $F(x)$.   **(4 marks)**

    e Find the median of X.   **(3 marks)**

E/P **14** The continuous random variable X has cumulative distribution function $F(x)$ given by

$$F(x) = \begin{cases} 0 & x < 2 \\ \frac{1}{81}(-2x^3 + 15x^2 - 44) & 2 \leqslant x \leqslant 5 \\ 1 & x > 5 \end{cases}$$

    a Find the probability density function $f(x)$.   **(3 marks)**

    b Find the mode of X.   **(2 marks)**

    c Sketch $f(x)$ for all values of x.   **(3 marks)**

    d Find the mean μ of X.   **(3 marks)**

    e Show that $F(\mu) > 0.5$.   **(1 mark)**

    f Show that the median of X lies between the mode and the mean.
       Hence describe the skewness of the distribution.   **(3 marks)**

E/P **15** A continuous random variable X has cumulative distribution function $F(x)$ given by

$$F(x) = \begin{cases} 0 & x < 0 \\ k(35x - 2x^2) & 0 \leqslant x \leqslant 5 \\ 1 & x > 5 \end{cases}$$

    a Show that $k = \frac{1}{125}$ **(1 mark)**

    b Find the median of X.   **(3 marks)**

    c Find the probability density function $f(x)$.   **(3 marks)**

    d Sketch $f(x)$ for all values of x.   **(3 marks)**

    e Write down the mode of X.   **(1 mark)**

    f Find $E(X)$.   **(3 marks)**

    g Comment on the skewness of this distribution.   **(2 marks)**

P **16** The continuous random variable X has probability density function $f(x)$ given by

$$f(x) = \begin{cases} ax + b & 0 \leqslant x \leqslant 2 \\ 0 & \text{otherwise} \end{cases}$$

    If $E(X) = \frac{9}{8}$, find the values of a and b.

Ⓔ **17** A continuous random variable X has probability density function $f(x)$ where

$$f(x) = \begin{cases} k(x + 1)^3 & -1 \leqslant x \leqslant 0 \\ 0 & \text{otherwise} \end{cases}$$

where k is a positive integer.

a Show that $k = 4$ **(3 marks)**

Find:

b $E(X)$ **(4 marks)**

c the cumulative distribution function $F(x)$ **(4 marks)**

d the median. **(3 marks)**

18 The continuous random variable X is uniformly distributed over the interval $[-2, 5]$.

a Sketch the probability density function $f(x)$ of X.

Find:

b $E(X)$ **c** $Var(X)$

d the cumulative distribution function of X, for all x

e $P(3.5 < X < 5.5)$ **f** $P(X = 4)$

g $P(X > 0 \mid X < 2)$ **h** $P(X > 3 \mid X > 0)$

19 The continuous random variable X has p.d.f. as shown in the diagram.

Find:

a the value of k **b** $P(-2 < X < -1)$

c $E(X)$ **d** $Var(X)$

e the cumulative distribution function of X, for all x.

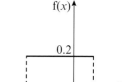

20 The continuous random variable Y is uniformly distributed on the interval $a \leqslant Y \leqslant b$. Given $E(Y) = 2$ and $Var(Y) = 3$, find:

a the value of a and the value of b **b** $P(Y > 1.8)$

Ⓔ/ℙ **21** The continuous random variable X has a continuous uniform distribution on the interval $[0, 2]$ The continuous random variable $Y = 10 - 5X$.

a Describe the distribution of Y. **(2 marks)**

b Find $P(Y < 3)$. **(2 marks)**

c Find $P(Y > 3 \mid X > 0.5)$. **(3 marks)**

Ⓔ/ℙ **22** A child has a pair of scissors and a piece of string 20 cm long which has a mark on one end. The child cuts the string, at a randomly chosen point, into two pieces. Let X represent the length of the piece of string with the mark on it.

a Write down the name of the probability distribution of X and sketch the graph of its probability density function. **(3 marks)**

b Find the values of $E(X)$ and $Var(X)$. **(4 marks)**

c Using your model, calculate the probability that the shorter piece of string is at least 8 cm long. **(3 marks)**

E/P **23** Joan records the temperature every day. The highest temperature she recorded was 29 °C to the nearest degree. Let X represent the error in the measured temperature.

    a Suggest a suitable model for the distribution of X. **(1 mark)**

    b Using your model, calculate the probability that the error will be less than 0.2 °C. **(3 marks)**

    c Find the variance of the error in the measured temperature. **(2 marks)**

E/P **24** Jameil catches a bus to work every morning. According to the timetable the bus is due at 9 am, but Jameil knows that the bus can arrive at a random time between three minutes early and ten minutes late. The random variable X represents the time, in minutes, after 9 am when the bus arrives.

    a Suggest a suitable model for the distribution of X and specify it fully. **(2 marks)**

    b Calculate the mean value of X. **(2 marks)**

    c Find the cumulative distribution function of X. **(4 marks)**

    Jameil will be late for work if the bus arrives after 9:05 am.

    d Find the probability that Jameil is late for work. **(2 marks)**

E/P **25** A plumber measures, to the nearest cm, the lengths of pipes.

    a Suggest a suitable model to represent the difference between the true lengths and the measured lengths. **(1 mark)**

    b Find the probability that for a randomly chosen rod the measured length will be within 0.2 cm of the true length. **(2 marks)**

    c Three pipes are selected at random. Find the probability that the measured lengths of all three pipes will be within 0.2 cm of the true length. **(2 marks)**

E/P **26** A coffee machine dispenses coffee into cups. It is electronically controlled to cut off the flow of coffee randomly between 190 ml and 210 ml. The random variable X is the volume of coffee dispensed into a cup.

    a Specify the probability density function of X and sketch its graph. **(3 marks)**

    b Find the probability that the machine dispenses

      i less than 198 ml

      ii exactly 198 ml. **(3 marks)**

    c Calculate the interquartile range of X. **(2 marks)**

    d Given that the machine has already dispensed 195 ml coffee into a cup, find the probability that it will dispense more than 200 ml into that cup. **(2 marks)**

27 Write down the name of the distribution you would recommend as a suitable model for each of the following situations:

    a the difference between the true height and the height measured, to the nearest cm, of randomly chosen people

    b the heights of randomly selected 18-year-old females.

28 The delay in departure, T hours, of a flight from *Statistics* airport is modelled by the probability density function

$$f(t) = \begin{cases} \frac{1}{72}(6-t)^2 & 0 \leqslant t \leqslant 6 \\ 0 & \text{otherwise} \end{cases}$$

a Find the cumulative distribution function $F(t)$.

b Find the median value of T.

c Find $E(T)$.

E/P **29** A continuous random variable X is uniform on the interval $[b, 5b]$.

a Write down the probability density function of X. **(2 marks)**

b Write down the value of $E(X)$. **(1 mark)**

c Show by integration that $\text{Var}(X) = \dfrac{4b^2}{3}$ **(3 marks)**

Given that $b = 3$

d find $P(X > 10)$ **(2 marks)**

Five observations are taken from this distribution.

e Find the probability that exactly three of them are bigger than 10. **(4 marks)**

A **30** The continuous random variable X has probability density function given by

E

$$f(x) = \begin{cases} \dfrac{2}{(2x-1)\ln 5} & 1 \leqslant x \leqslant 3 \\ 0 & \text{otherwise} \end{cases}$$

Find $F(x)$. **(3 marks)**

E **31** As part of its employee selection process, *ROBU Bank* sets an aptitude test. Over the years it has found that the percentage scored X (measured in 100s) by prospective employees can be modelled by the probability density function $f(x)$, where

$$f(x) = \begin{cases} kx\sin(\pi x) & 0 \leqslant x \leqslant 1 \\ 0 & \text{otherwise} \end{cases}$$

Find:

a the value of k **(5 marks)**

b $E(X)$ **(5 marks)**

32 A random variable X has probability density function given by

E

$$f(x) = \begin{cases} k & 0 \leqslant x \leqslant 1 \\ \dfrac{k}{x^2} & 1 \leqslant x \leqslant 2 \\ 0 & \text{otherwise} \end{cases}$$

a Find the value of k. **(3 marks)**

b Calculate the value of $E(X)$. **(4 marks)**

c Calculate the value of $\text{Var}(X)$. **(4 marks)**

Challenge

1 A spinner is made using a circle of radius r, and a pointer of length r which is pivoted at the centre of the circle. The pointer is spun and allowed to come to rest. The random variable θ represents the angle between the vertical and the resting position of the spinner, and the random variable X represents the horizontal distance of the end of the spinner from the centre of the circle.

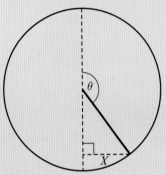

   a Describe a suitable distribution to model θ.

   b Hence, or otherwise, find $E(X)$.

   c Briefly explain how this spinner could be used as part of an experiment to estimate the value of π.

2 A continuous random variable X having a probability density function $f(x)$, where

$$f(x) = \begin{cases} \lambda e^{-\lambda x} & x \geqslant 0 \\ 0 & \text{otherwise} \end{cases}$$

where λ is a positive constant, is said to follow an **exponential** distribution.

Show that:

   a $E(X) = \dfrac{1}{\lambda}$ and $\text{Var}(X) = \dfrac{1}{\lambda^2}$

   b $P(X > a + b \mid X > a) = P(X > b)$

Summary of key points

1 If X is a continuous random variable with **probability density function** $f(x)$, then

   • $f(x) \geqslant 0$ for all $x \in \mathbb{R}$

   • $P(a < X < b) = \displaystyle\int_a^b f(x)\,dx$

   • $\displaystyle\int_{-\infty}^{\infty} f(x)\,dx = 1$

2 For a random variable X, the **cumulative distribution function** $F(x) = P(X \leqslant x)$.

3 If X is a continuous random variable with c.d.f. $F(x)$ and p.d.f. $f(x)$:

$$f(x) = \frac{d}{dx}F(x) \quad \text{and} \quad F(x) = \int_{-\infty}^{x} f(t)\,dt$$

4 If X is a continuous random variable with probability density function f(x):

- the mean or expected value of X is given by

$$E(X) = \mu = \int_{-\infty}^{\infty} x\text{f}(x)\,\text{d}x$$

- the variance of X is given by

$$Var(X) = \sigma^2 = \int_{-\infty}^{\infty} (x - \mu)^2\,\text{f}(x)\,\text{d}x$$

$$= \int_{-\infty}^{\infty} x^2\,\text{f}(x)\,\text{d}x - \mu^2$$

5 If X is a continuous random variable, then $E(\text{g}(X)) = \int_{-\infty}^{\infty} \text{g}(x)\,\text{f}(x)\,\text{d}x$

6 You can calculate Var(X) using

$$Var(X) = E(X^2) - (E(X))^2$$

7 In the case where g(X) is a linear function of the form $aX + b$,

- $E(aX + b) = aE(X) + b$
- $Var(aX + b) = a^2Var(X)$

8 The mode of a continuous random variable is the value of x for which the p.d.f. is a maximum.

9 If X is a continuous random variable with c.d.f. F(x):

- the median of X is the value m such that $F(m) = 0.5$
- the lower quartile of X is the value Q_1 such that $F(Q_1) = 0.25$
- the upper quartile of X is the value Q_3 such that $F(Q_3) = 0.75$
- the nth percentile of X is the value P_n such that $F(P_n) = \dfrac{n}{100}$

10 - Positive skew: mode < median < mean
- Negative skew: mean < median < mode

11 A random variable having a continuous uniform distribution U[a, b] has p.d.f.

$$\text{f}(x) = \begin{cases} \dfrac{1}{b-a} & a \leqslant x \leqslant b \\ 0 & \text{otherwise} \end{cases}$$

12 For a continuous uniform distribution U[a, b]

$$E(X) = \frac{a+b}{2}$$

$$Var(X) = \frac{(b-a)^2}{12}$$

$$F(x) = \begin{cases} 0 & x < a \\ \dfrac{x-a}{b-a} & a \leqslant x \leqslant b \\ 1 & x > b \end{cases}$$

Combinations of random variables

Prior knowledge check

1 A random variable $Y \sim N(20, 4^2)$. Find:

  a $P(Y > 15)$ **b** $P(19 < Y < 22)$

                      ← SM2, Chapter 3

2 Given that $X \sim N(0, 1^2)$, find:

  a p such that $P(X > p) = 0.25$

  b q such that $P(|X| < q) = 0.7$

                    ← SM2, Chapter 3

3 X is a random variable with $E(X) = 7$ and
$Var(X) = 3$. Find:

  a $E(2X)$ **b** $Var(3X - 5)$

  c $E(1 - 6X)$ ← FS1, Chapter 1

If the times of the first- and second-place
sprinters are normally distributed, then
the winning margin will also be normally
distributed.   → Exercise 4A, Q9

4.1 Combinations of random variables

Two random variables are **independent** if the outcome of one does not affect the distribution of the other. You need to be able to combine random variables with different distributions. You will use these two results:

- **If X and Y are two random variables, then**
  - **$E(X + Y) = E(X) + E(Y)$**
  - **$E(X - Y) = E(X) - E(Y)$**

Note You do not need to be able to prove these results for your exam.

- **If X and Y are two independent random variables, then**
  - **$\text{Var}(X + Y) = \text{Var}(X) + \text{Var}(Y)$**
  - **$\text{Var}(X - Y) = \text{Var}(X) + \text{Var}(Y)$**

Watch out You add the variances even when you subtract the random variables.

Example 1

If X is a random variable with $E(X) = \mu_1$ and $\text{Var}(X) = \sigma_1^2$ and Y is an independent random variable with $E(Y) = \mu_2$ and $\text{Var}(Y) = \sigma_2^2$, find the mean and variance of:

a $X + Y$ **b** $X - Y$

a $\quad E(X + Y) = E(X) + E(Y)$
$$= \mu_1 + \mu_2$$
$\quad \text{Var}(X + Y) = \text{Var}(X) + \text{Var}(Y)$
$$= \sigma_1^2 + \sigma_2^2$$

b $\quad E(X - Y) = E(X) - E(Y)$
$$= \mu_1 - \mu_2$$
$\quad \text{Var}(X - Y) = \text{Var}(X) + \text{Var}(Y)$
$$= \sigma_1^2 + \sigma_2^2$$

Variances are always added.

You can combine the above result with standard results about expectations and variances of multiples of a random variable to analyse linear combinations of independent random variables.

- **If X and Y are two random variables, then**
  - **$E(aX + bY) = aE(X) + bE(Y)$**
  - **$E(aX - bY) = aE(X) - bE(Y)$**

- **If X and Y are two independent random variables, then**
  - **$\text{Var}(aX + bY) = a^2\text{Var}(X) + b^2\text{Var}(Y)$**
  - **$\text{Var}(aX - bY) = a^2\text{Var}(X) + b^2\text{Var}(Y)$**

In this chapter you will apply these results to analyse **normally distributed** random variables.

- **A linear combination of normally distributed random variables is also normally distributed.**

This result allows you to fully define the distribution of a linear combination of independent normal random variables.

A ■ If X and Y are independent random variables with $X \sim \mathbf{N}(\mu_1, \sigma_1{}^2)$ and $Y \sim \mathbf{N}(\mu_2, \sigma_2{}^2)$, then

- $aX + bY \sim \mathbf{N}(a\mu_1 + b\mu_2, a^2\sigma_1{}^2 + b^2\sigma_2{}^2)$
- $aX - bY \sim \mathbf{N}(a\mu_1 - b\mu_2, a^2\sigma_1{}^2 + b^2\sigma_2{}^2)$

You can also use this result to find the distribution of sums of identically distributed independent normal random variables.

■ If $X_1, X_2, \ldots X_n$ are independent identically distributed random variables with

$$X_i \sim \mathbf{N}(\mu, \sigma^2), \text{ then } \sum_{i=1}^{n} X_i \sim \mathbf{N}(n\mu, n\sigma^2)$$

> **Watch out** The random variable $\sum_{i=1}^{n} X_i$ is **not** the same as the random variable nX_1.
>
> For example, $\mathrm{Var}(X_1 + X_2) = \sigma^2 + \sigma^2 = 2\,\mathrm{Var}(X_1)$, but $\mathrm{Var}(2X_1) = 4\,\mathrm{Var}(X_1)$.

Example 2

The independent random variables X and Y have distributions $X \sim N(5, 2^2)$ and $Y \sim N(10, 3^2)$.

a Find the distribution of:
  i $A = X + Y$ **ii** $B = 9X - 2Y$

b Find $P(B > 30)$.

The independent random variables Y_1, Y_2, Y_3 and Y_4 all have the same distribution as Y. The random variable Z is defined as:

$$Z = \sum_{i=1}^{4} Y_i$$

c Find the mean and standard deviation of Z.

---

a i $A \sim N(5 + 10, 2^2 + 3^2)$

  So $A \sim N(15, 13)$

ii $B \sim N(9 \times 5 - 2 \times 10, 9^2 \times 2^2 + 2^2 \times 3^2)$

  So $B \sim N(25, 360)$

b $P(B > 30) = 0.3961$

c $Z \sim N(4 \times 10, 4 \times 3^2)$

  $E(X) = 40$, $\mathrm{Var}(Z) = 36$

  So Z has mean 40 and standard deviation 6.

> **Watch out** When you subtract random variables you still **add** the variances.

> Use your calculator to find normal distribution probabilities. ← **SM2, Chapter 3**

> Be careful with means and standard deviations. The random variable Z has variance 36, so it has standard deviation 6.

Example 3

The independent random variables X and Y have distributions $X \sim N(25, 6)$ and $Y \sim N(22, 10)$. Find $P(X > Y)$.

Let $C = X - Y$.

Then $P(X > Y) = P(C > 0)$

$C \sim N(25 - 22, 6 + 10)$ so $C \sim N(3, 4^2)$

$P(C > 0) = 0.7734$ so $P(X > Y) = 0.7734$ (4 s.f.)

> **Problem-solving**
>
> You can **compare** independent normal random variables by defining a new random variable to be the difference between them. If $X - Y > 0$ then $X > Y$, and if $X - Y < 0$ then $Y > X$.

Example 4

A

Bottles of mineral water are delivered to shops in crates containing 12 bottles each. The weights of bottles are normally distributed with mean weight 2 kg and standard deviation 0.05 kg. The weights of empty crates are normally distributed with mean 2.5 kg and standard deviation 0.3 kg.

a Assuming that all random variables are independent, find the probability that a full crate will weigh between 26 kg and 27 kg.

b Two bottles are selected at random from a crate. Find the probability that they differ in weight by more than 0.1 kg.

c Find the weight, M, that a full crate should have on its label so that there is a 1% chance that it will weigh more than M.

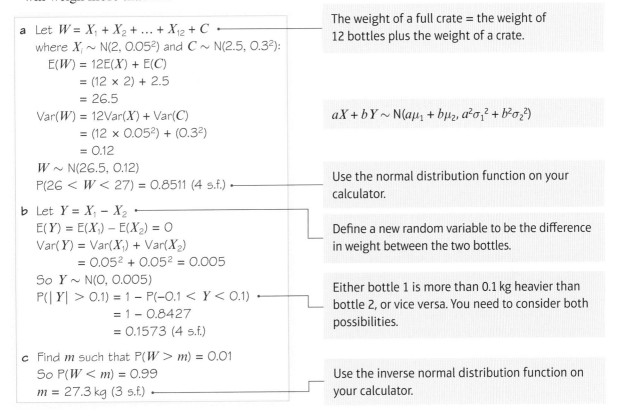

a Let $W = X_1 + X_2 + \ldots + X_{12} + C$ ———
     where $X_i \sim N(2, 0.05^2)$ and $C \sim N(2.5, 0.3^2)$:

     $E(W) = 12E(X) + E(C)$
            $= (12 \times 2) + 2.5$
            $= 26.5$
     $Var(W) = 12Var(X) + Var(C)$
            $= (12 \times 0.05^2) + (0.3^2)$
            $= 0.12$
     $W \sim N(26.5, 0.12)$
     $P(26 < W < 27) = 0.8511$ (4 s.f.) ———

b Let $Y = X_1 - X_2$ ———
     $E(Y) = E(X_1) - E(X_2) = 0$
     $Var(Y) = Var(X_1) + Var(X_2)$
            $= 0.05^2 + 0.05^2 = 0.005$
     So $Y \sim N(0, 0.005)$
     $P(|Y| > 0.1) = 1 - P(-0.1 < Y < 0.1)$ ———
                 $= 1 - 0.8427$
                 $= 0.1573$ (4 s.f.)

c Find m such that $P(W > m) = 0.01$
     So $P(W < m) = 0.99$
     $m = 27.3$ kg (3 s.f.) ———

The weight of a full crate = the weight of 12 bottles plus the weight of a crate.

$aX + bY \sim N(a\mu_1 + b\mu_2, a^2\sigma_1^2 + b^2\sigma_2^2)$

Use the normal distribution function on your calculator.

Define a new random variable to be the difference in weight between the two bottles.

Either bottle 1 is more than 0.1 kg heavier than bottle 2, or vice versa. You need to consider both possibilities.

Use the inverse normal distribution function on your calculator.

Exercise 4A

1 Given the random variables $X \sim N(80, 3^2)$ and $Y \sim N(50, 2^2)$ where X and Y are independent, find the distribution of W where:

  a $W = X + Y$ **b** $W = X - Y$

2 Given the random variables $X \sim N(45, 6)$, $Y \sim N(54, 4)$ and $W \sim N(49, 8)$ where X, Y and W are independent, find the distribution of R where $R = X + Y + W$.

3 X and Y are independent normal random variables. $X \sim N(60, 25)$ and $Y \sim N(50, 16)$. Find the distribution of T where:

  a $T = 3X$ **b** $T = 7Y$ **c** $T = 3X + 7Y$ **d** $T = X - 2Y$

A **4** X, Y and W are independent normal random variables. $X \sim N(8, 2)$, $Y \sim N(12, 3)$ and $W \sim N(15, 4)$. Find the distribution of A where:

a $A = X + Y + W$ **b** $A = W - X$ **c** $A = X - Y + 3W$

d $A = 3X + 4W$ **e** $A = 2X - Y + W$

5 A, B and C are independent normal random variables. $A \sim N(50, 6)$, $B \sim N(60, 8)$ and $C \sim N(80, 10)$. Find:

a $P(A + B < 115)$ **b** $P(A + B + C > 198)$ **c** $P(B + C < 138)$

d $P(2A + B - C < 70)$ **e** $P(A + 3B - C > 140)$ **f** $P(105 < A + B < 116)$

E **6** Given the random variables $X \sim N(20, 5)$ and $Y \sim N(10, 4)$ where X and Y are independent, find:

a $E(X - Y)$ **(2 marks)**

b $Var(X - Y)$ **(2 marks)**

c $P(13 < X - Y < 16)$ **(2 marks)**

P **7** X and Y are independent random variables with $X \sim N(76, 15)$ and $Y \sim N(80, 10)$. Find:

a $P(Y > X)$ **b** $P(X > Y)$

c the probability that X and Y differ by   **i** less than 3   **ii** more than 7

E/P **8** The random variable R is defined as $R = X + 4Y$ where $X \sim N(8, 2^2)$, $Y \sim N(14, 3^2)$ and X and Y are independent. Find:

a $E(R)$ **(2 marks)**

b $Var(R)$ **(2 marks)**

c $P(R < 41)$ **(2 marks)**

The random variables Y_1, Y_2 and Y_3 are independent and each has the same distribution as Y.

The random variable S is defined as $S = \sum_{i=1}^{3} Y_i - \frac{1}{2}X$

d Find $Var(S)$.   **(2 marks)**

E/P **9** Two runners recorded the mean and standard deviation of their 100 m sprint times in a table.

	Mean	Standard deviation
Runner A	13.2 seconds	0.9 seconds
Runner B	12.9 seconds	1.3 seconds

a Assuming that each runner's times are normally distributed, find the probability that in a head-to-head race, runner A will win by more than 0.5 seconds.   **(5 marks)**

A 'photo finish' occurs if the winning margin is less than 0.1 seconds.

b Find the probability of a 'photo finish'.   **(2 marks)**

E/P **10** A factory makes steel rods and steel tubes. The diameter of a steel rod is normally distributed with mean 3.55 cm and standard deviation 0.02 cm. The internal diameter of a steel tube is normally distributed with mean 3.60 cm and standard deviation 0.02 cm.

A rod and a tube are selected at random. Find the probability that the rod cannot pass through the tube.   **(6 marks)**

E/P **11** The mass of a randomly selected jar of jam is normally distributed with a mean mass of 1 kg and a standard deviation of 12 g. The jars are packed in boxes of 6 and the mass of the box is normally distributed with mean mass 250 g and standard deviation 10 g. Find the probability that a randomly chosen box of 6 jars will have a mass less than 6.2 kg.   **(6 marks)**

A **12** The thickness of paperback books can be modelled as a normal random variable with mean
E/P 2.1 cm and variance 0.39 cm². The thickness of hardback books can be modelled as a normal
random variable with mean 4.0 cm and variance 1.56 cm². A small bookshelf is 30 cm long.

   a Find the probability that a random sample of

     i 15 paperback books can be placed side-by-side on the bookshelf

     ii 5 hardback and 5 paperback books can be placed side-by-side on the bookshelf. **(8 marks)**

   b Find the shortest length of bookshelf needed so that there is at least a 99% chance
that it will hold a random sample of 15 paperback books. **(3 marks)**

E/P **13** A sweet manufacturer produces two varieties of fruit sweet, Xtras and Yummies. The masses, X
and Y in grams, of randomly selected Xtras and Yummies are such that

$$X \sim N(30, 25) \text{ and } Y \sim N(32, 16)$$

   a Find the probability that the mass of two randomly selected Yummies will differ by more
than 5 g. **(5 marks)**

One sweet of each variety is selected at random.

   b Find the probability that the Yummy sweet has a greater mass than the Xtra. **(5 marks)**

A packet contains 6 Xtras and 4 Yummies.

   c Find the probability that the average mass of the sweets in the packet lies between
280 g and 330 g. **(6 marks)**

E/P **14** A certain brand of biscuit is individually wrapped. The mass of a biscuit can be taken to be
normally distributed with mean 75 g and standard deviation 5 g. The mass of an individual
wrapping is normally distributed with mean 10 g and standard deviation 2 g. Six of these
individually wrapped biscuits are then packed together. The mass of the packing material is a
normal random variable with mean 40 g and standard deviation 3 g. Find, to 3 decimal places,
the probability that the total mass of the packet lies between 535 g and 565 g. **(7 marks)**

E/P **15** The independent normal random variables X and Y have distributions $N(10, 2^2)$ and
$N(40, 3^2)$ respectively. The random variable Q is defined as

$$Q = 2X + Y$$

   a Find:

     i $E(Q)$ **(2 marks)**

     ii $\text{Var}(Q)$ **(3 marks)**

The random variables X_1, X_2, X_3, X_4, X_5 are independent and all share the same distribution
as X. The random variable R is defined as

$$R = \sum_{i=1}^{5} X_i$$

   b **i** Find the distribution of R.   **ii** Find $P(Q > R)$. **(7 marks)**

E/P **16** The usable capacity of the hard drive on a games console is normally distributed with mean
60 GB and standard deviation 2.5 GB. The amounts of storage required by games are modelled
as being identically normally distributed with mean 5.5 GB and standard deviation 1.2 GB.

   a Chloe wants to save 10 randomly chosen games onto her empty hard drive.
Find the probability that they will fit. **(8 marks)**

   b State one assumption you have made in your calculations, and comment on its
validity. **(1 mark)**

A 17 X_1, X_2, X_3 and X_4 are independent random variables, each with distribution N(4, 0.03).
E/P The random variables Y and Z are defined as

$$Y = X_1 + X_2 + X_3 \qquad Z = 3X_4$$

Find the probability that Y and Z differ by no more than 1. **(5 marks)**

E/P **18** A builder purchases bags of sand in two sizes, large and small. Large bags have mass L kg and small bags have mass S kg. L and S are independent normally distributed random variables with distributions N(75, 5^2) and N(40, 3^2) respectively.

A large and a small bag of sand are chosen at random.

a Find the probability that the mass of the small bag is more than half the mass of the large bag. **(6 marks)**

The builder purchases 10 small bags of sand. The total mass of these bags is represented by the random variable M.

b Find P($|M - 400| < 5$). **(5 marks),**

Challenge

For independent random variables X and Y, E(XY) = E(X)E(Y).

Use this result to prove that if X and Y are independent random variables, then Var($X + Y$) = Var(X) + Var(Y).

Hint You may make use of the fact that for any two random variables E($X + Y$) = E(X) + E(Y).

Summary of key points

1 If X and Y are two random variables, then
- E($X + Y$) = E(X) + E(Y)
- E($X - Y$) = E(X) - E(Y)

2 If X and Y are two independent random variables, then
- Var($X + Y$) = Var(X) + Var(Y)
- Var($X - Y$) = Var(X) + Var(Y)

3 If X and Y are two random variables, then
- E($aX + bY$) = aE(X) + bE(Y)
- E($aX - bY$) = aE(X) - bE(Y)

4 If X and Y are two independent random variables, then
- Var($aX + bY$) = a^2Var(X) + b^2Var(Y)
- Var($aX - bY$) = a^2Var(X) + b^2Var(Y)

5 A linear combination of normally distributed random variables is also normally distributed.

6 If X and Y are independent random variables with $X \sim N(\mu_1, \sigma_1^2)$ and $Y \sim N(\mu_2, \sigma_2^2)$, then
- $aX + bY \sim N(a\mu_1 + b\mu_2, a^2\sigma_1^2 + b^2\sigma_2^2)$
- $aX - bY \sim N(a\mu_1 - b\mu_2, a^2\sigma_1^2 + b^2\sigma_2^2)$

7 If X_1, X_2, ... X_n are independent identically distributed random variables with $X_i \sim N(\mu, \sigma^2)$, then
$$\sum_{i=1}^{n} X_i \sim N(n\mu, n\sigma^2)$$

Review exercise

(E) **1** A long distance lorry driver recorded the distance travelled, m miles, and the amount of fuel used, f litres, each day. Summarised below are data from the driver's records for a random sample of 8 days.

The data are coded such that $x = m - 250$ and $y = f - 100$.

The data collected can be summarised as follows:

$$\sum x = 130 \qquad \sum y = 48$$
$$\sum xy = 8880 \qquad S_{xx} = 20\,487.5$$

a Find the equation of the regression line of y on x in the form $y = a + bx$. **(2)**

b Hence find the equation of the regression line of f on m. **(3)**

c Predict the amount of fuel used on a journey of 235 miles. **(1)**

← Section 1.1

(E/P) **2** A manufacturer stores drums of chemicals. During storage, evaporation takes place. A random sample of 10 drums was taken and the time in storage, x weeks, and the evaporation loss, y ml, are shown in the table below.

x	3	5	6	8	10	12	13	15	16	18
y	36	50	53	61	69	79	82	90	88	96

a On graph paper, draw a scatter diagram to represent these data. **(2)**

b Give a reason to support fitting a regression model of the form $y = a + bx$ to these data. **(1)**

c Find, to 2 decimal places, the value of a and the value of b.

(You may use $\sum x^2 = 1352$, $\sum y^2 = 53\,112$ and $\sum xy = 8354$.) **(2)**

d Give an interpretation of the value of b. **(1)**

e Using your model, predict the amount of evaporation that would take place after:
i 19 weeks
ii 35 weeks. **(2)**

f Comment, with a reason, on the reliability of each of your predictions. **(2)**

← Section 1.1

(E/P) **3** A metallurgist measured the length, l mm, of a copper rod at various temperatures, $t\,°C$, and recorded the following results.

t	l
20.4	2461.12
27.3	2461.41
32.1	2461.73
39.0	2461.88
42.9	2462.03
49.7	2462.37
58.3	2462.69
67.4	2463.05

The results were then coded such that $x = t$ and $y = l - 2460$.

a Calculate S_{xy} and S_{xx}.
(You may use $\sum x^2 = 15965.01$ and $\sum xy = 757.467$) **(2)**

b Find the equation of the regression line of y on x in the form $y = a + bx$. **(2)**

c Estimate the length of the rod at $40\,°C$. **(1)**

d Find the equation of the regression line of l on t. **(2)**

e Estimate the length of the rod at 90 °C. **(1)**

f Comment on the reliability of your estimate in part **e**. **(1)**

← Section 1.1

(E/P) **4** A student is investigating the relationship between the price (y pence) of 100 g of chocolate and the percentage ($x\%$) of the cocoa solids in the chocolate.

The following data are obtained

Chocolate brand	x (% cocoa)	y (pence)
A	10	35
B	20	55
C	30	40
D	35	100
E	40	60
F	50	90
G	60	110
H	70	130

(You may use: $\sum x = 315$, $\sum x^2 = 15\,225$, $\sum y = 620$, $\sum y^2 = 56\,550$, $\sum xy = 28\,750$)

a Draw a scatter diagram to represent these data. **(2)**

b Show that $S_{xy} = 4337.5$ and find S_{xx}. **(2)**

The student believes that a linear relationship of the form $y = a + bx$ could be used to describe these data.

c Use linear regression to find the value of a and the value of b, giving your answers to 3 significant figures. **(2)**

d Draw the regression line on your diagram. **(1)**

The student believes that one brand of chocolate is overpriced.

e Use the scatter diagram to:
  i state which brand is overpriced
  ii suggest a fair price for this brand.

Give reasons for both your answers. **(3)**

← Section 1.1

(E/P) **5** A mobile phone operator recorded the number of minutes used, x, and the number of text messages sent, y, for a random sample of 8 customers during a single month.

x	250	300	340	360	385	400	450	475
y	72	81	90	94	102	106	115	124

The operator suggests that the data can be described using the linear regression model $y = 12.476 + 0.2311x$.

a Calculate the residual values for this model. **(2)**

b By considering the residuals, explain whether a linear model is a suitable model for these data. **(1)**

c Given that $S_{xx} = 38\,850$, $S_{yy} = 2090$ and $S_{xy} = 8980$, calculate the residual sum of squares (RSS). **(2)**

A second mobile phone operator records the number of minutes used and number of text messages sent for a random sample of its customers. Using a linear regression model, the RSS for this sample is 18.254.

d State, with a reason, which sample is more closely modelled by a linear regression model. **(1)**

← Sections 1.1, 1.2

(E/P) **6** A sociologist recorded the marriage rates per 1000 people, m, and divorce rates per 1000 people, d, from a random sample of 6 US states.

The data are summarised below:

$$\sum m = 101.9 \qquad \sum m^2 = 1768.47$$
$$\sum d = 49.7 \qquad \sum d^2 = 430.63$$
$$\sum md = 868.06$$

a Calculate S_{mm} and S_{md}. **(2)**

b Find the equation of the regression line of d on m. **(3)**

c Use your equation to estimate the divorce rate if the marriage rate is 15.5. **(1)**

d Calculate the residual sum of squares (RSS). **(3)**

The table shows the residual for each value of m.

m	Residual
13.7	−0.703 95
14.4	−0.347 4
16.9	1.468 85
17.1	0.642 15
18.6	x
21.2	−0.552

e Find the value of x. **(2)**

f By considering the signs of the residuals, explain whether or not the linear regression model is suitable for these data. **(1)**

← Sections 1.1, 1.2

 7 A random sample of 7 online companies was taken. The monthly amount of advertising expenditure, x, in £1000s, and the monthly sales, y, in £1000, was recorded. The results were as follows:

x	1.4	1.5	2.5	3.4	1.3	2.2	1.8
y	370	440	660	950	330	550	720

(You may use $S_{xx} = 3.388\,571$, $S_{yy} = 289\,771.4$, $S_{xy} = 895.571\,4$)

a Find the equation of the regression line of y on x in the form $y = a + bx$ as a model for these results. **(2)**

b Show that the residual sum of squares is 53 100, correct to three significant figures. **(2)**

c Calculate the residual values. **(2)**

d Write down the outlier. **(1)**

e Comment on the validity of ignoring this outlier. **(1)**

f Ignoring the outlier, produce another model. **(2)**

g Use this model to estimate the monthly sales for a company with advertising expenditure of £1800. **(1)**

h Comment, giving a reason, on the reliability of your estimate. **(1)**

← Sections 1.1, 1.2

(E/P) **8** An anthropologist uses a linear regression model to predict the height, h cm, of a male humanoid from the length, l cm, of the femur bone. She collects data from 8 skeletons.

l	h
50.2	178.6
48.4	173.7
45.3	164.9
44.8	163.8
44.6	168.4
42.8	165.1
39.6	155.5
38.1	155.1

The data collected can be summarised as follows:

$\sum l = 353.8$ $\sum h = 1325.1$
$\sum l^2 = 15\,762.5$ $\sum h^2 = 219\,944.9$
$\sum lh = 58\,825.04$

a Calculate the equation of the regression line of h on l, giving your answer in the form $h = a + bl$. **(3)**

b Use your regression line to predict the height of a male humanoid with femur length 45.1 cm. **(2)**

The table shows the residuals for each value of l.

l	h
50.2	1.47138
48.4	0.03296
45.3	−2.80543
44.8	−2.94388
44.6	2.04074
42.8	p
39.6	−1.24376
38.1	1.24089

c Find the value of p. **(2)**

d By considering the signs of the residuals, or otherwise, comment on the suitability of the linear regression model for these data. **(1)**

e Calculate the residual sum of squares (RSS). **(2)**

An equivalent random sample of female humanoids is taken and the residual sum of squares is found to be 25.467.

f State, with a reason, which sample is likely to have the best linear fit. **(1)**

← **Sections 1.1, 1.2**

(E) **9** The scatter diagrams below were drawn by a student.

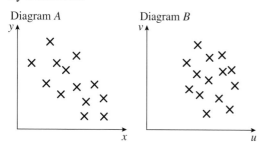

Diagram A

Diagram B

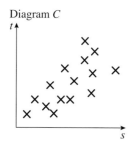

Diagram C

The student calculated the value of the product moment correlation coefficient for each of the sets of data.

The values were:

   0.68   −0.79   0.08

Write down, with a reason, which value corresponds to which scatter diagram. **(3)**

← **Section 2.1**

(E/P) **10** A young family were looking for a new three-bedroom semi-detached house. A local survey recorded the price, x, in £1000s, and the distance, y, in miles, from the nearest railway station of such houses. The following summary statistics were provided:

$S_{xx} = 113\,573$ $S_{yy} = 8.657$
$S_{xy} = -808.917$

a Use these values to calculate the product moment correlation coefficient. **(2)**

b Give an interpretation of your answer to part **a**. **(1)**

In another survey, the data for the same houses were supplied in km rather than miles.

c State the value of the product moment correlation coefficient in this case. **(1)**

← **Section 2.1**

(E/P) **11** As part of a statistics project, Gill collected data relating to the length of time, to the nearest minute, spent by shoppers in a supermarket and the amount of money they spent. Her data for a random sample of 10 shoppers are summarised in the table below, where t represents time and m represents the amount spent over £20.

t (minutes)	m (£)
15	−3
23	17
5	−19
16	4
30	12
6	−9
32	27
23	6
35	20
27	6

a Write down the actual amount spent by the shopper who was in the supermarket for 15 minutes. **(1)**

b Calculate S_{tt}, S_{mm} and S_{tm}.

(You may use $\sum t^2 = 5478$, $\sum m^2 = 2101$, and $\sum tm = 2485$) **(3)**

c Calculate the value of the product moment correlation coefficient between t and m. **(2)**

d Write down the value of the product moment correlation coefficient between t and the actual amount spent. Give a reason to justify your value. **(1)**

On another day Gill collected similar data. For these data, the product moment correlation coefficient was 0.178.

e Give an interpretation to both of these coefficients. **(2)**

f Suggest a practical reason why these two values are so different. **(1)**

← Section 2.1

(E) **12** During a village show, two judges, P and Q, had to award a mark out of 30 to some flower displays. The marks they awarded to a random sample of 8 displays are shown in the table below.

Display	A	B	C	D	E	F	G	H
Judge P	25	19	21	23	28	17	16	20
Judge Q	20	9	21	13	17	14	11	15

a Calculate Spearman's rank correlation coefficient for the marks awarded by the two judges. **(4)**

After the show, one competitor complained about the judges. She claimed that there was no positive correlation between their marks.

b Stating your hypotheses clearly, test whether this sample provides support for the competitor's claim.
Use a 5% level of significance. **(4)**

← Sections 2.2, 2.3

(E) **13** The table below shows the price of the same ice lolly at different stands on a beach, and the distance of each stand from the pier.

Stand	Distance from pier (m)	Price (£)
A	50	1.75
B	175	1.20
C	270	2.00
D	375	1.05
E	425	0.95
F	580	1.25
G	710	0.80
H	790	0.75
I	890	1.00
J	980	0.85

a Find, to 3 decimal places, the Spearman rank correlation coefficient between the distance of the stand from the pier and the price of the ice lolly. **(4)**

b Stating your hypotheses clearly and using a 5% significance level, test for negative rank correlation between price and distance. **(4)**

← Sections 2.2, 2.3

(E/P) **14** The numbers of deaths in one year from pneumoconiosis and lung cancer in a developing country are given in the table below.

Age group (years)	Deaths from pneumoconiosis (1000s)	Deaths from lung cancer (1000s)
20–29	12.5	3.7
30–39	5.9	9
40–49	18.5	10.2
50–59	19.4	19
60–69	31.2	13
70 and over	31	18

A charity claims that the relative vulnerabilities of different age groups are similar for both diseases.

a Give **one** reason to support the use Spearman's rank correlation coefficient in this instance. **(1)**

b Calculate Spearman's rank correlation coefficient for these data. **(4)**

c Test the charity's claim at the 5% significance level. State your hypotheses clearly. **(4)**

← **Sections 2.2, 2.3**

(E/P) **15** The product moment correlation coefficient is denoted by r and Spearman's rank correlation coefficient is denoted by r_s.

a Sketch separate scatter diagrams, with five points on each diagram, to show:
  i $r = 1$
  ii $r_s = -1$ but $r > -1$

Two judges rank seven collie dogs in a competition. The collie dogs are labelled A to G and the rankings are as follows.

Rank	1	2	3	4	5	6	7
Judge 1	A	C	D	B	E	F	G
Judge 2	A	B	D	C	E	G	F

b i Calculate Spearman's rank correlation coefficient for these data.
  ii Stating your hypotheses clearly, test, at the 5% level of significance, whether or not the judges are generally in agreement. **(8)**

← **Sections 2.1, 2.2, 2.3**

(E) **16** The masses of a reactant t mg and a product p mg in ten different instances of a chemistry experiment were recorded in a table.

t	p
1.2	3.8
1.9	7
3.2	11
3.9	12
2.5	9
4.5	12
5.7	13.5
4	12.2
1.1	2
5.9	13.9

(You may use $\sum t^2 = 141.51$, $\sum p^2 = 1081.74$ and $\sum tp = 386.32$)

a Draw a scatter diagram to represent these data. **(2)**

b State what is measured by the product moment correlation coefficient. **(1)**

c Calculate S_{tt}, S_{pp} and S_{tp}. **(3)**

d Calculate the value of the product moment correlation coefficient r between t and p. **(2)**

e Stating your hypotheses clearly, test, at the 1% significance level, whether or not the correlation coefficient is greater than zero. **(4)**

f With reference to your scatter diagram, comment on your result in part **e**. **(1)**

← **Sections 2.1, 2.3**

(E) **17** A geographer claims that the speed of flow of water in a river gets slower the wider the river is. He measures the width of the river, w metres, at seven points and records the rate of flow, f m s^{-1}.

Point	A	B	C	D	E	F	G
w	1.3	1.8	2.2	3.1	4.8	5.2	7.3
f	5.4	4.8	4.9	4.4	3.8	3.9	2.5

The Spearman's rank correlation coefficient between w and f is -0.93.

a Stating your hypotheses clearly, test whether or not the data provides support for the geographer's claim. Test at the 1% level of significance. **(4)**

b Without recalculating the correlation coefficient, explain how the Spearman's rank correlation coefficient would change if:
  i the speed of flow at G was actually 2.6 m s^{-1}
  ii an extra measurement, H was taken with a width of 0.8 m and a speed of flow of 6.2 m s^{-1}. **(3)**

The geographer collected data from a further 10 locations and found that there were now many tied ranks.

c Describe how you could find Spearman's rank correlation coefficient in this situation. **(2)**

← Sections 2.2, 2.3

(E) 18 A continuous random variable X has probability density function

$$f(x) = \begin{cases} k(4x - x^3) & 0 \leqslant x \leqslant 2 \\ 0 & \text{otherwise} \end{cases}$$

where k is a positive constant.

a Show that $k = \frac{1}{4}$ **(3)**

b Sketch $f(x)$. **(2)**

Find:

c $E(X)$ **(3)**

d the mode of X **(2)**

e the median of X **(3)**

f Comment on the skewness of the distribution. **(2)**

← Sections 3.1, 3.2, 3.3, 3.4

(E) 19 A continuous random variable X has probability density function $f(x)$ where

$$f(x) = \begin{cases} kx(x - 2) & 2 \leqslant x \leqslant 3 \\ 0 & \text{otherwise} \end{cases}$$

and k is a positive integer.

a Show that $k = \frac{3}{4}$ **(3)**

b Given that $E(X) = \frac{43}{16}$, find $Var(X)$. **(4)**

c Find the cumulative distribution function $F(x)$. **(4)**

d Show that the median value of X lies between 2.70 and 2.75. **(3)**

← Sections 3.1, 3.2, 3.3, 3.4

(E) 20 Ben attempts to model the continuous random variable Y with the cumulative distribution function:

$$F_1(y) = \begin{cases} 0 & y < 1 \\ 13y - 4y^2 - 9 & 1 \leqslant y \leqslant 2 \\ 1 & y > 2 \end{cases}$$

a Explain what is wrong with Ben's model.

Ben adapts his model to use the following cumulative distribution function.

$$F_2(y) = \begin{cases} 0 & y < 1 \\ k(y^4 + y^2 - 2) & 1 \leqslant y \leqslant 2 \\ 1 & y > 2 \end{cases}$$

Using Ben's second model,

b show that $k = \frac{1}{18}$ **(3)**

c find $P(Y > 1.5)$ **(2)**

d specify fully the probability density function $f(y)$. **(3)**

← Sections 3.1, 3.2

(E) 21 The continuous random variable X has probability density function $f(x)$ given by

$$f(x) = \begin{cases} 2(x - 2) & 2 \leqslant x \leqslant 3 \\ 0 & \text{otherwise} \end{cases}$$

a Sketch $f(x)$ for all values of x. **(2)**

b Write down the mode of X. **(1)**

c Given that $E(X) = \frac{8}{3}$, find $Var(X)$. **(3)**

d Find the median of X. **(3)**

e Comment on the skewness of this distribution. Give a reason for your answer. **(2)**

← Sections 3.1, 3.2, 3.3, 3.4

(E) 22 The continuous random variable X has probability density function given by

$$f(x) = \begin{cases} \frac{1}{6}x & 0 \leqslant x < 3 \\ 2 - \frac{1}{2}x & 3 \leqslant x \leqslant 4 \\ 0 & \text{otherwise} \end{cases}$$

a Sketch the probability density function of X. **(3)**

b Find the mode of X. **(1)**

c Specify fully the cumulative distribution function of X. **(4)**

d Using your answer to part c, find the median of X. **(3)**

e Find the 10th to 90th percentile range, giving your answer correct to three decimal places. **(4)**

← Sections 3.1, 3.2, 3.4

105

E **23** The continuous random variable X has cumulative distribution function

$$F(x) = \begin{cases} 0 & x < 0 \\ 2x^2 - x^3 & 0 \leqslant x \leqslant 1 \\ 1 & x > 1 \end{cases}$$

   a Find $P(X > 0.3)$. **(2)**

   b Verify that the median value of X lies between $x = 0.59$ and $x = 0.60$. **(3)**

   c Find the probability density function $f(x)$. **(3)**

   d Evaluate $E(X)$. **(3)**

   e Find the mode of X. **(2)**

   f Comment on the skewness of X. Justify your answer. **(2)**

   ← **Sections 3.1, 3.2, 3.3, 3.4**

A **E** **24** The continuous random variable X has probability density function given by

$$f(x) = \begin{cases} k & 0 \leqslant x \leqslant 2 \\ \dfrac{k}{x} & 2 < x \leqslant 4 \\ 0 & \text{otherwise} \end{cases}$$

   a Show that $k = \dfrac{1}{2 + \ln 2}$ **(4)**

   b Find $E(X)$. **(3)**

   ← **Sections 3.1, 3.3**

E **25** The random variable X is uniformly distributed over the interval $[-1, 5]$.

   a Sketch the probability density function $f(x)$ of X. **(2)**

   Find:

   b $E(X)$ **(2)**

   c $\text{Var}(X)$ **(2)**

   d $P(-0.3 < X < 3.3)$ **(2)**

   ← **Section 3.5**

E **26** The continuous random variable X is uniformly distributed over the interval $[2, 6]$.

   a Write down the probability density function $f(x)$. **(2)**

   Find:

   b $E(X)$ **(2)**

   c $\text{Var}(X)$ **(2)**

   d the cumulative distribution function of X, for all x **(2)**

   e $P(2.3 < X < 3.4)$ **(2)**

   ← **Section 3.5**

E **27** A string AB of length 5 cm is cut, in a random place C, into two pieces. The random variable X is the length of AC.

   a Write down the name of the probability distribution of X and sketch the graph of its probability density function. **(3)**

   b Find the values of $E(X)$ and $\text{Var}(X)$. **(4)**

   c Find $P(X > 3)$. **(2)**

   d Write down the probability that AC is exactly 3 cm long. **(1)**

   ← **Sections 3.5, 3.6**

E **28** The continuous random variable X is uniformly distributed over the interval $\alpha < x < \beta$.

   a Write down the probability density function of X, for all x. **(2)**

   b Given that $E(X) = 2$ and $P(X < 3) = \frac{5}{8}$, find the value of α and the value of β. **(3)**

   ← **Sections 3.5, 3.6**

E **29** A gardener has wire cutters and a piece of wire 150 cm long which has a ring attached at one end. The gardener cuts the wire, at a randomly chosen point, into 2 pieces. The length, in cm, of the piece of wire with the ring on it is represented by the random variable X.

   Find:

   a $E(X)$ **(2)**

   b the standard deviation of X **(2)**

   c the probability that the shorter piece of wire is at most 30 cm long. **(2)**

   ← **Sections 3.5, 3.6**

A
E/P
30 At a funfair, the duration B seconds of a ride on the Big Dipper has the normal distribution $N(82, 3^2)$. The duration F of a ride on the Ferris Wheel has the normal distribution $N(238, 7^2)$. Alice rides on the Big Dipper and the Ferris Wheel.

  a Find the probability that her ride on the Ferris Wheel is less than three times as long as her ride on the Big Dipper. **(6)**

  b State one assumption you have made and comment on its validity. **(1)**

Paul rides on the Big Dipper three times in a row. The random variable D represents the total duration of the three rides.

  c Find the distribution of D. **(3)**

Given that Alice starts a ride on the Ferris Wheel at the same time as Paul starts his three rides on the Big Dipper,

  d find the probability that Alice and Paul's rides finish within 10 seconds of one another. **(5)**

  ← Section 4.1

E **31** A workshop makes two types of electrical resistor.

The resistance, X ohms, of resistors of Type A is such that $X \sim N(20, 4)$.

The resistance, Y ohms, of resistors of Type B is such that $Y \sim N(10, 0.84)$.

When a resistor of each type is connected into a circuit, the resistance R ohms of the circuit is given by $R = X + Y$ where X and Y are independent.

Find:

  a $E(R)$ **(2)**

  b $Var(R)$ **(2)**

  c $P(28.90 < R < 32.64)$ **(3)**

  ← Section 4.1

E **32** The weights of adult men are normally distributed with a mean of 84 kg and a standard deviation of 11 kg.

A
  a Find the probability that the total weight of 4 randomly chosen adult men is less than 350 kg. **(3)**

The weights of adult women are normally distributed with a mean of 62 kg and a standard deviation of 10 kg.

  b Find the probability that the weight of a randomly chosen adult man is less than one and a half times the weight of a randomly chosen adult woman. **(4)**

  ← Section 4.1

E **33** The random variable D is defined as

$$D = A - 3B + 4C$$

where $A \sim N(5, 2^2)$, $B \sim N(7, 3^2)$ and $C \sim N(9, 4^2)$, and A, B and C are independent.

  a Find $P(D < 44)$. **(4)**

The random variables B_1, B_2 and B_3 are independent and each has the same distribution as B.

The random variable X is defined as

$$X = A - \sum_{i=1}^{3} B_i + 4C.$$

  b Find $P(X > 0)$. **(4)**

  ← Section 4.1

E **34** A manufacturer produces two flavours of soft drink, cola and lemonade. The weights, C and L, in grams, of randomly selected cola and lemonade cans are such that $C \sim N(350, 8)$ and $L \sim N(345, 17)$.

  a Find the probability that the weights of two randomly selected cans of cola will differ by more than 6 g. **(4)**

One can of each flavour is selected at random.

  b Find the probability that the can of cola weighs more than the can of lemonade. **(3)**

Cans are delivered to shops in boxes of 24 cans. The weights of empty boxes are normally distributed with mean 100 g and standard deviation 2 g.

c Find the probability that a full box of cola cans weighs between 8.51 kg and 8.52 kg. **(4)**

d State an assumption you made in your calculation in part **c**. **(1)**

← Sections 4.1

← Sections 4.1

Challenge

1 The table shows data that was collected from a scientific experiment:

x	1	3	4	5	7	8
y	1.5	3.3	5.3	7.5	13.8	16.8

a Use your calculator to find the following regression models for these data:
  i Linear ($y = a + bx$)
  ii Quadratic ($y = a + bx + cx^2$)
  iii Exponential ($y = ae^{bx}$ or $y = ab^x$)

b By calculating the residuals for each model, determine which model is most suitable.

← Sections 1.1, 1.2

2 A continuous random variable X has probability density function

$$f(x) = \begin{cases} ke^{-x} & x > 0 \\ 0 & \text{otherwise} \end{cases}$$

a Show that $k = 1$.

b Find the cumulative distribution function, $F(X)$.

c Hence, or otherwise, find the exact value of $P(1 < X < 4)$.

← Sections 3.1, 3.2

3 Let $X_1, X_2, X_3, \ldots X_n$ be identically distributed independent random variables, each with the continuous uniform distribution on $[0, 1]$. Define the random variable Y as the maximum value taken by each of the X_i.

a Show that $E(Y) = \dfrac{n}{n + 1}$

b Find an expression for the median of Y in terms of n.

If X and Y are independent continuous random variables with probability density functions $f(x)$ and $g(y)$ respectively, then the probability density function of $Z = X + Y$ is given by

$$h(z) = \int_{-\infty}^{\infty} f(z - t)g(t)\, dt$$

Find and sketch the probability density function of

c $X_1 + X_2$

d $X_1 + X_2 + X_3$.

← Section 3.5

Estimation, confidence intervals and tests using a normal distribution

5

In large-scale production processes it might be impossible to test every component. Engineers use samples to determine ranges of values that are likely to contain population parameters such as the mean or variance.

→ **Mixed exercise Q13**

Prior knowledge check

1 The independent normal random variables A and B have distributions $N(6, 2^2)$ and $N(7, 3^2)$ respectively.

  a Find $P(A > B)$.

  The random variable X is defined as $X = 3A + B$.

  b Find the distribution of X.   ← Chapter 4

2 A random sample of size 20 is taken from a population having a normal distribution with mean μ and standard deviation 1.5. The sample mean is 21.

  Test the following hypotheses at the 1% level of significance:

    $H_0: \mu = 20$, $H_1: \mu > 20$ ← SM2, Chapter 3

3 A sample of size 8 is taken from a population with distribution $N(6, 3^2)$, and the sample mean, $\overline{X}$, is calculated.

  Find $P(\overline{X} > 7)$.   ← FS1, Chapter 5

5.1 Estimators, bias and standard error

A

If you have a large population, for example the population of students in a school, or the population of trees in a forest, it is too time consuming and often costly to carry out a census to record, for example, the heights of all pupils or trees. In cases like these, **population parameters** such as the mean, μ, or the standard deviation, σ, are likely to be unknown.

In your A level course, you looked at methods of sampling that allow you to take a representative sample that can be used to estimate various population parameters.

> **Links** A census observes every member of a population, whereas a sample is a selection of observations taken from a subset of the population. There are various ways of selecting a random sample in practice. ← **SM1, Chapter 1**

A common way of estimating population parameters is to take a **random sample** from the population.

- **If X is a random variable then a random sample of size n will consist of n observations of the random variable X which are referred to as $X_1, X_2, X_3, \ldots, X_n$, where the X_i**
  - **are independent random variables**
  - **each have the same distribution as X.**

- **A statistic, T, is defined as a random variable consisting of any function of the X_i that involves no other quantities, such as unknown population parameters.**

> **Notation** X_i represents the ith observation of a sample. The specific value of the observation is denoted by x_i.

For example, $\overline{X}$, the sample mean, is a statistic whereas $\sum_{i=1}^{n} \dfrac{X_i^2}{n} - \mu^2$ is not a statistic since it involves the unknown population parameter μ.

Example 1

A sample, $X_1, X_2, \ldots, X_n$, is taken from a population with unknown population parameters μ and σ. State whether or not each of the following are statistics.

a $\dfrac{X_1 + X_3 + X_5}{3}$
b $\max(X_1, X_2, \ldots, X_n)$
c $\sum_{i=1}^{n} \left(\dfrac{X_i - \mu}{\sigma} \right)^2$

a $\dfrac{X_1 + X_3 + X_5}{3}$ is a statistic. ⟶ It is only a function of the sample $X_1, X_2, \ldots, X_n$. A statistic need not involve all members of the sample.

b $\max(X_1, X_2, \ldots, X_n)$ is a statistic. ⟶ It is only a function of the sample $X_1, X_2, \ldots, X_n$.

c $\sum_{i=1}^{n} \left(\dfrac{X_i - \mu}{\sigma} \right)^2$ is not a statistic. ⟶ The function contains μ and σ.

A Since it is possible to repeat the process of taking a sample, the specific value of a statistic T, namely t, will be different for each sample. If all possible samples are taken, these values will form a probability distribution called the **sampling distribution of** T.

- **The sampling distribution of a statistic** T **is the probability distribution of** T.

If the distribution of the population is known, then the sampling distribution of a statistic can sometimes be found.

Example 2

The weights, in grams, of a consignment of apples are normally distributed with a mean μ and standard deviation 4. A random sample of size 25 is taken and the statistics R and T are calculated as follows:

$$R = X_{25} - X_1 \text{ and } T = X_1 + X_2 + \dots + X_{25}$$

Find the distributions of R and T.

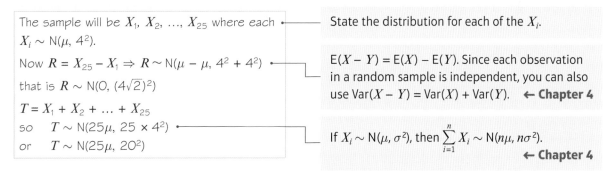

The sample will be $X_1, X_2, \dots, X_{25}$ where each —— State the distribution for each of the X_i.
$X_i \sim N(\mu, 4^2)$.

Now $R = X_{25} - X_1 \Rightarrow R \sim N(\mu - \mu, 4^2 + 4^2)$ —— $E(X - Y) = E(X) - E(Y)$. Since each observation
that is $R \sim N(0, (4\sqrt{2})^2)$ in a random sample is independent, you can also
use $\text{Var}(X - Y) = \text{Var}(X) + \text{Var}(Y)$. **← Chapter 4**

$T = X_1 + X_2 + \dots + X_{25}$

so $\quad T \sim N(25\mu, 25 \times 4^2)$ —— If $X_i \sim N(\mu, \sigma^2)$, then $\displaystyle\sum_{i=1}^{n} X_i \sim N(n\mu, n\sigma^2)$.

or $\quad T \sim N(25\mu, 20^2)$ **← Chapter 4**

Example 3

A large bag contains counters. 60% of the counters have the number 0 on them and 40% have the number 1.

a Find the population mean μ and population variance σ^2 of the values shown on the counters.

A simple random sample of size 3 is taken from this population.

b List all the possible observations from this sample.

c Find the sampling distribution for the mean

$$\overline{X} = \frac{X_1 + X_2 + X_3}{3}$$

where X_1, X_2 and X_3 are the values shown on the three counters in the sample.

d Hence find $E(\overline{X})$ and $\text{Var}(\overline{X})$.

e Find the sampling distribution for the sample mode, M.

f Hence find $E(M)$ and $\text{Var}(M)$.

A

a If X represents the value shown on a randomly chosen counter, then X has distribution

x	0	1
$P(X = x)$	$\frac{3}{5}$	$\frac{2}{5}$

$$\mu = E(X) = \sum_{\forall x} xP(X = x) = 0 + \frac{2}{5} \Rightarrow \mu = \frac{2}{5}$$

$$\sigma^2 = \text{Var}(X) = \sum_{\forall x} x^2P(X = x) - \mu^2 = 0 + 1^2 \times \frac{2}{5} - \frac{4}{25} \Rightarrow \sigma^2 = \frac{6}{25}$$

> Find μ and σ^2.
> ← FS1, Chapter 1

b The possible observations are

(0, 0, 0)

(1, 0, 0) (0, 1, 0) (0, 0, 1)

(1, 1, 0) (1, 0, 1) (0, 1, 1)

(1, 1, 1)

> List these systematically.

c $P(\overline{X} = 0) = \left(\frac{3}{5}\right)^3 = \frac{27}{125}$ i.e. the (0, 0, 0) case

$P\left(\overline{X} = \frac{1}{3}\right) = 3 \times \frac{2}{5} \times \left(\frac{3}{5}\right)^2 = \frac{54}{125}$ i.e. the (1, 0, 0), (0, 1, 0), (0, 0, 1) cases

$P\left(\overline{X} = \frac{2}{3}\right) = 3 \times \left(\frac{2}{5}\right)^2 \times \frac{3}{5} = \frac{36}{125}$ i.e. the (1, 1, 0), (1, 0, 1), (0, 1, 1) cases

$P(\overline{X} = 1) = \left(\frac{2}{5}\right)^3 = \frac{8}{125}$ i.e. the (1, 1, 1) case.

> Since the sample is random the observations are independent. So to find the probability of case (1, 0, 0) you can multiply the probabilities $P(X_1 = 1) \times P(X_2 = 0) \times P(X_3 = 0)$. Remember that each X_i has the same distribution as X.

So the distribution for $\overline{X}$ is

$\overline{x}$	0	$\frac{1}{3}$	$\frac{2}{3}$	1
$p(\overline{x})$	$\frac{27}{125}$	$\frac{54}{125}$	$\frac{36}{125}$	$\frac{8}{125}$

d $E(\overline{X}) = 0 + \frac{1}{3} \times \frac{54}{125} + \frac{2}{3} \times \frac{36}{125} + 1 \times \frac{8}{125} = \frac{18 + 24 + 8}{125} = \frac{2}{5}$

$\text{Var}(\overline{X}) = 0 + \frac{1}{9} \times \frac{54}{125} + \frac{4}{9} \times \frac{36}{125} + 1 \times \frac{8}{125} - \frac{4}{25} = \frac{6 + 16 + 8}{125} - \frac{20}{125} = \frac{2}{25}$

e The sample mode can take values 0 or 1.

$P(M = 0) = \frac{27}{125} + \frac{54}{125} = \frac{81}{125}$ i.e. cases (0, 0, 0), (1, 0, 0), (0, 1, 0), (0, 0, 1)

and $P(M = 1) = \frac{44}{125}$ i.e. the other cases.

so the distribution of M is

m	0	1
$p(m)$	$\frac{81}{125}$	$\frac{44}{125}$

> **Note** Notice that $E(\overline{X}) = \mu$ but $E(M) \neq \mu$ and that neither $E(\overline{X})$ nor $E(M)$ is equal to the population mode, which is of course zero as 60% of the counters have a zero on them.

f $E(M) = 0 + 1 \times \frac{44}{125} = \frac{44}{125}$

and $\text{Var}(M) = 0 + 1 \times \frac{44}{125} - \left(\frac{44}{125}\right)^2 = 0.228$

A In your A level course, you calculated the mean and variance of sets of sample data and used these as estimates for the equivalent population parameters.

- **A statistic that is used to estimate a population parameter is called an estimator and the particular value of the estimator generated from the sample taken is called an estimate.**

You need to be able to determine how **reliable** these sample statistics are as estimators for the corresponding population parameters.

Since all the X_i are random variables having the same mean and variance as the population, you can sometimes find expected values of a statistic T, $E(T)$, and this will tell you what the 'average' value of the statistic should be.

Example 4

A random sample $X_1, X_2, \ldots, X_n$ is taken from a population with $X \sim N(\mu, \sigma^2)$.
Show that $E(\overline{X}) = \mu$.

$$\overline{X} = \frac{1}{n}(X_1 + \ldots + X_n)$$

$$E(\overline{X}) = \frac{1}{n}E(X_1 + \ldots + X_n) \quad\longleftarrow\quad \text{Use } E(aX) = aE(X).$$

$$= \frac{1}{n}(E(X_1) + \ldots + E(X_n)) \quad\longleftarrow\quad E(X + Y) = E(X) + E(Y).$$

$$= \frac{1}{n}(\mu + \ldots + \mu)$$

$$= \frac{n\mu}{n}$$

$$E(\overline{X}) = \mu$$

This example shows that if you use the sample mean as an estimator of the population mean then 'on average' it will give the correct value.

This is an important property for an estimator to have. You say that $\overline{X}$ is an **unbiased estimator** of μ. A specific value of $\overline{x}$ will be an **unbiased estimate** for μ.

- **If a statistic T is used as an estimator for a population parameter θ and $E(T) = \theta$, then T is an unbiased estimator for θ.**

When selecting suitable estimators for population parameters, bias is one consideration. In Example 3, you found two statistics based on samples of size 3 from a population of counters of which 60% had the number 0 on them and 40% had the number 1. The population mean μ was $\frac{2}{5}$ and the population mode was 0 (since 60% of the counters had 0 on them). The two statistics that you calculated were the sample mean $\overline{X}$ and the sample mode M. You could use either of them as estimators for μ, the population mean, but you saw that $E(\overline{X}) = \mu$ and $E(M) \neq \mu$. So in this case, if you wanted an unbiased estimator for μ, you would choose the sample mean $\overline{X}$ rather than the sample mode M. How about an estimator for the population mode? Neither of the statistics that you calculated had the property of being unbiased since $E(\overline{X}) = \mu = \frac{2}{5}$ and $E(M) = \frac{44}{125}$ whereas the population mode was 0.

A Intuitively, you might prefer the estimator M since it is, after all, a mode and is also slightly closer to the population mode. In this case you refer to M as a **biased estimator** for the population mode. The **bias** is simply the expected value of the estimator minus the parameter of the population it is estimating.

- **If a statistic T is used as an estimator for a population parameter θ then the bias = E(T) − θ.**

In this case the bias is $\frac{44}{125}$

- **For an unbiased estimator, the bias is 0.**

You saw in Example 4 that the mean of a sample is an unbiased estimator for the population mean. If you take a sample X_1 of size one from a population with mean μ and variance σ^2, then the sample mean is $\overline{X} = X_1$, because there is only one value. So $E(\overline{X}) = E(X_1) = \mu$.

If you wanted to find an estimator for the **population variance**, you might try using the variance of the sample, $V = \dfrac{\sum(X_i - \overline{X})^2}{n}$.

For our sample X_1 of size one, the variance of the sample will be $\dfrac{(X_1 - \overline{X})^2}{1} = (X_1 - X_1)^2 = 0$.

> **Note** In general, the variance of the sample will be an **underestimate** for the variance of the population. This is because the statistic $\dfrac{\sum(X_i - \overline{X})^2}{n}$ uses the sample mean $\overline{X}$ rather than the population mean μ, and on average the sample observations will be closer to $\overline{X}$ than to μ.

So for a sample of size one, $E(V) = 0 \neq \sigma^2$. This illustrates that the variance of the sample is not an unbiased estimator for the variance of the population.

You can use a slightly different statistic, called the **sample variance**, as an unbiased estimator for the population variance.

> **Online** Explore biased and unbiased estimators using GeoGebra.

- **An unbiased estimator for σ^2 is given by the sample variance S^2 where**

$$S^2 = \frac{1}{n-1}\sum_{i=1}^{n}(X_i - \overline{X})^2$$

> **Notation** S^2 is the estimator (a random variable), and s^2 is the estimate (an observation from this random variable).

There are several ways to calculate the value of s^2 for a particular sample:

$$s^2 = \frac{1}{n-1}\sum_{i=1}^{n}(x_i - \overline{x})^2$$

$$= \frac{S_{xx}}{n-1}$$

$$= \frac{n}{n-1}\left(\frac{\sum x^2}{n} - \overline{x}^2\right)$$

$$= \frac{1}{n-1}\left(\sum x^2 - n\overline{x}^2\right)$$

> **Links** You can use the equivalence of these forms to show that s^2 is an unbiased estimate for σ^2. → **Exercise 5A Challenge**

The form that you use will depend on the information that you are given in the question.

Although a sample of size one can be used as an unbiased estimator of μ, it is clear that, in practice, a single observation from a population will not provide a useful estimate of the population mean. You need some way of differentiating between the **quality** of different unbiased estimators.

Example 5

A

A random sample X_1, X_2, ..., X_n is taken from a population with $X \sim N(\mu, \sigma^2)$.

Show that $\text{Var}(\overline{X}) = \dfrac{\sigma^2}{n}$

$$\overline{X} = \frac{1}{n}(X_1 + \dots + X_n)$$

$$\text{Var}(\overline{X}) = \frac{1}{n^2}\text{Var}(X_1 + \dots + X_n)$$ ——— Use $\text{Var}(aX) = a^2\text{Var}(X)$.

$$= \frac{1}{n^2}(\text{Var}(X_1) + \dots + \text{Var}(X_n))$$ ——— Use $\text{Var}(X + Y) = \text{Var}(X) + \text{Var}(Y)$.

$$= \frac{1}{n^2}(\sigma^2 + \dots + \sigma^2)$$

$$= \frac{n\sigma^2}{n^2}$$

$$\text{Var}(\overline{X}) = \frac{\sigma^2}{n}$$

One reason that the **sample mean** is used as an estimator for μ is that the variance of the estimator $\text{Var}(\overline{X}) = \dfrac{\sigma^2}{n}$ decreases as n increases. For larger values of n, the value of an estimate is more likely to be close to the population mean. So the greater the value of n, the **better** the estimator is.

- **The standard deviation of an estimator is called the standard error of the estimator.**

When you are using the sample mean, $\overline{X}$, you can use the following result for the standard error.

- **Standard error of $\overline{X} = \dfrac{\sigma}{\sqrt{n}}$ or $\dfrac{s}{\sqrt{n}}$**

Watch out Although in general $\sigma \neq s$, you can use the second version of this standard error in situations where you **do not know** the population standard deviation.

Example 6

The table below summarises the number of breakdowns, X, on a town's bypass on 30 randomly chosen days.

Number of breakdowns	2	3	4	5	6	7	8	9
Number of days	3	5	4	3	5	4	4	2

a Calculate unbiased estimates of the mean and variance of the number of breakdowns.

20 more days were randomly sampled and this sample had $\overline{x} = 6.0$ days and $s^2 = 5.0$.

b Treating the 50 results as a single sample, obtain further unbiased estimates of the population mean and variance.

c Find the standard error of this new estimate of the mean.

d Estimate the size of sample required to achieve a standard error of less than 0.25.

Notation 'Hat' notation is used to describe an estimate of a parameter.
For example:
$\hat{\sigma}^2$ represents an estimate for the population variance σ^2.
$\hat{\mu}$ represents an estimate for the population mean μ.

115

A

a By calculator,

$$\sum x = 160 \text{ and } \sum x^2 = 990$$

So $\quad \hat{\mu} = \bar{x} = \dfrac{160}{30} = 5.33$

and $\quad \hat{\sigma}^2 = s_x^2 = \dfrac{990 - 30\bar{x}^2}{29}$

$\qquad\qquad = 4.7126 = 4.71$ (3 s.f.)

> Use $s^2 = \dfrac{1}{n-1}\left(\sum x^2 - n\bar{x}^2\right)$ since you have values for $\sum x^2$ and $\bar{x}$.

> You can use your calculator to find unbiased estimates of the mean and variance but you should show your working out in the exam.

b New sample: $\bar{y} = 6.0 \Rightarrow \sum y = 20 \times 6.0 = 120$

$$s_y^2 = 5.0 \Rightarrow \dfrac{\sum y^2 - 20 \times 6^2}{19} = 5$$

So $\quad \sum y^2 = 5 \times 19 + 20 \times 36$

$\Rightarrow \quad \sum y^2 = 815$

So the combined sample (w) of size 50 has

$$\sum w = 160 + 120 = 280$$
$$\sum w^2 = 990 + 815 = 1805$$

Then the combined estimate of μ is

$$\bar{w} = \dfrac{280}{50} = 5.6$$

and the estimate for σ^2 is

$$s_w^2 = \dfrac{1805 - 50 \times 5.6^2}{49}$$

$\Rightarrow s_w^2 = 4.8367\ldots = 4.84$ (3 s.f.)

> **Problem-solving**
>
> First you need to 'unwrap' the formulae for $\bar{y}$ and s_y^2 to find $\sum y$ and $\sum y^2$.

> Now combine with $\sum x$ and $\sum x^2$. Let the combined variable be w.

c The best estimate of σ^2 will be s_w^2 since it is based on a larger sample than s_x^2 or s_y^2.

So the standard error is $\dfrac{s_w}{\sqrt{50}} = \sqrt{\dfrac{4.836\ldots}{50}}$

$\qquad\qquad\qquad\qquad = 0.311$ (3 s.f.)

> Use the $\dfrac{s}{\sqrt{n}}$ formula for standard error.

d To achieve a standard error < 0.25 you require

$$\sqrt{\dfrac{4.836\ldots}{n}} < 0.25$$

$\Rightarrow \qquad \sqrt{n} > \dfrac{\sqrt{4.836\ldots}}{0.25}$

$\qquad\qquad \sqrt{n} > 8.797\ldots$

$\Rightarrow \qquad\quad n > 77.38\ldots$

So we need a sample size of at least 78.

> You do not know the value for σ so you will have to use your best estimate of it, namely s_w.

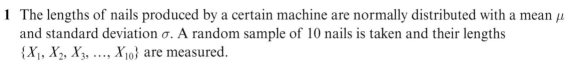
Exercise 5A

A

1 The lengths of nails produced by a certain machine are normally distributed with a mean μ and standard deviation σ. A random sample of 10 nails is taken and their lengths $\{X_1, X_2, X_3, ..., X_{10}\}$ are measured.

  i Write down the distributions of the following:

    a $\displaystyle\sum_{1}^{10} X_i$ **b** $\dfrac{2X_1 + 3X_{10}}{5}$ **c** $\displaystyle\sum_{1}^{10}(X_i - \mu)$

    d $\overline{X}$ **e** $\displaystyle\sum_{1}^{5} X_i - \sum_{6}^{10} X_i$ **f** $\displaystyle\sum_{1}^{10}\left(\dfrac{X_i - \mu}{\sigma}\right)$

  ii State which of the above are statistics.

2 A large bag of coins contains 1p, 5p and 10p coins in the ratio $2:2:1$.

  a Find the mean μ and the variance σ^2 for the value of coins in this population.

  A random sample of two coins is taken and their values X_1 and X_2 are recorded.

  b List all the possible observations from this sample.

  c Find the sampling distribution for the mean $\overline{X} = \dfrac{X_1 + X_2}{2}$

  d Hence show that $E(\overline{X}) = \mu$ and $\text{Var}(\overline{X}) = \dfrac{\sigma^2}{2}$

3 Find unbiased estimates of the mean and variance of the populations from which the following random samples have been taken.

  a 21.3; 19.6; 18.5; 22.3; 17.4; 16.3; 18.9; 17.6; 18.7; 16.5; 19.3; 21.8; 20.1; 22.0

  b 1; 2; 5; 1; 6; 4; 1; 3; 2; 8; 5; 6; 2; 4; 3; 1

  c 120.4; 230.6; 356.1; 129.8; 185.6; 147.6; 258.3; 329.7; 249.3

  d 0.862; 0.754; 0.459; 0.473; 0.493; 0.681; 0.743; 0.469; 0.538; 0.361

4 Find unbiased estimates of the mean and the variance of the populations from which random samples with the following summaries have been taken.

  a $n = 120$ $\sum x = 4368$ $\sum x^2 = 162\,466$

  b $n = 30$ $\sum x = 270$ $\sum x^2 = 2546$

  c $n = 1037$ $\sum x = 1140.7$ $\sum x^2 = 1278.08$

  d $n = 15$ $\sum x = 168$ $\sum x^2 = 1913$

E

5 The concentrations, in mg per litre, of a trace element in 7 randomly chosen samples of water from a spring were:

      240.8   237.3   236.7   236.6   234.2   233.9   232.5.

  a Explain what is meant by an unbiased estimator.   **(1 mark)**

  b Determine unbiased estimates of the mean and the variance of the concentration of the trace element per litre of water from the spring.   **(4 marks)**

6 Cartons of orange juice are filled by a machine. A sample of 10 cartons selected at random from the production contained the following quantities of orange juice (in ml).

201.2   205.0   209.1   202.3   204.6   206.4   210.1   201.9   203.7   207.3

Calculate unbiased estimates of the mean and variance of the population from which this sample was taken. **(4 marks)**

7 A manufacturer of self-assembly furniture required bolts of two lengths, 5 cm and 10 cm, in the ratio 2 : 1 respectively.

a Find the mean μ and the variance σ^2 for the lengths of bolts in this population.

A random sample of three bolts is selected from a large box containing bolts in the required ratio.

b List all the possible observations from this sample.

c Find the sampling distribution for the mean $\overline{X}$.

d Hence find $E(\overline{X})$ and $Var(\overline{X})$.

e Find the sampling distribution for the mode M.

f Hence find $E(M)$ and $Var(M)$.

g Find the bias when M is used as an estimator of the population mode.

8 A biased six-sided dice has probability p of landing on a six.

Every day, for a period of 25 days, the dice is rolled 10 times and the number of sixes X is recorded, giving rise to a sample X_1 , X_2 , ..., X_{25}.

a Write down $E(X)$ in terms of p.

b Show that the sample mean $\overline{X}$ is a biased estimator of p and find the bias.

c Suggest a suitable unbiased estimator of p.

9 The random variable $X \sim U[-\alpha, \alpha]$.

a Find $E(X)$ and $E(X^2)$.

A random sample X_1, X_2, X_3 is taken and the statistic $Y = X_1^2 + X_2^2 + X_3^2$ is calculated.

b Show that Y is an unbiased estimator of α^2.

10 John and Mary each independently took a random sample of sixth-formers in their college and asked them how much money, in pounds, they earned last week. John used his sample of size 20 to obtain unbiased estimates of the mean and variance of the amount earned by a sixth-former at their college last week. He obtained values of $\overline{x} = 15.5$ and $s_x^2 = 8.0$.

Mary's sample of size 30 can be summarised as $\sum y = 486$ and $\sum y^2 = 8222$.

a Use Mary's sample to find unbiased estimates of μ and σ^2. **(2 marks)**

b Combine the samples and use all 50 observations to obtain further unbiased estimates of μ and σ^2. **(4 marks)**

c Explain what is meant by standard error. **(1 mark)**

d Find the standard error of the mean for each of these estimates of μ. **(2 marks)**

e Comment on which estimate of μ you would prefer to use. **(1 mark)**

A **11** A machine operator checks a random sample of 20 bottles from a production line in order to estimate the mean volume of bottles (in cm³) from this production run. The 20 values can be summarised as $\sum x = 1300$ and $\sum x^2 = 84\,685$.

E/P

a Use this sample to find unbiased estimates of μ and σ^2. **(2 marks)**

A supervisor knows from experience that the standard deviation of volumes on this process, σ, should be 3 cm³ and he wishes to have an estimate of μ that has a standard error of less than 0.5 cm³.

b Recommend a sample size for the supervisor, showing working to support your recommendation. **(2 marks)**

c Does your recommended sample size guarantee a standard error of less than 0.5 cm³? Give a reason for your answer. **(1 mark)**

The supervisor takes a further sample of size 16 and finds $\sum x = 1060$.

d Combine the two samples to obtain a revised estimate of μ. **(2 marks)**

E **12** The heights of certain seedlings after growing for 10 weeks in a greenhouse have a standard deviation of 2.6 cm. Find the smallest sample that must be taken for the standard error of the mean to be less than 0.5 cm. **(3 marks)**

E **13** The hardness of a plastic compound was determined by measuring the indentation produced by a heavy pointed device.

The following observations in tenths of a millimetre were obtained:

4.7, 5.2, 5.4, 4.8, 4.5, 4.9, 4.5, 5.1, 5.0, 4.8.

a Estimate the mean indentation for this compound. **(1 mark)**

b Find the standard error for your estimate. **(2 marks)**

c Estimate the size of sample required in order that in future the standard error of the mean should be just less than 0.05. **(3 marks)**

P **14** Prospective army recruits receive a medical test. The probability of each recruit passing the test is p, independent of any other recruit. The medicals are carried out over two days and on the first day n recruits are seen and on the next day $2n$ are seen. Let X_1 be the number of recruits who pass the test on the first day and let X_2 be the number who pass on the second day.

a Write down $E(X_1)$, $E(X_2)$, $Var(X_1)$ and $Var(X_2)$.

b Show that $\dfrac{X_1}{n}$ and $\dfrac{X_2}{2n}$ are both unbiased estimates of p and state, giving a reason, which you would prefer to use.

c Show that $X = \dfrac{1}{2}\left(\dfrac{X_1}{n} + \dfrac{X_2}{2n}\right)$ is an unbiased estimator of p.

d Show that $Y = \left(\dfrac{X_1 + X_2}{3n}\right)$ is an unbiased estimator of p.

e Which of the statistics $\dfrac{X_1}{n}, \dfrac{X_2}{2n}, X$ or Y is the best estimator of p?

The statistic $T = \left(\dfrac{2X_1 + X_2}{3n}\right)$ is proposed as an estimator of p.

f Find the bias.

15 A large bag of counters has 40% with the number 0 on, 40% with the number 2 and 20% with the number 1.

a Find the mean μ and the variance σ^2 for this population of counters.

A random sample of size 3 is taken from the bag.

b List all the possible observations from this sample.

c Find the sampling distribution for the mean $\overline{X}$.

d Find $E(\overline{X})$ and $Var(\overline{X})$.

e Find the sampling distribution for the median N.

f Hence find $E(N)$ and $Var(N)$.

g Show that N is an unbiased estimator of μ.

h Explain which estimator, $\overline{X}$ or N, you would choose as an estimator of μ.

> **Challenge**
>
> **a** Show that $\dfrac{1}{n-1}\displaystyle\sum_{i=1}^{n}(x_i - \overline{x})^2 = \dfrac{1}{n-1}\left(\sum x^2 - n\overline{x}^2\right)$.
>
> **b** Hence, or otherwise, show that s^2 is an unbiased estimate for the population variance σ^2.

5.2 Confidence intervals

The value of $\hat{\theta}$ which is found from a sample and used as an unbiased estimate for the population parameter θ is very unlikely to be exactly equal to θ.

There is no way of establishing, from the sample data only, how close the estimate is.

You can instead form a **confidence interval** for θ.

- **A confidence interval (C.I.) for a population parameter θ is a range of values defined so that there is a specific probability that the true value of the parameter lies within that range.**

For example, you could establish a 90% confidence interval, or a 95% confidence interval.

A 95% confidence interval is an interval such that there is a 0.95 probability that the interval contains θ.

> **Watch out** The population parameter, θ, is fixed, so you cannot talk about its value in probabilistic terms.

Different samples will generate different confidence intervals since estimates for the parameter will change based on the data i n the sample and the sample size.

For large n, you know from the central limit theorem that whatever the distribution of the random variable X, the sample mean will be approximately normally distributed:

> **Links** This approximation improves for larger samples and is exact if the population is normally distributed.
> ← FS1, Chapter 5

$$\overline{X} \sim N\left(\mu, \frac{\sigma^2}{n}\right)$$

Hence, if you know the population standard deviation, you can establish a confidence interval for the population mean, μ, using the standardised normal distribution.

Example 7

A

Show that a 95% confidence interval for μ, based on a sample of size n, is given by

$$\left(\overline{x} - 1.96 \times \frac{\sigma}{\sqrt{n}}, \overline{x} + 1.96 \times \frac{\sigma}{\sqrt{n}}\right)$$

$\overline{X}$ is approximately $\sim N\left(\mu, \frac{\sigma^2}{\sqrt{n}}\right)$

and therefore

$$Z = \frac{\overline{X} - \mu}{\frac{\sigma}{\sqrt{n}}} \sim N(0, 1^2)$$

Using tables, or your calculator, you can see that for the $N(0, 1^2)$ distribution

$$P(Z > 1.9600) = P(Z < -1.9600) = 0.025$$

and so 95% of the distribution is between −1.9600 and 1.9600.

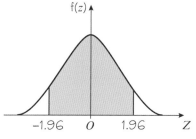

So $P(-1.96 < Z < 1.96) = 0.95$

$$\Rightarrow P\left(-1.96 < \frac{\overline{X} - \mu}{\frac{\sigma}{\sqrt{n}}} < 1.96\right) = 0.95$$

Look at the inequality inside the probability statement:

$$-1.96 \times \frac{\sigma}{\sqrt{n}} < \overline{X} - \mu < 1.96 \times \frac{\sigma}{\sqrt{n}}$$

$$\overline{X} + 1.96 \times \frac{\sigma}{\sqrt{n}} > \mu > \overline{X} - 1.96 \times \frac{\sigma}{\sqrt{n}}$$

$$\overline{X} - 1.96 \times \frac{\sigma}{\sqrt{n}} < \mu < \overline{X} + 1.96 \times \frac{\sigma}{\sqrt{n}}$$

So the 95% confidence interval for μ is

$$\left(\overline{x} - 1.96 \times \frac{\sigma}{\sqrt{n}}, \overline{x} + 1.96 \times \frac{\sigma}{\sqrt{n}}\right)$$

Problem-solving

You will need to use the **standardised** normal distribution $N(0, 1^2)$ to tackle problems like this.

If $X \sim N(\mu, \sigma^2)$ then $Z = \frac{X - \mu}{\sigma} \sim N(0, 1^2)$.

← **SM2, Chapter 3**

Whatever the distribution of the population, you know by the central limit theorem that $\overline{X}$ will be approximately normal.

Start to isolate μ.

Multiply by −1 and change the inequalities.

Notation The upper and lower values of a confidence interval are sometimes called the **confidence limits**.

- **The 95% confidence interval for μ is $\left(\overline{x} - 1.96 \times \dfrac{\sigma}{\sqrt{n}}, \overline{x} + 1.96 \times \dfrac{\sigma}{\sqrt{n}}\right)$.**

The 1.96 in the formula above is determined by the percentage points of the standardised normal distribution. By changing this value you can formulate confidence intervals with different levels of confidence.

A For example, a 99% confidence interval would see 1.96 replaced by 2.5758, since that is the value of z such that $P(-z < Z < z) = 0.99$.

Note The choice of what level of confidence to use in a particular situation will depend on the problem involved but a value of 95% is commonly used if no other value is specified.

Interpreting confidence intervals

- First, it is important to remember that μ is a fixed, but unknown, number and as such it cannot vary and does not have a distribution.
- Secondly, it is worth remembering that you base a 95% confidence interval on a probability statement about the normal distribution $Z \sim N(0, 1^2)$.
- However, although you start by considering probabilities associated with the random variable Z, the final confidence interval does not tell you the probability that μ lies inside a fixed interval. Rather, since μ is fixed, it is the confidence interval that varies (according to the value of $\bar{x}$).
- What a 95% confidence interval tells you is that the probability that **the interval contains** μ is 0.95.

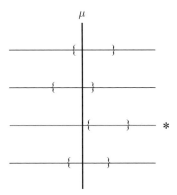

The diagram opposite illustrates the 95% confidence intervals calculated from different samples and also shows the position of μ. Suppose 20 samples of size 100 were taken and 95% confidence intervals for μ were calculated for each sample. This would give 20 different confidence intervals, each based on one of the 20 different values of $\bar{x}$. If you imagine for a moment that you actually do know what the value of μ is then you can plot each of these confidence intervals on a diagram similar to the one here; you would expect that 95% of these confidence intervals would contain the value μ but about once in every 20 times you would get an interval which did not contain μ (like the one marked * here). The problem for the statistician is that they never know whether the confidence interval they have just calculated is one that contains μ or not.

However, 95% (or 90% or 99%, depending on the degree of confidence required) of the time the interval will contain μ.

Example 8

The breaking strains of reels of string produced at a certain factory have a standard deviation of 1.5 kg. A sample of 100 reels from a certain batch were tested and their mean breaking strain was 5.30 kg.

a Find a 95% confidence interval for the mean breaking strain of string in this batch.

The manufacturer becomes concerned if the lower 95% confidence limit falls below 5 kg. A sample of 80 reels from another batch gave a mean breaking strain of 5.31 kg.

b Will the manufacturer be concerned?

The distribution for breaking strains is not known but the sample is quite large and by the central limit theorem $\bar{X}$ will be approximately normally distributed.

a 95% confidence limits are

$$\bar{x} \pm 1.96 \times \frac{\sigma}{\sqrt{n}} = 5.30 \pm 1.96 \times \frac{1.5}{\sqrt{100}}$$

So a 95% confidence interval is (5.006, 5.594).

b Lower 95% confidence limit is

$$\bar{x} - 1.96 \times \frac{\sigma}{\sqrt{n}} = 5.31 - \frac{1.96 \times 1.5}{\sqrt{80}}$$
$$= 4.98$$

so the manufacturer will be concerned.

> Use the $x \pm 1.96 \times \frac{\sigma}{\sqrt{n}}$ formula.

Notation In your exam you can define a confidence interval
- by giving the confidence limits,
  e.g. 5.30 ± 0.294
- using interval notation,
  e.g. (5.006, 5.594) or [5.006, 5.594]
- using inequalities,
  e.g. $5.006 < \mu < 5.594$.

- **The width of a confidence interval is the difference between the upper confidence limit and the lower confidence limit. This is $2 \times z \times \frac{\sigma}{\sqrt{n}}$, where z is the relevant percentage point from the standardised normal distribution, for example 1.96, 1.6449, etc.**

The greater the width, the less information you have about the population mean. There are three factors that affect the width: the value of σ, the size of the sample n and the degree of confidence required. In a particular example where σ and n are determined, the only factor you can change to alter the width is the degree of confidence. A high level of confidence (e.g. 99%) will give a greater width than a lower level of confidence (e.g. 90%), and the statistician has to weigh up the advantages of high confidence against greater width when calculating a confidence interval.

Example 9

A random sample of size 25 is taken from a normal population with standard deviation 2.5. The mean of the sample is 17.8.

a Find a 99% C.I. for the population mean μ.

b What size sample is required to obtain a 99% C.I. of width of at most 1.5?

c What confidence level would be associated with the interval based on the above sample of 25 but of width 1.5, i.e. (17.05, 18.55)?

a 99% confidence limits are

$$x \pm 2.5758 \times \frac{\sigma}{\sqrt{n}} = 17.8 \pm 2.5758 \times \frac{2.5}{\sqrt{25}}$$

So a 99% confidence interval is (16.51, 19.09).

> Use the table on page 214 to find 2.5758.

b Width of 99% C.I. is $2 \times 2.5758 \times \frac{2.5}{\sqrt{n}}$

so you require $1.5 > \frac{12.879...}{\sqrt{n}}$

i.e. $n > 73.719...$

so you need $n = 74$

> Use the $2 \times z \times \frac{\sigma}{\sqrt{n}}$ formula or the definition for the width.

A

c A width of 1.5 $\Rightarrow 1.5 = 2 \times z \times \dfrac{2.5}{\sqrt{25}}$

$$z = 1.5$$

From the table on page 214 you find that

$$P(Z < 1.5) = 0.9332$$

and so $P(Z > 1.5) = P(Z < -1.5)$
$$= 1 - 0.9332$$
$$= 0.0668$$

So the confidence level is
$$100 \times (1 - 2 \times 0.0668) = 86.6\%.$$

The percentage of the confidence interval is given by the area between $z = \pm1.5$.

Exercise 5B

1 A random sample of size 9 is taken from a normal distribution with variance 36. The sample mean is 128.
 a Find a 95% confidence interval for the mean μ of the distribution.
 b Find a 99% confidence interval for the mean μ of the distribution.

2 A random sample of size 25 is taken from a normal distribution with standard deviation 4. The sample mean is 85.
 a Find a 90% confidence interval for the mean μ of the distribution.
 b Find a 95% confidence interval for the mean μ of the distribution.

E/P 3 A random sample is taken from a distribution with mean μ and variance 4.41. The sample has the following values:

23.1, 21.8, 24.6, 22.5.

Niall says that even though the sample is small, he can still use the normal distribution to obtain confidence limits for μ.
 a State what assumption Niall must make in order for this to be true. **(1 mark)**
 b Given that this assumption is true, use the sample to find 98% confidence limits for the mean μ. **(3 marks)**

4 A normal distribution has standard deviation 15. Estimate the sample size required if the following confidence intervals for the mean should have width of less than 2.
 a 90% **b** 95% **c** 99%

A **5** An experienced poultry farmer knows that the mean mass μ kg for a large population of
E chickens will vary from season to season but the standard deviation of the masses should
remain at 0.70 kg. A random sample of 100 chickens is taken from the population and the mass
X kg of each chicken in the sample is recorded, giving $\sum x = 190.2$.

    a Explain what is meant by a 95% confidence interval for a population parameter θ.   **(1 mark)**

    b Find a 95% confidence interval for μ.   **(3 marks)**

E/P **6** A railway watchdog is studying the number of seconds that express trains are late to arrive.
Previous surveys have shown that the standard deviation is 50. A random sample of 200 trains
was selected and gave rise to a mean of 310 seconds late.

    a Find a 90% confidence interval for μ, the mean number of seconds that express
    trains are late.   **(3 marks)**

    Five different independent random samples of 200 trains are selected, and each sample is used
    to generate a different 90% confidence interval for μ.

    b Find the probability that exactly three of these | **Hint** Use a suitable binomial distribution.
    confidence intervals contain μ.   **(2 marks)**

E/P **7** Amy is investigating the total distance travelled by lorries in current use. The standard deviation
can be assumed to be 15 000 km. A random sample of 80 lorries was stopped and their mean
distance travelled was found to be 75 872 km.

    Amy suspects that the population is not normally distributed, but claims that she can still use
    the normal distribution to find a confidence interval for μ.

    a State, with a reason, whether Amy is correct.   **(2 marks)**

    b Find a 90% confidence interval for the mean distance travelled by lorries in
    current use.   **(3 marks)**

E/P **8** It is known that each year the standard deviation of the marks in a certain examination is 13.5
but the mean mark μ will fluctuate. An examiner wants to estimate the mean mark of all the
candidates on the examination but he only has the marks of a sample of 250 candidates, which
gives a sample mean of 68.4.

    a What assumption about the candidates must the examiner make in order to use this
    sample mean to calculate a confidence interval for μ?   **(1 mark)**

    b Assuming that the above assumption is justified, calculate a 95% confidence interval
    for μ.   **(3 marks)**

    Later, the examiner discovers that the actual value of μ was 65.3.

    c What conclusions might the examiner draw about his sample?   **(2 marks)**

E/P **9** The battery life in hours of a new model of mobile phone in standby mode is modelled as
having a uniform distribution on $[\mu - 10, \mu + 10]$, where the value of μ is not known.

    a Show that the variance of the battery life is $\frac{100}{3}$ **(3 marks)**

    A random sample of 120 phones was tested and the mean battery life on standby was
    78.7 hours.

    b Find a 95% confidence interval for μ.   **(3 marks)**

A **10** A statistics student calculated 95% and 99% confidence intervals for the mean μ of a certain

E/P population but failed to label them. The two intervals were (22.7, 27.3) and (23.2, 26.8).

    a State, with a reason, which interval is the 95% one. **(1 mark)**

    b Estimate the standard error of the mean in this case. **(2 marks)**

    c What was the student's unbiased estimate of the mean μ in this case? **(2 marks)**

E **11** The managing director of a firm has commissioned a survey to estimate the mean expenditure
of customers on electrical appliances. A random sample of 100 people were questioned
and the research team presented the managing director with a 95% confidence interval of
(£128.14, £141.86).

    The director says that this interval is too wide and wants a confidence interval of total width £10.

    a Using the same value of $\bar{x}$, find the confidence limits in this case. **(3 marks)**

    b Find the level of confidence for the interval in part **a**. **(2 marks)**

    The managing director is still not happy and now wishes to know how large a sample would be
required to obtain a 95% confidence interval of total width no greater than £10.

    c Find the smallest size of sample that will satisfy this request. **(3 marks)**

E **12** A plant produces steel sheets whose masses are known to be normally distributed with a
standard deviation of 2.4 kg. A random sample of 36 sheets had a mean mass of 31.4 kg.
Find 99% confidence limits for the population mean. **(3 marks)**

E **13** A machine is regulated to dispense liquid into cartons in such a way that the amount of liquid
dispensed on each occasion is normally distributed with a standard deviation of 20 ml.

    Find 99% confidence limits for the mean amount of liquid dispensed if a random sample of
40 cartons had an average content of 266 ml. **(3 marks)**

E/P **14** **a** The error made when a certain instrument is used to measure the body length of a butterfly
of a particular species is known to be normally distributed with mean 0 and standard
deviation 1 mm. Calculate, to 3 decimal places, the probability that the size of the
error made when the instrument is used once is less than 0.4 mm. **(2 marks)**

    b Given that the body length of a butterfly is measured 9 times with the instrument,
calculate, to 3 decimal places, the probability that the mean of the 9 readings will be
within 0.5 mm of the true length. **(3 marks)**

    c Given that the mean of the 9 readings was 22.53 mm, determine a 98% confidence
interval for the true body length of the butterfly. **(3 marks)**

5.3 Hypothesis testing for the difference between means

A

You need to be able to carry out a hypothesis test for the **difference** between the means of two normal distributions with known variances.

Links In your A level course you used the sample mean to carry out hypothesis tests for the mean of a single normal distribution. ← SM2, Section 3.7

If, instead of one population, you now have two independent populations then you can test hypotheses about the difference between the population means.

In Chapter 4 you saw tha,t if X and Y are two independent normal distributions with means of μ_x and μ_y and standard deviations σ_x and σ_y respectively, then

$$X - Y \sim N(\mu_x - \mu_y, \sigma_x^2 + \sigma_y^2)$$

Now, if $\overline{X}$ and $\overline{Y}$ are sample means based on samples of size n_x and n_y respectively from the above two normal populations, then

$$\overline{X} - \overline{Y} \sim N\left(\mu_x - \mu_y, \frac{\sigma_x^2}{n_x} + \frac{\sigma_y^2}{n_y}\right)$$

and the statistic $\overline{X} - \overline{Y}$ can be used to test hypotheses about the values of μ_x and μ_y.

The central limit theorem tells you that, provided the sample sizes n_x and n_y are large, $\overline{X} - \overline{Y}$ will have a normal distribution whatever the distributions of X and Y. You can therefore use this to test whether there is a significant difference between the means of any two populations. The usual null hypothesis is that the values of μ_x and μ_y are equal, but other situations are possible provided that the null hypothesis gives you a value for $\mu_x - \mu_y$.

The test statistic you will need to use is based upon the distribution of $\overline{X} - \overline{Y}$ and is

$$Z = \frac{\overline{X} - \overline{Y} - (\mu_x - \mu_y)}{\sqrt{\dfrac{\sigma_x^2}{n_x} + \dfrac{\sigma_y^2}{n_y}}}$$

Note The formula for the test statistic is in the formula book.

- **Test for difference between two means:**

  - **If $X \sim N(\mu_x, \sigma_x^2)$ and the independent random variable $Y \sim N(\mu_y, \sigma_y^2)$, then a test of the null hypothesis H_0: $\mu_x = \mu_y$ can be carried out using the test statistic**

    $$Z = \frac{\overline{X} - \overline{Y} - (\mu_x - \mu_y)}{\sqrt{\dfrac{\sigma_x^2}{n_x} + \dfrac{\sigma_y^2}{n_y}}}$$

  - **If the sample sizes n_x and n_y are large, then the result can be extended, by the central limit theorem, to include cases where the distributions of X and Y are not normal.**

Example 10

A

The weights of boys and girls in a certain school are known to be normally distributed with standard deviations of 5 kg and 8 kg respectively. A random sample of 25 boys had a mean weight of 48 kg and a random sample of 30 girls had a mean weight of 45 kg.

Stating your hypotheses clearly, test, at the 5% level of significance, whether there is evidence that the mean weight of the boys in the school is greater than the mean weight of the girls.

$H_0: \mu_{boy} = \mu_{girl}$ $H_1: \mu_{boy} > \mu_{girl}$

The alternative hypothesis is $\mu_{boy} > \mu_{girl}$, since this is what you are testing for. The null hypothesis is that the two population means are the same.

$\sigma_1 = 5, n_1 = 25, \sigma_2 = 8$ and $n_2 = 30$

The value of the test statistic is

$$z = \frac{\bar{x} - \bar{y} - (\mu_x - \mu_y)}{\sqrt{\dfrac{\sigma_x^2}{n_x} + \dfrac{\sigma_y^2}{n_y}}}$$

$\mu_x - \mu_y = 0$ from the null hypothesis.

$$= \frac{48 - 45}{\sqrt{\dfrac{25}{25} + \dfrac{64}{30}}}$$

$$= \frac{3}{\sqrt{3.1333\ldots}}$$

$$= 1.6947\ldots$$

The 5% (one-tailed) critical value for Z is $z = 1.6449$ (table on page 214) so this value is significant and you can reject H_0 and conclude that there is evidence that the mean weight of the boys is greater than the mean weight of the girls.

Watch out   Always quote the critical value from the tables in full and give your conclusion in context.

Sometimes you may be asked to test, for example, whether the mean weight of the boys exceeds the mean weight of the girls by more than 2 kg. The test would be similar to the above but the hypotheses would be slightly different and this will affect the test statistic.

Example 11

The weights of boys and girls in a certain school are known to be normally distributed with standard deviations of 5 kg and 8 kg respectively. A random sample of 25 boys had a mean weight of 48 kg and a random sample of 30 girls had a mean weight of 45 kg.

Stating your hypotheses clearly, test, at the 5% level of significance, whether there is evidence that the mean weight of the boys in the school is more than 2 kg greater than the mean weight of the girls.

A

$H_0: \mu_{boy} - \mu_{girl} = 2 \qquad H_1: \mu_{boy} - \mu_{girl} > 2$

$\sigma_1 = 5$, $n_1 = 25$, $s_2 = 8$ and $n_2 = 30$

The value of the test statistic is

$$z = \frac{\bar{x} - \bar{y} - (\mu_x - \mu_y)}{\sqrt{\dfrac{\sigma_x^2}{n_x} + \dfrac{\sigma_y^2}{n_y}}}$$

$$= \frac{48 - 45 - 2}{\sqrt{\dfrac{25}{25} + \dfrac{64}{30}}}$$

$$= 0.565$$

The 5% (one-tailed) critical value for Z is $z = 1.6449$ (table on page 214) so this value is not significant.

There is insufficient evidence that the mean weight of the boys is more than 2 kg greater than the mean weight of the girls.

The null hypothesis is that the difference between the means is 2. The alternative hypothesis is that the difference is *greater* than 2.

Notice how the test statistic calculation has changed. This 2 comes from $\mu_x - \mu_y$.

Example 12

A manufacturer of personal stereos can use batteries made by two different manufacturers. The standard deviation of lifetimes for *Never Die* batteries is 3.1 hours and for *Everlasting* batteries is 2.9 hours. A random sample of 80 *Never Die* batteries and a random sample of 90 *Everlasting* batteries were tested and their mean lifetimes were 7.9 hours and 8.2 hours respectively.

Stating your hypotheses clearly, test, at the 5% level of significance, whether there is evidence of a difference between the mean lifetimes of the two makes of batteries.

Let μ_x be the mean lifetime of *Never Die* batteries and let μ_y be the mean lifetime of *Everlasting* batteries.

$H_0: \mu_x = \mu_y \qquad H_1: \mu_x \neq \mu_y$

$\sigma_x = 3.1$, $n_x = 80$, $\sigma_y = 2.9$ and $n_y = 90$

$\bar{x} - \bar{y} = 7.9 - 8.2 = -0.3$

So $z = \dfrac{\bar{x} - \bar{y} - (\mu_x - \mu_y)}{\sqrt{\dfrac{\sigma_x^2}{n_x} + \dfrac{\sigma_y^2}{n_y}}}$

$$= \frac{-0.3}{\sqrt{\dfrac{(3.1)^2}{80} + \dfrac{(2.9)^2}{90}}}$$

$$= -0.649...$$

Watch out You are testing for a **difference** (in either direction) between the means, so use a **two-tailed test**.

You are not told that the distributions of lifetimes of the batteries are normally distributed, but the sample sizes are both quite large, so by the central limit theorem you can proceed with $\bar{X} - \bar{Y}$ approximately normally distributed.

From the null hypothesis you know that $\mu_x - \mu_y = 0$.

A The 5% (two-tailed) critical values for Z are $z = \pm 1.9600$.

So this value is not significant and you do not reject H_0. You can conclude that there is no significant evidence of a difference in the mean lifetimes of the two makes of batteries.

Exercise 5C

In Questions 1–3 carry out a test on the given hypotheses at the given level of significance. The populations from which the random samples are drawn are normally distributed.

1 $H_0: \mu_1 = \mu_2$, $H_1: \mu_1 > \mu_2$, $n_1 = 15$, $\sigma_1 = 5.0$, $n_2 = 20$, $\sigma_2 = 4.8$, $\bar{x}_1 = 23.8$ and $\bar{x}_2 = 21.5$ using a 5% level.

2 $H_0: \mu_1 = \mu_2$, $H_1: \mu_1 \neq \mu_2$, $n_1 = 30$, $\sigma_1 = 4.2$, $n_2 = 25$, $\sigma_2 = 3.6$, $\bar{x}_1 = 49.6$ and $\bar{x}_2 = 51.7$ using a 5% level.

3 $H_0: \mu_1 = \mu_2$, $H_1: \mu_1 < \mu_2$, $n_1 = 25$, $\sigma_1 = 0.81$, $n_2 = 36$, $\sigma_2 = 0.75$, $\bar{x}_1 = 3.62$ and $\bar{x}_2 = 4.11$ using a 1% level.

In Questions 4–6 carry out a test on the given hypotheses at the given level of significance. Given that the distributions of the populations are unknown, explain the significance of the central limit theorem in these tests.

4 $H_0: \mu_1 = \mu_2$, $H_1: \mu_1 \neq \mu_2$, $n_1 = 85$, $\sigma_1 = 8.2$, $n_2 = 100$, $\sigma_2 = 11.3$, $\bar{x}_1 = 112.0$ and $\bar{x}_2 = 108.1$ using a 1% level.

5 $H_0: \mu_1 = \mu_2$, $H_1: \mu_1 > \mu_2$, $n_1 = 100$, $\sigma_1 = 18.3$, $n_2 = 150$, $\sigma_2 = 15.4$, $\bar{x}_1 = 72.6$ and $\bar{x}_2 = 69.5$ using a 5% level.

6 $H_0: \mu_1 = \mu_2$, $H_1: \mu_1 < \mu_2$, $n_1 = 120$, $\sigma_1 = 0.013$, $n_2 = 90$, $\sigma_2 = 0.015$, $\bar{x}_1 = 0.863$ and $\bar{x}_2 = 0.868$ using a 1% level.

(E/P) 7 A factory has two machines designed to cut piping. The first machine works to a standard deviation of 0.011 cm and the second machine has a standard deviation of 0.015 cm. A random sample of 10 pieces of piping from the first machine has a mean length of 6.531 cm and a random sample of 15 pieces from the second machine has a mean length of 6.524 cm. Assuming that the lengths of piping follow a normal distribution, test, at the 5% level, whether the machines are producing piping of the same mean length. **(7 marks)**

(E/P) 8 A farmer grows wheat. He wants to improve his yield per acre by at least 1 tonne by buying a different variety of seed. The variance of the yield of the old seed is 0.6 tonnes² and the variance of the yield of the new seed is 0.8 tonnes². A random sample of 70 acres of wheat planted with the old seed has a mean yield of 5 tonnes and a random sample of 80 acres of wheat planted with the new seed has mean yield of 6.5 tonnes.

A **a** Test, at the 5% level of significance, whether there is evidence that the mean yield of the new seed is more than 1 tonne greater than the mean yield of the old seed. State your hypotheses clearly. **(9 marks)**

b Explain the relevance of the central limit theorem to the test in part **a**. **(2 marks)**

(E/P) **9** An agricultural research scientist investigated whether the diet of cows influenced the amount of fat in their milk. The fat content of 60 litres of milk selected at random from cows fed entirely on grain had a mean value of 4.1 g per litre and a random sample of 50 litres of milk from cows fed on a combination of grain and grass had a mean value of 3.7 g per litre.

It is known that the variance of fat content for a grain-only diet is $0.8\,(\text{g/l})^2$ and that the variance of fat content for a grain and grass diet is $0.75\,(\text{g/l})^2$.

a Stating your hypotheses clearly and using a 5% level of significance, test whether there is a difference between the mean fat content of milk from cows fed on these two diets. **(7 marks)**

b State, in the context of this question, an assumption you have made in carrying out the test in part **a**. **(1 mark)**

Challenge

Two independent random variables, X and Y, have unknown means μ_x and μ_y respectively. The variances, σ_x^2 and σ_y^2, are both known. Random samples of n_x observations from the random variable X and n_y observations from the random variable Y are taken. The sample means are $\bar{x}$ and $\bar{y}$.

A hypothesis test is carried out to see if there is a difference in the means of the two samples and it is found that the null hypothesis is accepted.

A confidence interval is to be found for the common mean of the two samples, μ.

$\hat{\mu}$ is used to denote the **pooled estimate of the population mean**, and is found by finding the mean of the combined samples.

a Show that $\hat{\mu} = \dfrac{n_x\bar{x} + n_y\bar{y}}{n_x + n_y}$.

The distribution of the corresponding random variable is given as

$$\text{N}\left(\mu, \frac{n_x\,\sigma_x^2 + n_y\,\sigma_y^2}{(n_x + n_y)^2}\right)$$

Given that $n_x = 100$, $n_y = 120$, $\bar{x} = 46.0$, $\bar{y} = 47.0$, $\sigma_x^2 = 16.0$ and $\sigma_y^2 = 24.0$,

b find the 99% confidence interval for μ.

5.4 Use of large sample results for an unknown population

One of the practical difficulties that you will encounter when carrying out hypothesis tests based on samples, and also in finding confidence intervals, is the need to know the value of the population standard deviation. In practice, if you do not know μ, it is unlikely that you will know σ. Sometimes it is reasonable to assume that similar processes or populations will have the same standard deviation but perhaps just have an altered mean. Occasionally you can look at historical data and see that over a period of time the standard deviation has been constant, and it may be reasonable to assume that it remains so but often it may be impossible to choose a reliable value of σ.

In situations such as this, you can use the central limit theorem and the sample variance, s^2, which is an unbiased estimator of the population variance.

- **If the population is normal, or can be assumed to be so, then, for large samples, $\dfrac{\overline{X} - \mu}{\dfrac{s}{\sqrt{n}}}$ has an approximate N(0, 1²) distribution.**

> **Watch out** Both of these tests rely on large sample sizes, and the second test also relies on the central limit theorem.

- **If the population is not normal, by assuming that s is a close approximation to σ, then for large samples, $\dfrac{\overline{X} - \mu}{\dfrac{s}{\sqrt{n}}}$ can be treated as having an approximate N(0, 1²) distribution.**

Example 13

As part of a study of the health of young schoolchildren, a random sample of 220 children from area A and a second, independent random sample of 180 children from area B were weighed. The results are given in the table below.

	n	$\overline{x}$	s
Area A	220	37.8	3.6
Area B	180	38.6	4.1

a Test, at the 5% level of significance, whether there is evidence of a difference in the mean weight of children in the two areas. State your hypotheses clearly.

b State an assumption you have made in carrying out this test.

c Explain the significance of the central limit theorem to this test.

A

a $H_0 : \mu_A = \mu_B \qquad H_1 : \mu_A \neq \mu_B$

Test statistic $\quad z = \dfrac{\bar{x}_A - \bar{x}_B - (\mu_A - \mu_B)}{\sqrt{\dfrac{\sigma_A^2}{n_A} + \dfrac{\sigma_B^2}{n_B}}}$

$z = \dfrac{38.6 - 37.8}{\sqrt{\dfrac{3.6^2}{220} + \dfrac{4.1^2}{180}}}$

$z = 2.0499...$

$\quad = 2.05$ (3 s.f.)

Two-tail 5% critical values are $z = \pm 1.96$

Since $2.05 > 1.96$, the result is significant so reject H_0.

There is evidence that the mean weight of children in the two areas is different.

b The test statistic requires σ so you have to assume that $s^2 = \sigma^2$ for both samples.

c You are not told that the populations are normally distributed but the samples are both large and so the central limit theorem enables us to assume that $\bar{X}_A$ and $\bar{X}_B$ are both normal.

State the hypotheses in terms of μ. The word 'difference' suggests a two-tailed test.

For a two-tailed test you can choose whether to use $\bar{x}_A - \bar{x}_B$ or $\bar{x}_B - \bar{x}_A$. It is usually easier to choose the case which gives a positive value to the test statistic, which is $\bar{x}_B - \bar{x}_A$ in this case.

Give the value to at least 3 s.f.

Always state the critical value(s).

Always give your conclusion in context.

Watch out Note that the central limit theorem is not an assumption – it is a theorem that can be invoked to enable you to use the normal distribution.

The assumption that $s^2 = \sigma^2$ is reasonable since both samples are large.

Exercise **5D**

E/P **1** An experiment was conducted to compare the drying properties of two paints, Quickdry and Speedicover. In the experiment, 200 similar pieces of metal were painted, 100 randomly allocated to Quickdry and the rest to Speedicover.

The table below summarises the times, in minutes, taken for these pieces of metal to become touch-dry.

	Quickdry	Speedicover
Mean	28.7	30.6
Standard deviation	7.32	3.51

Using a 5% significance level, test whether the mean time for Quickdry to become touch-dry is less than that for Speedicover. State your hypotheses clearly. **(6 marks)**

E/P **2** A supermarket examined a random sample of 80 weekend shoppers' purchases and an independent random sample of 120 weekday shoppers' purchases. The results are summarised in the table below.

	n	$\bar{x}$	s
Weekend	80	38.64	6.59
Weekday	120	40.13	8.23

A **a** Stating your hypotheses clearly, test, at the 5% level of significance, whether there is evidence that the mean expenditure in the week is more than at weekends. **(6 marks)**

b State an assumption you have made in carrying out this test. **(1 mark)**

E/P **3** It is claimed that the components produced in a small factory have a mean mass of 10 g. A random sample of 250 of these components is tested and the sample mean, $\bar{x}$, is 9.88 g and the standard deviation, s, is 1.12 g.

a Test, at the 5% level, whether there has been a change in the mean mass of a component. **(6 marks)**

b State any assumptions you would make to carry out this test. **(1 mark)**

E/P **4** Two independent samples are taken from population A and population B. Carry out the following tests using the information given.

a $H_0 : \mu_A = \mu_B$, $H_1 : \mu_A < \mu_B$ using a 1% level of significance
$n_A = 90$, $n_B = 110$, $\bar{x}_A = 84.1$, $\bar{x}_B = 87.9$, $s_A = 12.5$, $s_B = 14.6$ **(5 marks)**

b $H_0 : \mu_A - \mu_B = 2$, $H_1 : \mu_A - \mu_B > 2$ using a 5% level of significance
$n_A = 150$, $n_B = 200$, $\bar{x}_A = 125.1$, $\bar{x}_B = 119.3$, $s_A = 23.2$, $s_B = 18.4$ **(5 marks)**

c State an assumption that you have made in carrying out these tests. **(1 mark)**

E/P **5** A shopkeeper complains that the average mass of chocolate bars of a certain type that he is buying from a wholesaler is less than the stated value of 85.0 g. The shopkeeper measured the mass of 100 bars from a large delivery and found that their masses had a mean of 83.6 g and a standard deviation of 7.2 g. Using a 5% significance level, determine whether the shopkeeper is justified in his complaint. State clearly the null and alternative hypotheses that you are using, and express your conclusion in words. **(6 marks)**

E/P **6** A health authority set up an investigation to examine the ages of mothers when they give birth to their first child.

A random sample of 250 first-time mothers from a certain year had a mean age of 22.45 years with a standard deviation of 2.9 years. A further random sample of 280 first-time mothers taken 10 years later had a mean age of 22.96 years with a standard deviation of 2.8 years.

a Test whether these figures suggest that there is a difference in the mean age of first-time mothers between these two dates. Use a 5% level of significance. **(6 marks)**

b State any assumptions you have made about the distribution of ages of first-time mothers. **(1 mark)**

Mixed exercise 5

1 The masses of bags of lentils, X kg, have a normal distribution with unknown mean μ kg and a known standard deviation σ kg. A random sample of 80 bags of lentils gave a 90% confidence interval for μ of (0.4533, 0.5227).

    a Without carrying out any further calculations, use this confidence interval to test whether $\mu = 0.48$. State your hypotheses clearly and write down the significance level you have used. **(3 marks)**

    A second random sample of 120 of these bags of lentils had a mean mass of 0.482 kg.

    b Calculate a 95% confidence interval for μ based on this second sample. **(6 marks)**

2 The lengths of the tails of mice in a pet shop are assumed to have unknown mean μ and unknown standard deviation σ.

    A random sample of 20 mice is taken and the length of their tails recorded.

    The sample is represented by $X_1, X_2, \ldots, X_{20}$.

    a State whether or not the following are statistics.

       Give reasons for your answers.

       i $\dfrac{2X_1 + X_{20}}{3}$ **ii** $\displaystyle\sum_1^{20}(X_i - \mu)^2$ **iii** $\dfrac{\displaystyle\sum_1^{20}X_i^2}{n}$ **(4 marks)**

    b Find the mean and variance of $\dfrac{4X_1 - X_{20}}{3}$ **(3 marks)**

3 The breaking stresses of rubber bands are normally distributed.

    A company uses bands with a mean breaking stress of 46.50 N.

    A new supplier claims that they can supply bands that are stronger and provides a sample of 100 bands for the company to test. The company checked the breaking stress, X, for each of these 100 bands and the results are summarised as follows:

       $n = 100$ $\sum X = 4715$ $\sum X^2 = 222\,910$

    a Test, at the 5% level, whether there is evidence that the new bands are stronger. **(6 marks)**

    b Find an approximate 95% confidence interval for the mean breaking stress of these new rubber bands. **(3 marks)**

4 On each of 100 days, a conservationist took a sample of 1 litre of water from a particular place along a river, and measured the amount, X mg, of chlorine in the sample. The results she obtained are shown in the table.

X	1	2	3	4	5	6	7	8	9
Number of days	4	8	20	22	16	13	10	6	1

    a Estimate the mean amount of chlorine present per litre of water, and estimate, to 3 decimal places, the standard error of this estimate. **(3 marks)**

A

b Obtain approximate 98% confidence limits for the mean amount of chlorine present per litre of water. **(3 marks)**

Given that measurements at the same point under the same conditions are taken for a further 100 days,

c estimate, to 3 decimal places, the probability that the mean of these measurements will be greater than 4.6 mg per litre of water. **(3 marks)**

E **5** The amount, to the nearest mg, of a certain chemical in particles in the atmosphere at a meteorological station was measured each day for 300 days. The results are shown in the table.

Amount of chemical (mg)	12	13	14	15	16
Number of days	5	42	210	31	12

Estimate the mean amount of this chemical in the atmosphere, and find, to 2 decimal places, the standard error of this estimate. **(3 marks)**

E/P **6** From time to time a firm manufacturing pre-packed furniture needs to check the mean distance between pairs of holes drilled by a machine in pieces of chipboard to ensure that no change has occurred. It is known from experience that the standard deviation of the distance is 0.43 mm. The firm intends to take a random sample of size n, and to calculate a 99% confidence interval for the mean of the population. The width of this interval must be no more than 0.60 mm. Calculate the minimum value of n. **(4 marks)**

E **7** The times taken by five-year-old children to complete a certain task are normally distributed with a standard deviation of 8.0 s. A random sample of 25 five-year-old children from school A were given this task and their mean time was 44.2 s.

a Find 95% confidence limits for the mean time taken by five-year-old children from school A to complete this task. **(3 marks)**

The mean time for a random sample of 20 five-year-old children from school B was 40.9 s. The headteacher of school B concluded that the overall mean for school B must be less than that of school A. Given that the two samples were independent,

b test the headteacher's conclusion using a 5% significance level. State your hypotheses clearly. **(6 marks)**

E/P **8** The random variable X is normally distributed with mean μ and variance σ^2.

a Write down the distribution of the sample mean $\overline{X}$ of a random sample of size n. **(1 mark)**

b State, with a reason, whether this distribution is exact or is an estimate. **(1 mark)**

An efficiency expert wishes to determine the mean time taken to drill a fixed number of holes in a metal sheet.

c Determine how large a random sample is needed so that the expert can be 95% certain that the sample mean time will differ from the true mean time by less than 15 seconds. Assume that it is known from previous studies that $\sigma = 40$ seconds. **(4 marks)**

E/P **9** A commuter regularly uses a train service which should arrive in London at 09:31. He decided to test this stated arrival time. Each working day for a period of 4 weeks, he recorded the number of minutes X that the train was late on arrival in London. If the train arrived early then the value of X was negative. His results are summarised as follows:

$$n = 20 \qquad \sum x = 15.0 \qquad \sum x^2 = 103.21$$

A

a Calculate unbiased estimates of the mean and variance of the number of minutes late of this train service. **(3 marks)**

The random variable X represents the number of minutes that the train is late on arriving in London. Records kept by the railway company show that over fairly short periods, the standard deviation of X is 2.5 minutes. The commuter made two assumptions about the distribution of X and the values obtained in the sample and went on to calculate a 95% confidence interval for the mean arrival time of this train service.

b State the two assumptions. **(2 marks)**

c Find the confidence interval. **(3 marks)**

d Given that the assumptions are reasonable, comment on the stated arrival time of the service. **(1 mark)**

E/P **10** The random variable X is normally distributed with mean μ and variance σ^2.

a Write down the distribution of the sample mean $\overline{X}$ of a random sample of size n. **(1 mark)**

b Explain what you understand by a 95% confidence interval. **(2 marks)**

A garage sells both leaded and unleaded petrol. The distribution of the values of sales for each type is normal. During 2010 the standard deviation of individual sales of each type of petrol was £3.25. The mean of the individual sales of leaded petrol during this time was £8.72. A random sample of 100 individual sales of unleaded petrol gave a mean of £9.71.

Calculate:

c an interval within which 90% of the sales of leaded petrol will lie, **(3 marks)**

d a 95% confidence interval for the mean sales of unleaded petrol. **(3 marks)**

The mean of the sales of unleaded petrol for 2009 was £9.10.

e Using a 5% significance level, investigate whether there is sufficient evidence to conclude that the mean of all the 2010 unleaded sales was greater than the mean of the 2009 sales. **(6 marks)**

f Find the size of the sample that should be taken so that the garage proprietor can be 95% certain that the sample mean of sales of unleaded petrol during 2010 will differ from the true mean by less than 50p. **(4 marks)**

E/P **11** **a** Explain what is meant by a 98% confidence interval for a population mean. **(2 marks)**

The lengths, in cm, of the leaves of willow trees are known to be normally distributed with variance $1.33\,\text{cm}^2$.

A sample of 40 willow tree leaves is found to have a mean of 10.20 cm.

b Estimate, giving your answer to 3 decimal places, the standard error of the mean. **(2 marks)**

c Use this value to estimate symmetrical 95% confidence limits for the mean length of the population of willow tree leaves, giving your answer to 2 decimal places. **(3 marks)**

d Find the minimum size of the sample of leaves which must be taken if the width of the symmetrical 98% confidence interval for the population mean is at most 1.50 cm. **(4 marks)**

A **12 a** Write down the mean and the variance of the distribution of the means of all possible
E/P samples of size n taken from an infinite population having mean μ and variance σ^2. **(2 marks)**

b Describe the form of this distribution of sample means when
  i n is large
  ii the distribution of the population is normal. **(2 marks)**

The standard deviation of all the till receipts of a supermarket during 2014 was £4.25.

c Given that the mean of a random sample of 100 of the till receipts is £18.50, obtain an
approximate 95% confidence interval for the mean of all the till receipts during 2014. **(3 marks)**

d Find the size of sample that should be taken so that the management can be 95% confident
that the sample mean will not differ from the true mean by more than 50p. **(3 marks)**

e The mean of all the till receipts of the supermarket during 2013 was £19.40. Using a 5%
significance level, investigate whether the sample in part **a** provides sufficient evidence to
conclude that the mean of all the 2014 till receipts is different from that in 2013. **(6 marks)**

E/P **13** Records of the diameters of spherical ball bearings produced on a certain machine indicate that
the diameters are normally distributed with mean 0.824 cm and standard deviation 0.046 cm.
Two hundred samples are chosen, each consisting of 100 ball bearings.

a Calculate the expected number of the 200 samples having a mean diameter less than
0.823 cm. **(2 marks)**

On a certain day it was suspected that the machine was malfunctioning. It may be assumed that
if the machine is malfunctioning it will change the mean of the diameters without changing
their standard deviation. On that day a random sample of 100 ball bearings had mean diameter
0.834 cm.

b Determine a 98% confidence interval for the mean diameter of the ball bearings being
produced that day. **(3 marks)**

c Hence state whether or not you would conclude that the machine is malfunctioning on that
day given that the significance level is 2%. **(3 marks)**

E/P **14** A cardiologist claims that there is a higher mean heart rate in people who always drive to
work compared to people who regularly walk to work. She measures the heart rates, X, of 30
people who always drive to work and 36 people who regularly walk to work. Her results are
summarised in the table below.

	n	$\bar{x}$	s^2
Drive to work	30	52	60.2
Walk to work	36	47	55.8

a Test, at the 5% level of significance, the cardiologist's claim. State your hypotheses clearly. **(6 marks)**

b State any assumptions you have made in testing the cardiologist's claim. **(2 marks)**

The cardiologist decides to add another person who drives to work to her data. She measures
the person's heart rate and finds $X = 55$.

c Find an unbiased estimate of the variance for the sample of 31 people who drive to work.
Give your answer to 3 significant figures. **(4 marks)**

Challenge

A Two independent random samples $X_1, X_2, \ldots, X_n$ and $Y_1, Y_2, \ldots, Y_m$ are taken from a population with mean μ and variance σ^2. The unbiased estimators $\overline{X}$ and $\overline{Y}$ of μ are calculated. A new unbiased estimator T of μ is sought of the form $T = r\overline{X} + s\overline{Y}$.

a Show that, since T is unbiased, $r + s = 1$.

b By writing $T = r\overline{X} + (1 - r)\overline{Y}$, show that

$$\text{Var}(T) = \left(\sigma^2 \frac{r^2}{n} + \frac{1 - r^2}{m} \right)$$

c Show that the minimum variance of T is when $r = \dfrac{n}{n + m}$

d Find the best (in the sense of minimum variance) unbiased estimator of μ of the form $r\overline{X} + s\overline{Y}$.

Summary of key points

1 If X is a random variable, then a random sample of size n will consist of n observations of the random variable X, which are referred to as $X_1, X_2, X_3, \ldots, X_n$, where the X_i
   - are independent random variables
   - each have the same distribution as X.

   A statistic, T, is defined as a random variable consisting of any function of the X_i that involves no other quantities, such as unknown population parameters.

2 The **sampling distribution** of a statistic T is the probability distribution of T.

3 A statistic that is used to estimate a population parameter is called an **estimator** and the particular value of the estimator generated from the sample taken is called an **estimate**.

4 If a statistic T is used as an estimator for a population parameter θ and $\text{E}(T) = \theta$ then T is an unbiased estimator for θ.

5 If a statistic T is used as an estimator for a population parameter θ then the bias $= \text{E}(T) - \theta$. For an unbiased estimator, the bias is 0.

6 An unbiased estimator for σ^2 is given by the **sample variance** S^2 where

$$S^2 = \frac{1}{n - 1} \sum_{i=1}^{n} (X_i - \overline{X})^2$$

7 The standard deviation of an estimator is called the **standard error** of the estimator.

8 When you are using the sample mean, $\overline{X}$, you can use the following result for the standard error:

   Standard error of $\overline{X} = \dfrac{\sigma}{\sqrt{n}}$ or $\dfrac{s}{\sqrt{n}}$

9 A **confidence interval** for a population parameter θ is a range of values defined so that there is a specific probability that the true value of the parameter lies within that range.

10 A 95% confidence interval for the population mean μ is $\left(\bar{x} - 1.96 \times \dfrac{\sigma}{\sqrt{n}}, \bar{x} + 1.96 \times \dfrac{\sigma}{\sqrt{n}}\right)$.

11 The width of a confidence interval is the difference between the upper confidence limit and the lower confidence limit. This is $2 \times z \times \dfrac{\sigma}{\sqrt{n}}$, where z is the relevant percentage point from the standard normal distribution, for example 1.96, 1.6449, etc.

12 Test for difference between two means:
- If $X \sim N(\mu_x, \sigma_x^2)$ and the independent random variable $Y \sim N(\mu_y, \sigma_y^2)$, then a test of the null hypothesis $H_0: \mu_x = \mu_y$ can be carried out using the test statistic

$$Z = \frac{\bar{X} - \bar{Y} - (\mu_x - \mu_y)}{\sqrt{\dfrac{\sigma_x^2}{n_x} + \dfrac{\sigma_y^2}{n_y}}}$$

- If the sample sizes n_x and n_y are large then the result can be extended, by the central limit theorem, to include cases where the distributions of X and Y are not normal.

13
- If the population is normal, or can be assumed to be so, then, for large samples, $\dfrac{\bar{X} - \mu}{\dfrac{s}{\sqrt{n}}}$ has an approximate $N(0, 1^2)$ distribution.

- If the population is not normal, by assuming that s is a close approximation to σ, then $\dfrac{\bar{X} - \mu}{\dfrac{s}{\sqrt{n}}}$ can be treated as having an approximate $N(0, 1^2)$ distribution.

Further hypothesis tests

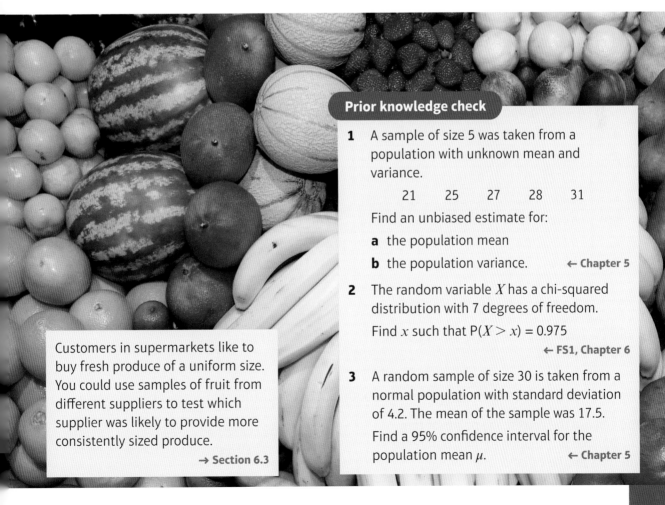

6

Objectives

After completing this chapter, you should be able to:

Prior knowledge check

1 A sample of size 5 was taken from a
   population with unknown mean and
   variance.

       21   25   27   28   31

   Find an unbiased estimate for:

   a the population mean

   b the population variance.   ← **Chapter 5**

2 The random variable X has a chi-squared
   distribution with 7 degrees of freedom.
   Find x such that $P(X > x) = 0.975$

   ← **FS1, Chapter 6**

3 A random sample of size 30 is taken from a
   normal population with standard deviation
   of 4.2. The mean of the sample was 17.5.
   Find a 95% confidence interval for the
   population mean μ.   ← **Chapter 5**

Customers in supermarkets like to
buy fresh produce of a uniform size.
You could use samples of fruit from
different suppliers to test which
supplier was likely to provide more
consistently sized produce.

→ **Section 6.3**

6.1 Variance of a normal distribution

A If you take a sample of n independent observations $X_1, X_2, \ldots, X_n$ with sample mean $\overline{X}$, then

$S^2 = \dfrac{1}{n-1} \displaystyle\sum_{i=1}^{n} (X_i - \overline{X})^2$ is an unbiased estimator of the population variance σ^2.

Different samples from the same population will give different estimates for σ^2, so in the same way that $\overline{x}$ is a particular value of the random variable $\overline{X}$, s^2 is a particular value of a random variable S^2.

In order to find a confidence interval for σ^2 you need to know something about the distribution of S^2.

> **Watch out** To find confidence intervals for the population mean, you used the central limit theorem to say that the sample mean $\overline{X}$ was approximately normally distributed. You cannot apply the central limit theorem to S^2.

The distribution of S^2 is not easy to find, but if X is normally distributed, then the distribution of $\dfrac{(n-1)S^2}{\sigma^2}$ is a **chi-squared** distribution with $n-1$ degrees of freedom.

> **Links** You have previously used the chi-squared family of distributions to model goodness of fit.
> ← FS1, Chapter 6

- **If a random sample of n observations $X_1, X_2, \ldots, X_n$ is selected from N(μ, σ^2), then**

$$\frac{(n-1)S^2}{\sigma^2} \sim \chi^2_{n-1}$$

Percentage points for the chi-squared distribution are given in the table in the formulae booklet, and on page 215. The number down the left-hand side is the number of degrees of freedom, which should be equal to $n-1$.

ν	0.995	0.990	0.975	0.950	0.900	0.100	0.050	0.025	0.010	0.005
1	0.000	0.000	0.001	0.004	0.016	2.705	3.841	5.024	6.635	7.879
2	0.010	0.020	0.051	0.103	0.211	4.605	5.991	7.378	9.210	10.597
3	0.072	0.115	0.216	0.352	0.584	6.251	7.815	9.348	11.345	12.838
4	0.207	0.297	0.484	0.711	1.064	7.779	9.488	11.143	13.277	14.860
5	0.412	0.554	0.831	1.145	1.610	9.236	11.070	12.832	15.086	16.750
6	0.676	0.872	1.237	1.635	2.204	10.645	12.592	14.449	16.812	18.548
7	0.989	1.239	1.690	2.167	2.833	12.017	14.067	16.013	18.475	20.278

For example, $\chi^2_6 (0.95) = 1.635$ so the probability of χ^2_6 exceeding 1.635 is 95%. You should note that the probability that χ^2_6 is less than 1.635 is 100% − 95% = 5%.

For example, if you wish to find $\chi^2_4 (0.05)$ you look down the 0.05 column until you meet the row $\nu = 4$, and read off 9.488

Because the χ^2 distribution is non-symmetric, both tails of the distribution are given in the table.

You can use the percentage points of the chi-squared distribution to find confidence intervals for the variance of a normally distributed population.

> **Note** You can also find the percentage points for the chi-squared distribution on some graphical calculators.

A If a random sample $X_1, X_2, \ldots, X_n$ is taken from an $N(\mu, \sigma)$ distribution, then $\dfrac{(n-1)\,S^2}{\sigma^2}$ has a χ^2_{n-1}

distribution. Using the table of percentage points of the distribution χ^2_{n-1}, you can find given values

of $\dfrac{(n-1)\,S^2}{\sigma^2}$ such that the interval between them will contain σ^2 with 95% probability.

Thus, assuming that the two tails are of equal size,

$$\chi^2_{n-1}(0.975) < \frac{(n-1)\,S^2}{\sigma^2} < \chi^2_{n-1}(0.025)$$

We want to get bounds for σ^2 so we isolate it by

$$\frac{1}{\chi^2_{n-1}(0.025)} < \frac{\sigma^2}{(n-1)\,S^2} < \frac{1}{\chi^2_{n-1}(0.975)}$$

1 turning the fractions upside down and reversing the inequality.

$$\frac{(n-1)\,S^2}{\chi^2_{n-1}(0.025)} < \sigma^2 < \frac{(n-1)\,S^2}{\chi^2_{n-1}(0.975)}$$

2 multiplying by $(n-1)\,S^2$

If you have a specific estimate s^2, this becomes

$$\frac{(n-1)\,s^2}{\chi^2_{n-1}(0.025)} < \sigma^2 < \frac{(n-1)\,s^2}{\chi^2_{n-1}(0.975)}$$

The values $\dfrac{(n-1)\,s^2}{\chi^2_{n-1}(0.025)}$ and $\dfrac{(n-1)\,s^2}{\chi^2_{n-1}(0.975)}$ are the lower and upper 95% confidence limits respectively.

In a similar way, the 90% confidence limits are $\dfrac{(n-1)\,s^2}{\chi^2_{n-1}(0.05)}$ and $\dfrac{(n-1)\,s^2}{\chi^2_{n-1}(0.95)}$

- **Generally, for a probability of α that the variance falls outside the limits,**
  - **the $100(1-\alpha)\%$ confidence limits are**

$$\frac{(n-1)\,s^2}{\chi^2_{n-1}\left(\dfrac{\alpha}{2}\right)} \quad \textbf{and} \quad \frac{(n-1)\,s^2}{\chi^2_{n-1}\left(1-\dfrac{\alpha}{2}\right)}$$

  - **the $100(1-\alpha)\%$ confidence interval for the variance of a normal distribution is**

$$\left(\frac{(n-1)\,s^2}{\chi^2_{n-1}\left(\dfrac{\alpha}{2}\right)}, \frac{(n-1)\,s^2}{\chi^2_{n-1}\left(1-\dfrac{\alpha}{2}\right)}\right)$$

Example 1

A In order to determine the accuracy of a new rifle, 8 marksmen were selected at random to fire the rifle at a target. The distances x, in mm, of the 8 shots from the centre of the target were as follows:

10, 14, 12, 8, 6, 11, 18, 14

Assuming that the distances are normally distributed, find a 95% confidence interval for the variance.

$\bar{x} = 11.625$ $s^2 = 14.2679$	You can use a calculator to find $\bar{x}$ and s^2.
$\chi_7^2(0.975) = 1.690$ $\chi_7^2(0.025) = 16.013$	
$\dfrac{(n-1)s^2}{\chi_{n-1}^2(0.025)} = \dfrac{7 \times 14.2679}{16.013} = 6.2371...$	There are 8 observations in the sample, so use χ_7^2 and find the percentage points from the table.
$\dfrac{(n-1)s^2}{\chi_{n-1}^2(0.975)} = \dfrac{7 \times 14.2679}{1.690} = 59.0976...$	Now find the critical points for the variance.
The 95% confidence interval for the variance is (6.237, 59.098).	Don't forget to write out the interval.

Example 2

A company manufactures 12 amp electrical fuses.

A random sample of 10 fuses was taken from a batch and the failure current, x, measured for each. The results are summarised below:

$$\sum x = 118.9 \quad \sum x^2 = 1414.89$$

Assume that the data can be regarded as a random sample from a normal population.

a Calculate an unbiased estimate for the variance of the batch based upon the sample.

b Use your estimate from part **a** to calculate a 95% confidence interval for the standard deviation.

a $s^2 = \dfrac{1}{n-1}\left(\sum x^2 - \dfrac{(\sum x)^2}{n}\right)$	Find s^2 and $\bar{x}$. Since the $\sum x$ and $\sum x^2$ are given you will have to use the formula.
$= \frac{1}{9}\left(1414.89 - \dfrac{118.9^2}{10}\right)$	
$= 0.1299$ (4 d.p.)	
b The percentage points are	
$\chi_9^2(0.975) = 2.700$ and $\chi_9^2(0.025) = 19.023$	This time we use the χ_9^2 distribution. Find the percentage points from the table.
The critical points are	
$\dfrac{(n-1)s^2}{\chi_{n-1}^2(0.975)} = \dfrac{9 \times 0.1299}{2.7} = 0.433$	
$\dfrac{(n-1)s^2}{\chi_{n-1}^2(0.025)} = \dfrac{9 \times 0.1299}{19.023} = 0.0615$	Calculate the critical points using the formula. Then write as a confidence interval.
The 95% confidence interval for the variance is (0.062, 0.433).	
Hence the 95% confidence interval for the standard deviation is (0.249, 0.658).	The confidence limits for the standard deviation are just the square root of the confidence limits for the variance.

Exercise 6A

A

1 A random sample of 15 observations of a normal population gave an unbiased estimate for the variance of the population of $s^2 = 4.8$. Calculate a 95% confidence interval for the population variance.

2 A random sample of 20 observations of a normally distributed variable X is summarised by $\sum x = 132.4$ and $\sum x^2 = 884.3$. Calculate a 90% confidence interval for the variance of X.

3 A random sample of 14 observations were taken from a population that is assumed to be normally distributed. The resulting values were:

   2.3, 3.9, 3.5, 2.2, 2.6, 2.5, 2.3, 3.9, 2.1, 3.6, 2.1, 2.7, 3.2, 3.4

   Calculate a 95% confidence interval for the population variance.

E 4 A random sample of female voles was trapped in a wood. Their lengths, in centimetres (excluding tails), were 7.5, 8.4, 10.1, 6.2 and 8.4.

   Assuming that this is a sample from a normal distribution, calculate a 95% confidence interval for the variance of the lengths of female voles. **(4 marks)**

E 5 A random sample of 10 is taken from the annual rainfall figures, x cm, in a certain district. The data can be summarised by $\sum x = 621$ and $\sum x^2 = 38\,938$.

   a Calculate 90% confidence limits for the variance of the annual rainfall. **(4 marks)**

   b What assumption have you made about the distribution of the annual rainfall in part **a**? **(1 mark)**

E 6 A new variety of small daffodil is grown in the trial ground of a nursery. During the flowering period, a random sample of 10 flowers was taken and the lengths, in millimetres, of their stalks were measured. The results were as follows:

   266, 254, 215, 220, 253, 230, 216, 248, 234, 244

   Assuming that the lengths are normally distributed, calculate a 95% confidence interval for the variance of the lengths. **(4 marks)**

E/P 7 A random sample of 8 cats was taken and the lengths of their tails, x cm, were measured. It was found that $\sum x = 234$.

   Assuming that the lengths of cats' tails are normally distributed and given that the lower limit for the 95% confidence interval for the population variance is 7.9623, find $\sum x^2$. **(5 marks)**

E/P 8 A hotel owner chooses 25 randomly selected dates and finds that the standard deviation of the number of rooms occupied is 6.21.

   Given that the number of rooms occupied is normally distributed, find, correct to 3 significant figures, a 90% confidence interval for the population standard deviation. **(5 marks)**

E/P 9 Francine works on a production line and is responsible for calibrating a machine that produces aluminium fastenings. She selects a random sample of 8 fastenings and records the diameter, in millimetres, of each one.

   6.7, 6.8, 6.9, 6.4, 6.3, 6.3, 6.7, 6.1

a Calculate a 99% confidence interval for the standard deviation of the diameters of the fastenings. **(6 marks)**

Given that the diameters of the fastenings are independent,

b state what further assumption is necessary for this confidence interval to be valid. **(1 mark)**

Francine requires the standard deviation of the diameters to be less than 0.15 mm, otherwise she must recalibrate the machine.

c Use your answer to part **a** to decide if the machine needs recalibrating. **(1 mark)**

6.2 Hypothesis testing for the variance of a normal distribution

You need to be able to carry out a hypothesis test for the variance of a normal distribution.

Suppose that a manufacturer of pistons for cars had a machine that finished the diameter of the piston to a specific size. The machine was set up so that it produced the pistons with a diameter that was normally distributed with mean 60 mm and variance 0.0009 mm². After the machine had been running for some time, a sample of 15 pistons was taken and the mean of the size of the pistons in the sample was still 60 mm, but the best estimate of the variance calculated from the sample was 0.002 mm². The manufacturer wishes to know whether the variance has increased.

Putting the manufacturer's question in the form of hypotheses you get:

> **Note** The null and alternative hypotheses are framed in terms of the population variance σ^2.

$H_0: \sigma^2 = 0.03^2$ $\qquad$ $H_1: \sigma^2 > 0.03^2$

If H_0 is assumed true, then $\dfrac{(n-1)S^2}{\sigma^2}$ will be a single observation from a χ^2 distribution.

Since in this case $s^2 > \sigma^2$, you need to consider how likely you would be to get the calculated value of $\dfrac{(n-1)S^2}{\sigma^2}$ if H_0 were true. The critical value separating the acceptance and rejection regions will be the relevant percentage point of the χ^2_{n-1} distribution.

In this case, $\nu = n - 1 = 15 - 1 = 14$ and we are using a 5% level of significance.

From the table on page 215, $\chi^2_{14}(0.05) = 23.685$ and the critical region is $\dfrac{(n-1)s^2}{\sigma^2} \geqslant 23.685$.

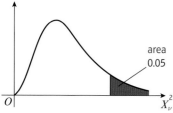

area
0.05

The value of the test statistic will be $\dfrac{(n-1)s^2}{\sigma^2}$

In the manufacturer's case $\dfrac{(n-1)s^2}{\sigma^2} = \dfrac{(15-1)0.002}{0.03^2} = 31.11$

31.11 is in the critical region so the result is significant and H_0 is rejected. There is evidence that the variance has increased.

A Note that there are seven steps to be followed:

1 Write down the null hypothesis (H_0).

2 Write down the alternative hypothesis (H_1).

3 Specify the significance level.

4 Write down the degrees of freedom ν.

5 Write down the critical region.

6 Identify the population variance, σ^2, and the unbiased estimate, s^2, and calculate the value of the test statistic $\dfrac{(n-1)s^2}{\sigma^2}$

7 Complete your test and state your conclusions, stating whether H_0 is accepted or rejected, and interpreting this in the context of the question.

Example 3

A random sample of 12 observations is taken from a normal distribution with a variance of σ^2. The unbiased estimate of the population variance is calculated as 0.015.

Test, at the 5% level, the null hypothesis that $\sigma^2 = 0.025$ against the alternative hypothesis that $\sigma^2 \neq 0.025$.

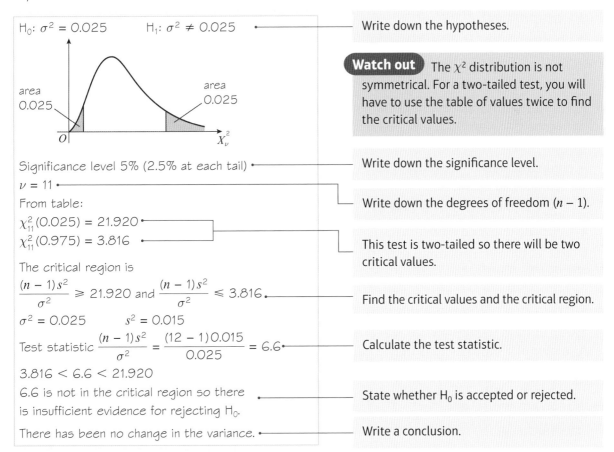

H_0: $\sigma^2 = 0.025$ H_1: $\sigma^2 \neq 0.025$ — Write down the hypotheses.

area 0.025   area 0.025

Watch out The χ^2 distribution is not symmetrical. For a two-tailed test, you will have to use the table of values twice to find the critical values.

Significance level 5% (2.5% at each tail) — Write down the significance level.

$\nu = 11$ — Write down the degrees of freedom $(n-1)$.

From table:

$\chi^2_{11}(0.025) = 21.920$

$\chi^2_{11}(0.975) = 3.816$ — This test is two-tailed so there will be two critical values.

The critical region is

$\dfrac{(n-1)s^2}{\sigma^2} \geq 21.920$ and $\dfrac{(n-1)s^2}{\sigma^2} \leq 3.816$. — Find the critical values and the critical region.

$\sigma^2 = 0.025$ $s^2 = 0.015$

Test statistic $\dfrac{(n-1)s^2}{\sigma^2} = \dfrac{(12-1)0.015}{0.025} = 6.6$ — Calculate the test statistic.

$3.816 < 6.6 < 21.920$

6.6 is not in the critical region so there is insufficient evidence for rejecting H_0. — State whether H_0 is accepted or rejected.

There has been no change in the variance. — Write a conclusion.

Exercise 6B

A

1 Twenty random observations (x) are taken from a normal distribution with variance σ^2. The results are summarised as follows:

$$\sum x = 332.1 \quad \sum x^2 = 5583.63$$

a Calculate an unbiased estimate for the population variance.

b Test, at the 5% significance level, $H_0: \sigma^2 = 1.5$ against $H_1: \sigma^2 > 1.5$.

2 A random sample of 10 observations is taken from a normal distribution with variance σ^2. The variance is thought to be equal to 0.09. The results were as follows:

0.35, 0.42, 0.30, 0.26, 0.31, 0.30, 0.40, 0.33, 0.30, 0.40

Test, at the 2.5% level of significance, $H_0: \sigma^2 = 0.09$ against $H_1: \sigma^2 < 0.09$.

P 3 The following random observations are taken from a normal distribution which is thought to have a variance of 4.1:

2.1, 2.3, 3.5, 4.6, 5.0, 6.4, 7.1, 8.6, 8.7, 9.1

Test, at the 5% significance level, $H_0: \sigma^2 = 4.1$ against $H_1: \sigma^2 \neq 4.1$.

P 4 It is claimed that the masses of a particular component produced in a small factory are normally distributed and have a mean mass of 10 g and a standard deviation of 1.12 g.

A random sample of 20 such components was found to have a variance of 1.15 g.

Test, at the 5% significance level, $H_0: \sigma^2 = 1.12^2$ against $H_1: \sigma^2 \neq 1.12^2$.

E/P 5 Rollers for use in roller bearings are produced on a certain machine. The rollers are supposed be normally distributed and to have a mean diameter (μ) of 10 mm with a variance (σ^2) of 0.04 mm^2.

A random sample of 15 rollers is taken and their diameters, x mm, were measured. The results are summarised below:

$$\sum x = 149.941 \quad \sum x^2 = 1498.83$$

a Calculate unbiased estimates for μ and σ^2. **(3 marks)**

b Test, at the 5% significance level, the hypothesis $\sigma^2 = 0.04$ against the hypothesis $\sigma^2 \neq 0.04$. **(5 marks)**

E/P 6 The diameters of the eggs of the little gull are approximately normally distributed with mean 4.11 cm with a variance of 0.19 cm^2.

A sample of 8 little gull eggs from a particular island which were measured had diameters in centimetres as follows:

4.4, 4.5, 4.1, 3.9, 4.4, 4.6, 4.5, 4.1

a Calculate an unbiased estimate for the variance of the population of little gull eggs on the island. **(2 marks)**

b Test, at the 10% significance level, the hypothesis $\sigma^2 = 0.19$ against the hypothesis $\sigma^2 \neq 0.19$. **(5 marks)**

A **7** Fishing line produced by a certain manufacturer is known to have a mean tensile breaking
strength of 170.2 kg and standard deviation 10.5 kg. The breaking strength of the fishing line is
normally distributed.

A new component is added to the material which will, it is claimed, decrease the standard
deviation without altering the tensile strength. A random sample of 20 pieces of the new fishing
line is selected. Each piece is tested to destruction and the tensile strength of each piece is noted.
The results are used to calculate unbiased estimates of the mean strength and standard deviation
of the population of new fishing line. These were found to be 172.4 kg and 8.5 kg.

   a Test, at the 5% level, whether the variance has been reduced. **(6 marks)**

   b What recommendation would you make to the manufacturer? **(1 mark)**

E/P **8** The masses of three-month-old Jack Russell puppies, X kg, are normally distributed
with $X \sim N(\mu, \sigma^2)$.

A random sample of 10 puppies is taken and their masses at three months old, x kg,
are recorded. The results are summarised as follows:

$$\sum x = 32.12 \quad \sum x^2 = 103.8592$$

   a Find an unbiased estimate for μ, and calculate the standard error of your estimate. **(3 marks)**

   b Stating your hypotheses clearly, test at the 5% level of significance whether the
   standard deviation of the masses of puppies is different from 0.25 kg. **(6 marks)**

6.3 The F-distribution

Customers in supermarkets like to buy produce of a uniform size. This means that the manager of a
supermarket is not only concerned with the mean size of, for example, apples, but also the variance.

Given two suppliers, the one likely to be chosen is the one with the lower variance.

Given also that the manager can only take a *sample* from each manufacturer, how can she tell
whether one variance is larger than the other?

The **F-test** is used to determine if two independent random samples from normal distributions have
equal variance. The F-test will be covered in Section 6.4. The F-test is based on the **F-distribution**,
which will be covered in this section.

Begin by assuming that the samples are independent and from normal distributions.

Suppose that you take a random sample of n_x observations from an $N(\mu_x, \sigma_x^2)$ distribution and,
independently, a random sample of n_y observations from an $N(\mu_y, \sigma_y^2)$ distribution. Unbiased
estimators for the two population variances are S_x^2 and S_y^2.

Using the result from Section 6.1,

$$\frac{(n_x - 1) S_x^2}{\sigma_x^2} \sim \chi_{n_x-1}^2 \text{ and } \frac{(n_y - 1) S_y^2}{\sigma_y^2} \sim \chi_{n_y-1}^2$$

A

It follows from this that $\dfrac{\dfrac{(n_x - 1)\,S_x^2}{\sigma_x^2}}{\dfrac{(n_y - 1)\,S_y^2}{\sigma_y^2}} \sim \dfrac{\chi^2_{n_x - 1}}{\chi^2_{n_y - 1}}$

so $\quad \dfrac{\dfrac{S_x^2}{\sigma_x^2}}{\dfrac{S_y^2}{\sigma_y^2}} \sim \dfrac{\dfrac{\chi^2_{n_x - 1}}{(n_x - 1)}}{\dfrac{\chi^2_{n_y - 1}}{(n_y - 1)}} \quad$ and $\quad \dfrac{\dfrac{S_y^2}{\sigma_y^2}}{\dfrac{S_x^2}{\sigma_x^2}} \sim \dfrac{\dfrac{\chi^2_{n_y - 1}}{(n_y - 1)}}{\dfrac{\chi^2_{n_x - 1}}{(n_x - 1)}}$

The distribution $\dfrac{\dfrac{\chi^2_{n_x - 1}}{(n_x - 1)}}{\dfrac{\chi^2_{n_y - 1}}{(n_y - 1)}}$ has two parameters, $(n_x - 1)$ and $(n_y - 1)$, and is usually denoted by $F_{n_x - 1,\, n_y - 1}$,

or by $F_{\nu_1,\, \nu_2}$ where $\nu_1 = (n_x - 1)$ and $\nu_2 = (n_y - 1)$, i.e. if $n_x = 13$ and $n_y = 9$ the distribution would be an $F_{12,\, 8}$-distribution.

Note Distributions of this type were first studied by Sir Ronald Fisher and hence are referred to as **F-distributions**.

- **For a random sample of n_x observations from an $N(\mu_x, \sigma_x^2)$ distribution and an independent random sample of n_y observations from an $N(\mu_y, \sigma_y^2)$ distribution,**

$$\frac{S_x^2/\sigma_x^2}{S_y^2/\sigma_y^2} \sim F_{n_x - 1,\, n_y - 1}$$

Note This distribution is given in the formulae booklet.

If you assume that $\sigma_x^2 = \sigma_y^2$, then $\dfrac{\sigma_y^2}{\sigma_x^2} = 1$ and

$$\frac{S_x^2}{S_y^2} \sim F_{n_x - 1,\, n_y - 1}$$

- **If a random sample of n_x observations is taken from a normal distribution with unknown variance σ^2 and an independent random sample of n_y observations is taken from a normal distribution with equal but unknown variance, then**

Watch out $F_{n_x - 1, n_y - 1} \neq F_{n_y - 1, n_x - 1}$, so the order of the parameters in the F-distribution is important. If S_x^2 is on top of the fraction, then $n_x - 1$ comes first, and if S_y^2 is on top of the fraction then $n_y - 1$ comes first.

$$\frac{S_x^2}{S_y^2} \sim F_{n_x - 1,\, n_y - 1}$$

The F-distribution has two parameters, $\nu_1 = n_x - 1$ and $\nu_2 = n_y - 1$, and to get all distributions relating to all possible combinations of ν_1 and ν_2 would require very extensive tables. However, the F-distribution is used mainly in hypothesis testing for variances, and so you are not really interested in all values of $F_{\nu_1,\, \nu_2}$.

Notation The numbers of degrees of freedom, ν_1 and ν_2, are used here because that is how they are described on the tables, but ν_x and ν_y could equally well be used.

A The values of F_{ν_1, ν_2} that are of interest are the critical values, which are exceeded with probabilities of 5%, 1% etc. These critical values are written $F_{\nu_1, \nu_2}(0.05)$, $F_{\nu_1, \nu_2}(0.01)$, etc.

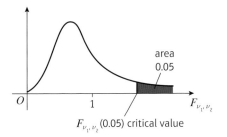

area 0.05

F_{ν_1, ν_2}

$F_{\nu_1, \nu_2}(0.05)$ critical value

> **Note** Some graphical calculators might also be able to give you critical values for an F-distribution.

A separate table is given for each significance level (the table on page 218 is for the 1% (0.01) and 5% (0.05) significance levels). The first row at the top gives values of ν_1 (remember that ν_1 comes first after the F), and the first column on the left gives values of ν_2. Where row and column meet gives the critical value corresponding to the significance level of that table. A short extract from the 0.05 significance level table is given below.

Probability ν_1 / ν_2	1	2	3	4	5	6	8	10	12	24	∞
1	161.4	199.5	215.7	224.6	230.2	234.0	238.9	241.9	243.9	249.1	254.3
2	18.51	19.00	19.16	19.25	19.30	19.33	19.37	19.40	19.41	19.46	19.50
3	10.13	9.55	9.28	9.12	9.01	8.94	8.85	8.79	8.74	8.64	8.53
4	7.71	6.94	6.59	6.39	6.26	6.16	6.04	5.96	5.91	5.77	5.63
5	6.61	5.79	5.41	5.19	5.05	4.95	4.82	4.74	4.68	4.53	4.37
6	5.99	5.14	4.76	4.53	4.39	4.28	4.15	4.06	4.00	3.84	3.67
7	5.59	4.74	4.35	4.12	3.97	3.87	3.73	3.64	3.57	3.41	3.23
8	5.32	4.46	4.07	3.84	3.69	3.58	3.44	3.35	3.28	3.12	2.93
9	5.12	4.26	3.86	3.63	3.48	3.37	3.23	3.14	3.07	2.90	2.71

(Probability **0.05** labels the left column block.)

To find the value of $F_{3,7}(0.05)$ you find the intersection of the 3rd column and 7th row.

> **Online** Expore the F-distribution using GeoGebra and use it to determine critical values of the sample variances.

Example 4

Use the table to find:

a the $F_{5, 8}(0.05)$ critical value

b the $F_{8, 5}(0.05)$ critical value.

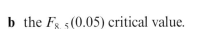

a $F_{5, 8}(0.05)$ critical value = 3.69

b $F_{8, 5}(0.05)$ critical value = 4.82

Using the table, the critical value at the 5% level for $F_{5, 8}$ is at the intersection of the 5th column and the 8th row.

The critical value of $F_{8, 5}$ is at the intersection of the 8th column and the 5th row.

A The F-distribution tables show you critical values which are **exceeded** with a given probability. So, for example, if f is an observation from a F_{ν_1, ν_2} distribution, then $P(f > F_{\nu_1, \nu_2}(0.05)) = 0.05$.

> **Notation** A critical value $F_{\nu_1, \nu_2}(p)$ which is exceeded with probability p is sometimes called an **upper critical value for p**. The value $F_{\nu_1, \nu_2}(1 - p)$ for which the observation is **less** than $F_{\nu_1, \nu_2}(1 - p)$ with probability p is sometimes called a **lower critical value for p**.

You are also interested in finding values at the other end of the distribution. For example, a value $F_{\nu_1, \nu_2}(0.95)$ for which $P(f > F_{\nu_1, \nu_2}(0.95)) = 0.95$, or equivalently $P(f < F_{\nu_1, \nu_2}(0.95)) = 0.05$. This is shown on the diagram below.

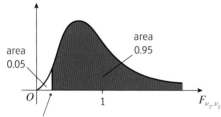

$F_{\nu_1, \nu_2}(0.95)$ upper critical value **or** 5% lower critical value.

You can use the table to find lower critical values such as this as follows:

$$P(F_{\nu_1, \nu_2} < x) = P\left(\frac{\frac{\chi^2_{\nu_1}}{\nu_1}}{\frac{\chi^2_{\nu_2}}{\nu_2}} < x\right)$$

$$= P\left(\frac{\frac{\chi^2_{\nu_2}}{\nu_2}}{\frac{\chi^2_{\nu_1}}{\nu_1}} > \frac{1}{x}\right) \qquad \text{If } a \text{ and } b \text{ are both positive then } a < b \Rightarrow \frac{1}{a} > \frac{1}{b}$$

$$= P\left(F_{\nu_2, \nu_1} > \frac{1}{x}\right)$$

So the lower critical value for p in an F_{ν_1, ν_2} distribution is the same as the **reciprocal** of the upper critical value for p in an F_{ν_2, ν_1} distribution. In other words,

■ $$F_{\nu_1, \nu_2}(p) = \frac{1}{F_{\nu_2, \nu_1}(1 - p)}$$

> **Watch out** The order of v_1 and v_2 have changed on the right-hand side.

Example 5

A Find upper critical values for:

a $F_{8,10}(0.95)$

b $F_{10,8}(0.95)$

a $F_{8,10}(0.95)$ critical value $= \dfrac{1}{3.35}$ ←— Use $\dfrac{1}{F_{10,8}(0.05) \text{ critical value}}$

$= 0.2985\ldots$

$= 0.30$ (2 d.p.)

b $F_{10,8}(0.95)$ critical value $= \dfrac{1}{3.07}$ ←— Use $\dfrac{1}{F_{8,10}(0.05) \text{ critical value}}$

$= 0.3257\ldots$

$= 0.33$ (2 d.p.)

Example 6

Find the lower and upper 5% critical values for an $F_{a,b}$-distribution in each of the following cases:

a $a = 6, b = 10$

b $a = 12, b = 8$

a The upper critical value is $F_{6,10}(0.05) = 3.22$

The lower 5% critical value is $F_{6,10}(0.95) = \dfrac{1}{F_{10,6}} = \dfrac{1}{4.06} = 0.25$ (2 d.p.)

b The upper critical value is $F_{12,8}(0.05) = 3.28$

The lower 5% critical value is $F_{12,8}(0.95) = \dfrac{1}{F_{8,12}} = \dfrac{1}{2.85} = 0.35$ (2 d.p.)

Example 7

The random variable X follows an F-distribution with 8 and 10 degrees of freedom.

Find $P\left(\dfrac{1}{5.81} < X < 5.06\right)$

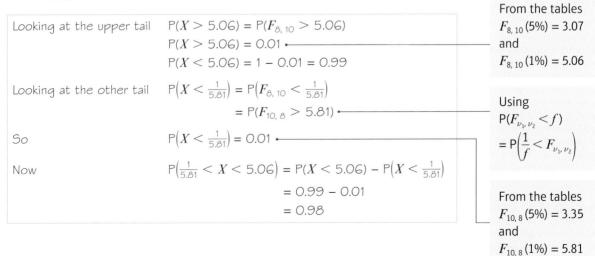

Looking at the upper tail	$P(X > 5.06) = P(F_{8,10} > 5.06)$	From the tables
	$P(X > 5.06) = 0.01$ ←	$F_{8,10}(5\%) = 3.07$
	$P(X < 5.06) = 1 - 0.01 = 0.99$	and
		$F_{8,10}(1\%) = 5.06$

Looking at the other tail $\quad P\left(X < \dfrac{1}{5.81}\right) = P\left(F_{8,10} < \dfrac{1}{5.81}\right)$

$= P(F_{10,8} > 5.81)$ ←

Using
$P(F_{\nu_1, \nu_2} < f)$
$= P\left(\dfrac{1}{f} < F_{\nu_1, \nu_2}\right)$

So $\quad P\left(X < \dfrac{1}{5.81}\right) = 0.01$ ←

Now $\quad P\left(\dfrac{1}{5.81} < X < 5.06\right) = P(X < 5.06) - P\left(X < \dfrac{1}{5.81}\right)$

$= 0.99 - 0.01$

$= 0.98$

From the tables
$F_{10,8}(5\%) = 3.35$
and
$F_{10,8}(1\%) = 5.81$

Example 8

A
The random variable X follows an $F_{6,12}$-distribution.
Find $P(X < 0.25)$.

$P(X < 0.25) = P(F_{6,12} < 0.25) = P\left(F_{12,6} > \frac{1}{0.25}\right)$

$\qquad\qquad\qquad\qquad\quad = P(F_{12,6} > 4)$

From the tables $\quad F_{12,6}(0.05) = 4$

So $\qquad\qquad\qquad P(F_{12,6} > 4) = P(F_{6,12} < 0.25) = 0.05$

Exercise 6C

1 Find the upper 5% critical value for an $F_{a,b}$-distribution in each of the following cases:
  a $a = 12, b = 18$ **b** $a = 4, b = 11$ **c** $a = 6, b = 9$

2 Find the lower 5% critical value for an $F_{a,b}$-distribution in each of the following cases:
  a $a = 6, b = 8$ **b** $a = 25, b = 12$ **c** $a = 5, b = 5$

3 Find the upper 1% critical value for an $F_{a,b}$-distribution in each of the following cases:
  a $a = 12, b = 18$ **b** $a = 6, b = 16$ **c** $a = 5, b = 9$

4 Find the lower 1% critical value for an $F_{a,b}$-distribution in each of the following cases:
  a $a = 3, b = 12$ **b** $a = 8, b = 12$ **c** $a = 5, b = 12$

5 Find the lower and upper 5% critical values for an $F_{a,b}$-distribution in each of the following cases:
  a $a = 8, b = 10$ **b** $a = 12, b = 10$ **c** $a = 3, b = 5$

(P) 6 The random variable X follows an $F_{40,12}$-distribution.
Find $P(X < 0.5)$.

(P) 7 The random variable X follows an $F_{12,8}$-distribution.
Find $P\left(\frac{1}{2.85} < X < 3.28\right)$.

(P) 8 The random variable X has an F-distribution with 2 and 7 degrees of freedom.
Find $P(X < 9.55)$.

9 The random variable X follows an F-distribution with 6 and 12 degrees of freedom.

a Show that $P(0.25 < X < 3.00) = 0.9$.

A large number of values are randomly selected from an F-distribution with 6 and 12 degrees of freedom.

b Find the probability that the seventh value to be selected will be the third value to lie between 0.25 and 3.00.

6.4 The F-test

From Section 6.3, you know that if a random sample of n_x observations is taken from a normal distribution with unknown variance σ^2 and an independent random sample of n_y observations is taken from a normal distribution with equal but unknown variance then

$$\frac{S_x^2}{S_y^2} \sim F_{n_x - 1,\, n_y - 1}$$

If $\sigma_x^2 = \sigma_y^2$, then you would expect $\dfrac{S_x^2}{S_y^2}$ to be close to 1, but if $\sigma_x^2 > \sigma_y^2$ then you would expect it to be greater than 1.

You wish to test the null hypothesis H_0: $\sigma_x^2 = \sigma_y^2$ against the alternative hypothesis H_1: $\sigma_x^2 > \sigma_y^2$.

As before when testing hypotheses, if your value of $\dfrac{S_x^2}{S_y^2}$ is such that it could only occur under the null hypothesis with a probability $\leqslant p$ (typically $p = 0.05$) then you reject the null hypothesis; otherwise you have to conclude that there is insufficient evidence to reject the null hypothesis.

To test whether two variances are the same, a simple set of rules can be followed.

1 Find s_l^2 and s_s^2, the larger and smaller variances respectively.

2 Write down the null hypothesis H_0: $\sigma_l^2 = \sigma_s^2$

3 Write down the alternative hypothesis H_1: $\sigma_l^2 > \sigma_s^2$ (one-tailed)

or H_1: $\sigma_l^2 \neq \sigma_s^2$ (two-tailed).

4 Look up the critical value of F_{ν_l, ν_s}, where ν_l is the number of degrees of freedom of the distribution with the larger variance and ν_s is the number of degrees of freedom of the distribution with the smaller variance. If a two-tailed test is used, p is halved (e.g. for a 10% significance level you would use $F_{\nu_l, \nu_s}(0.05)$ as the critical value).

5 Write down the critical region.

6 Calculate $F_{\text{test}} = \dfrac{s_l^2}{s_s^2}$

7 See whether F_{test} lies in the critical region and draw your conclusions. Relate these to the original problem.

Example 9

A

Two samples of sizes 13 and 9 are taken from normal distributions X and Y with variances σ_x^2 and σ_y^2 respectively. The two samples give values $s_x^2 = 24$ and $s_y^2 = 18$. Test, at the 5% level, H_0: $\sigma_x^2 = \sigma_y^2$ against H_1: $\sigma_x^2 > \sigma_y^2$

$\nu_l = 13 - 1 = 12,$ $\qquad$ $\nu_s = 9 - 1 = 8$

$s_l^2 = 24$ and $s_s^2 = 18$

The critical value is $F_{12,8}(0.05) = 3.28$

The test statistic is $\dfrac{s_l^2}{s_s^2} = \dfrac{24}{18} = 1.33$

$1.33 < 3.28$

There is insufficient evidence to reject H_0; the two populations have equal variances.

> The sample from X has the largest value of s^2, so use this sample size to calculate ν_1.

> This is the intersection of the 12th column and the 8th row in the table.

> The largest variance is divided by the smallest.

> Always state whether you accept or reject H_0 and draw a conclusion (in context if possible).

Example 10

Two samples of sizes 7 and 11 are taken from normal distributions X and Y with variances σ_x^2 and σ_y^2 respectively. The two samples give values $s_x^2 = 5$ and $s_y^2 = 25$. Test, at the 5% level of significance, H_0: $\sigma_x^2 = \sigma_y^2$ against H_1: $\sigma_x^2 < \sigma_y^2$

$\nu_l = 11 - 1 = 10,$ $\qquad$ $\nu_s = 7 - 1 = 6$

$s_l^2 = 25$ and $s_s^2 = 5$

The critical value is $F_{10,6}(0.05) = 4.06$

The test statistic is $\dfrac{s_l^2}{s_s^2} = \dfrac{25}{5} = 5$

$5 > 4.06$

There is sufficient evidence to reject H_0.

So $\sigma_x^2 < \sigma_y^2$

> The sample from Y has the largest value of s^2, so use this sample size to calculate ν_1.

> The largest variance is divided by the smallest.

Example 11

Two samples of sizes 11 and 13 are taken from normal distributions X and Y with variances σ_x^2 and σ_y^2 respectively. The two samples give values $s_x^2 = 1.6$ and $s_y^2 = 2.4$. Test, at the 10% level of significance, H_0: $\sigma_x^2 = \sigma_y^2$ against H_1: $\sigma_x^2 \neq \sigma_y^2$

$\nu_l = 13 - 1 = 12,$ $\qquad$ $\nu_s = 11 - 1 = 10$

$s_l^2 = 2.4$ and $s_s^2 = 1.6$

The critical value is $F_{12,10}(0.05) = 2.91$

> **Watch out** For a two-tailed F-test, there is still only **one critical value**. The only thing that changes is that the p-value is halved.

> Two-tailed test so use $F_{\nu_l, \nu_s}\left(\dfrac{p}{2}\right)$.

A

The test statistic is $\dfrac{s_f^2}{s_s^2} = \dfrac{2.4}{1.6} = 1.5$

$1.5 < 2.91$

There is insufficient evidence to reject H_0; the two variances are equal.

> Always state whether you accept or reject H_0 and draw a conclusion (in context if possible).

Example 12

A manufacturer of wooden furniture stores some of its wood outside and some inside a special store. It is believed that the wood stored inside should have less variable hardness properties than that stored outside. A random sample of 25 pieces of wood stored outside was taken and compared to a random sample of 21 similar pieces taken from the inside store, with the following results:

	Outside	Inside
Sample size	25	21
Mean hardness (coded units)	110	122
Sum of squares about the mean	5190	3972

a Test, at the 0.05 level of significance, whether the manufacturer's belief is correct.

b State an assumption you made in order to do this test.

a $s^2_{\text{outside}} = \dfrac{5190}{24} = 216.25$ and $s^2_{\text{inside}} = \dfrac{3972}{20} = 198.6$

> $s^2 = \dfrac{\sum(x_i - x)^2}{n - 1}$

$\nu_o = 25 - 1 = 24, \qquad \nu_i = 21 - 1 = 20$

$H_0: \sigma_o^2 = \sigma_i^2 \qquad\qquad H_1: \sigma_o^2 > \sigma_i^2$

Critical value $= F_{24,\,20}(0.05) = 2.08$

> Outside has the largest variance.

$F_{\text{test}} = \dfrac{216.25}{198.6} = 1.089$

> One-tailed test.

$1.089 < 2.08$, so there is insufficient evidence to reject H_0; wood stored inside is just as variable in hardness as wood stored outside.

b The assumption made is that the populations are normally distributed.

> Always state whether you accept or reject H_0 and draw a conclusion (in context if possible).

Exercise 6D

1 Random samples are taken from two normally distributed populations. There are 11 observations from the first population and the best estimate for the population variance is $s^2 = 7.6$. There are 7 observations from the second population and the best estimate for the population variance is $s^2 = 6.4$.

Test, at the 5% level of significance, $H_0: \sigma_1^2 = \sigma_2^2$ against $H_1: \sigma_1^2 > \sigma_2^2$.

A **2** Random samples are taken from two normally distributed populations. There are 25 observations from the first population and the best estimate for the population variance is $s^2 = 0.42$. There are 41 observations from the second population and the best estimate for the population variance is $s^2 = 0.17$.

Test, at the 1% significance level, $H_0: \sigma_1^2 = \sigma_2^2$ against $H_1: \sigma_1^2 > \sigma_2^2$

E **3** The sample variance of the lengths of a sample of 9 tent-poles produced by a machine was $63\ mm^2$. A second machine produced a sample of 13 tent-poles with a sample variance of $225\ mm^2$. Both these values are unbiased estimates of the population variances.

   a Test, at the 10% level, whether there is evidence that the machines differ in variability, stating the null and alternative hypotheses. **(6 marks)**

   b State the assumption you have made about the distribution of the populations in order to carry out the test in part **a**. **(1 mark)**

4 Random samples are taken from two normally distributed populations. The size of the sample from the first population is $n_1 = 13$ and this gives an unbiased estimate for the population variance $s_1^2 = 36.4$. The figures for the second population are $n_2 = 9$ and $s_2^2 = 52.6$.

Test, at the 5% significance level, whether $\sigma_1^2 = \sigma_2^2$ or if $\sigma_2^2 > \sigma_1^2$

E/P **5** Dining Chairs Ltd are in the process of selecting a make of glue for using on the joints of their furniture. There are two possible contenders – Goodstick, which is the more expensive, and Holdtight, the cheaper of the two.

The company are concerned that, while both glues are said to have the same adhesive power, one might be more variable than the other.

A series of trials are carried out with each glue and the joints tested to destruction. The force in newtons at which each joint failed is recorded. The results are as follows:

Goodstick: 10.3, 8.2, 9.5, 9.9, 11.4
Holdtight: 9.6, 10.8, 9.9, 10.8, 10.0, 10.2

   a Test, at the 10% significance level, whether the variances are equal. **(6 marks)**

   b Which glue would you recommend and why? **(1 mark)**

E/P **6** The closing balances, £x, of a number of randomly chosen bank current accounts of two different types, Chegrit and Dicabalk, are analysed by a statistician. The summary statistics are given in the table below.

	Sample size	Σx	Σx^2
Chegrit	7	276	143 742
Dicabalk	15	394	102 341

Stating your hypotheses clearly, test, at the 10% significance level, whether the two distributions have the same variance. (You may assume that the closing balances of each type of account are normally distributed.) **(6 marks)**

7 Bigborough Council wants to change the bulbs in their traffic lights at regular intervals so that there is a very small probability that any light bulb will fail in service.

The council wants the length of time between light bulb changes to be as long as possible. They have obtained a sample of bulbs from another manufacturer, who claims the same bulb life as their present manufacturer. The council wishes therefore to select the manufacturer whose bulbs have the smallest variance.

When they last tested a random sample of 9 bulbs from their present supplier the summary results were $\Sigma x = 9415$ hours, $\Sigma x^2 = 9\,863\,681$, where x represents the lifetime of a bulb.

A random sample of 8 bulbs from the prospective new supplier gave the following bulb lifetimes in hours:

$\qquad$ 1002, 1018, 943, 1030, 984, 963, 1048, 994

a Calculate unbiased estimates for the means and variances of the two populations. **(2 marks)**

Assuming that the lifetimes of bulbs are normally distributed,

b test, at the 10% significance level, whether the two variances are equal. **(5 marks)**

c State your recommendation to the council, giving reasons for your choice. **(1 mark)**

1 A random sample of 14 observations was taken of a random variable X which was normally distributed. The sample had mean $\bar{x} = 23.8$ and variance $s^2 = 1.8$.

Calculate:

a a 95% confidence interval for the variance of the population

b a 90% confidence interval for the variance of the population.

 2 A woollen mill produces scarves. The mill has several machines, each operated by a different person. Jane has recently started working at the mill and the supervisor wishes to check the lengths of the scarves Jane is producing. A random sample of 20 scarves is taken and the length, x cm, of each scarf is recorded. The results are summarised as:

$\qquad \sum x = 1428 \quad \sum x^2 = 102\,286$

Assuming that the lengths of scarves produced by any individual follow a normal distribution,

a calculate a 95% confidence interval for the variance σ^2 of the lengths of scarves produced by Jane. **(5 marks)**

The mill's owners require that 90% of scarves should be within 10 cm of the mean length.

b Find the value of σ that would satisfy this condition. **(3 marks)**

c Explain whether the supervisor should be concerned about the scarves Jane is producing. **(1 mark)**

 3 a Define a confidence interval. **(1 mark)**

A car rental company owner chooses 20 randomly selected dates and finds that the standard deviation of the number of cars rented is 3.75.

b Given that the number of cars rented is normally distributed, find, correct to 3 significant figures, a 90% confidence interval for the population standard deviation. **(5 marks)**

A **4** The lengths of adult slow-worms, X cm, are normally distributed with $X \sim N(\mu, \sigma^2)$.

E A random sample of 7 slow-worms is taken and their lengths, x cm, are recorded. The results are summarised as follows:

$$\sum x = 225.9 \quad \sum x^2 = 7338.07$$

Stating your hypotheses clearly, test, at the 5% level of significance, whether the standard deviation of the lengths of slow-worms is different from 2.7 cm. **(6 marks)**

E/P **5** Giovanna is a pharmacist and is testing the amount of time it takes for a particular dosage of a drug to take effect on patients. She selects a random sample of 10 patients and records the time, in minutes, for the drug to take effect on each patient.

10, 12, 13, 15, 17, 17, 19, 21, 22, 25

a Calculate a 95% confidence interval for the standard deviation of the time taken for the drug to take effect. **(6 marks)**

Given that the times taken are independent,

b state what further assumption is necessary for this confidence interval to be valid. **(1 mark)**

Giovanna requires the standard deviation of the times taken to be less than 3.1 minutes otherwise she must change the dosage.

c Use your answer to part **a** to decide if the dosage needs changing. **(1 mark)**

E **6** The maximum weight that 50 cm lengths of a certain make of string can hold before breaking (the breaking strain) has a normal distribution with mean 40 kg and standard deviation 5 kg. The manufacturer of the string has developed a new process which should increase the mean breaking strain of the string but should not alter the standard deviation. Ten randomly selected pieces of string are tested and their breaking strains, in kg, are:

51, 48, 37, 46, 36, 53, 34, 49, 47, 50

Stating your hypotheses clearly, test, at the 5% level of significance, whether the new process has altered the variance. **(6 marks)**

7 The random variable X has an F-distribution with 5 and 10 degrees of freedom.
Find values of a and b such that $P(a \leqslant X \leqslant b) = 0.90$.

E **8** The standard deviation of the length of a random sample of 8 fence posts produced by a timber yard was 8 mm. A second timber yard produced a random sample of 13 fence posts with a standard deviation of 14 mm.

a Test, at the 10% significance level, whether there is evidence that the lengths of fence posts produced by these timber yards differ in variability. State your hypotheses clearly. **(6 marks)**

b State an assumption you have made in order to carry out the test in part **a**. **(1 mark)**

9 The lengths, x mm, of the forewings of a random sample of male and female adult butterflies are measured. The following statistics are obtained from the data.

	Number of butterflies	Sample mean $\bar{x}$	$\sum x^2$
Females	7	50.6	17 956.5
Males	10	53.2	28 335.1

Assuming the lengths of the forewings are normally distributed, test, at the 10% level of significance, whether the variances of the two distributions are the same. State your hypotheses clearly. **(6 marks)**

Challenge

Hint Use $\dfrac{(n-1)S^2}{\sigma^2} \sim \chi^2_{n-1}$

a Given that $\text{Var}(\chi^2_v) = 2v$, show that $\text{Var}(S^2) = \dfrac{2\sigma^4}{n-1}$ where S^2 is the sample variance and σ^2 is the population variance.

b Hence comment on the quality of S^2 as an estimator for σ^2 as the sample size increases.

Summary of key points

1. If a random sample of n observations $X_1, X_2, \ldots, X_n$ is selected from $N(\mu, \sigma^2)$, then
$$\frac{(n-1)S^2}{\sigma^2} \sim \chi^2_{n-1}$$

2. Generally, for a probability of α that the variance falls outside the limits,
   - the $100(1-\alpha)\%$ **confidence limits** are
   $$\frac{(n-1)s^2}{\chi^2_{n-1}\left(\frac{\alpha}{2}\right)} \text{ and } \frac{(n-1)s^2}{\chi^2_{n-1}\left(1-\frac{\alpha}{2}\right)}$$
   - the $100(1-\alpha)\%$ **confidence interval** for the variance of a normal distribution is
   $$\left(\frac{(n-1)s^2}{\chi^2_{n-1}\left(\frac{\alpha}{2}\right)}, \frac{(n-1)s^2}{\chi^2_{n-1}\left(1-\frac{\alpha}{2}\right)}\right)$$

3. To carry out a hypothesis test for for the variance of a normal distribution, follow these steps.
   - Write down the null hypothesis (H_0).
   - Write down the alternative hypothesis (H_1).
   - Specify the significance level.
   - Write down the degrees of freedom ν.
   - Write down the critical region.
   - Identify the population variance, σ^2, and the unbiased estimate, s^2, and calculate the value of the test statistic $\dfrac{(n-1)s^2}{\sigma^2}$
   - Complete your test and state your conclusions, stating whether H_0 is accepted or rejected, and interpreting this in the context of the question.

4 The F-distribution has two parameters, $(n_x - 1)$ and $(n_y - 1)$ and is usually denoted by F_{n_x-1, n_y-1}. For a random sample of n_x observations from an $N(\mu_x, \sigma_x^2)$ distribution and an independent random sample of n_y observations from an $N(\mu_y, \sigma_y^2)$ distribution,

$$\frac{S_x^2/\sigma_x^2}{S_y^2/\sigma_y^2} \sim F_{n_x-1, n_y-1}$$

5 If a random sample of n_x observations is taken from a normal distribution with unknown variance σ^2 and an independent random sample of n_y observations is taken from a normal distribution with equal but unknown variance, then

$$\frac{S_x^2}{S_y^2} \sim F_{n_x-1, n_y-1}$$

6 A critical value $F_{\nu_1, \nu_2}(p)$ which is exceeded with probability p is called an **upper critical value** for p. The value $F_{\nu_1, \nu_2}(1-p)$ for which the observation is less than $F_{\nu_1, \nu_2}(1-p)$ with probability p is called a **lower critical value** for p.

$$F_{\nu_1, \nu_2}(p) = \frac{1}{F_{\nu_2, \nu_1}(1-p)}$$

7 To test whether two variances are the same, a simple set of rules can be followed:

- Find s_l^2 and s_s^2, the larger and smaller variances respectively.
- Write down the null hypothesis H_0: $\sigma_l^2 = \sigma_s^2$
- Write down the alternative hypothesis H_1: $\sigma_l^2 > \sigma_s^2$ (one-tailed), or H_1: $\sigma_l^2 \neq \sigma_s^2$ (two-tailed).
- Look up the critical value of F_{ν_l, ν_s}, where ν_l is the number of degrees of freedom of the distribution with the larger variance and ν_s is the number of degrees of freedom of the distribution with the smaller variance. If a two-tailed test is used, p is halved (e.g. for a 10% significance level you would use $F_{\nu_l, \nu_s}(5\%)$ as the critical value).
- Write down the critical region.
- Calculate $F_{\text{test}} = \dfrac{s_l^2}{s_s^2}$
- See whether F_{test} lies in the critical region and draw your conclusions. Relate these to the original problem.

Confidence intervals and tests using the *t*-distribution

7

Prior knowledge check

1. A random sample of size 20 is taken from a normally distributed population with a standard deviation of 2. The mean of the sample was 16.

   Find a 95% confidence interval for the mean μ. ← **Section 5.2**

2. A researcher is comparing the heights of children in two towns. A random sample of 100 children from town A is taken and the sample mean and standard deviation are 145 cm and 4 cm respectively. An independent random sample of 120 children from town B is taken and the sample mean and standard deviation are 146 cm and 3.5 cm respectively.

   Test, at the 5% level of significance, whether there is evidence of a difference in the mean heights of the children in the two towns. ← **Section 5.3**

Farmers often try out different diets to see which is the most effective at producing high-yield animals. They can compare the effectiveness of two diets using a paired *t*-test. → **Mixed exercise Q13**

7.1 Mean of a normal distribution with unknown variance

A You know that for a normally distributed random variable X, the sample mean will also be distributed normally:

$$\overline{X} \sim N\left(\mu, \frac{\sigma^2}{n}\right)$$

This means that you can calculate a confidence interval for the population mean, μ, **as long as you know** the population variance. In most cases, however, if you are taking a sample from a large population you will not already know the population variance.

If the sample size, n, is **large**, then you can use the sample variance as an approximation of the population variance.

However, if n is **small**, S is unlikely to be very close to σ and $\dfrac{\overline{X} - \mu}{\frac{S}{\sqrt{n}}}$ can no longer be

modelled by the normal distribution N(0, 1²).

Links For a large sample of size n from a normal population, $\dfrac{\overline{X} - \mu}{\frac{S}{\sqrt{n}}}$ is approximately normal with distribution N(0, 1²). ← **Section 5.4**

When n is small we usually use the symbol t to denote the quantity $\dfrac{\overline{X} - \mu}{\frac{S}{\sqrt{n}}}$

- **If a random sample X_1, X_2, ... , X_n is selected from a normal distribution with mean μ and unknown variance σ^2 then**

$$t = \frac{\overline{X} - \mu}{\frac{S}{\sqrt{n}}}$$

has a t_{n-1}-distribution where S^2 is an unbiased estimator of σ^2.

Links $S^2 = \dfrac{1}{n-1}\left(\sum_{i=1}^{n} X_i^2 - n\overline{X}^2\right)$ ← **Section 5.1**

There are a family of t-distributions determined by the value of n. This establishes the number of degrees of freedom, similar to the chi-squared and F-distributions.

The number of degrees of freedom, ν, is equal to $n - 1$ and as $\nu \to \infty$, the t-distribution approaches the distribution N(0, 1²). For this reason the t-distribution is usually used when the **sample size, n, is small**. For larger sample sizes it is more convenient to approximate t with a normal distribution. The diagram below shows two examples of the t-distribution for different values of ν, together with the standardised normal distribution.

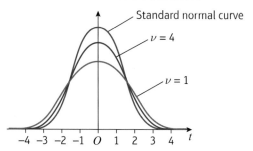

Notation W. S. Gosset, who published his works under the pseudonym 'the student', first investigated the probability distribution of $\dfrac{\overline{X} - \mu}{\frac{S}{\sqrt{n}}}$ for a sample taken from a normal distribution. The resulting distribution is known as 'Student's t-distribution', or more commonly just the t-distribution.

A As with the F-distribution and the χ^2 distribution, the critical values of the *t*-distribution depend on the number of degrees of freedom. The table of values in the formulae booklet, and on page 217, gives percentage points for the *t*-distribution for certain values of ν up to 120.

Note The *t*-distribution is symmetric in the same way as the normal distribution.

The values in the table are those which a random variable with Student's *t*-distribution on ν degrees of freedom exceeds with the probability shown.

ν	0.10	0.05	0.025	0.01	0.005
1	3.078	6.314	12.706	31.821	63.657
2	1.886	2.920	4.303	6.965	9.925
3	1.638	2.353	3.182	4.541	5.841
4	1.533	2.132	2.776	3.747	4.604
5	1.476	2.015	2.571	3.365	4.032
6	1.440	1.943	2.447	3.143	3.707
7	1.415	1.895	2.365	2.998	3.499
8	1.397	1.860	2.306	2.896	3.355

For example, if X has the t_7-distribution with 7 degrees of freedom ($n = 8$):
- $P(X > 1.895) = 0.05$
- $P(X < 1.895) = 0.95$

and by the symmetry of the *t*-distribution:
- $P(X < -1.895) = 0.05$

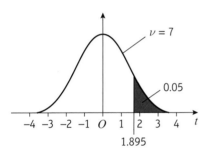

Note You may be able to use your calculator to find the percentage points for the *t*-distribution rather than the tables.

Watch out When working with the *t*-distribution you are strongly advised to **draw appropriate diagrams** so that you are sure in your own mind which areas under the *t*-distribution you are dealing with.

Online Expore the *t*-distribution using GeoGebra and use it to determine critical values of the sample variances.

Example **1**

The random variable X has a *t*-distribution with 10 degrees of freedom. Determine values of t for which:

a $P(X > t) = 0.025$

b $P(X < t) = 0.95$

c $P(X < t) = 0.025$

d $P(|X| > t) = 0.05$

e $P(|X| < t) = 0.98$

Notation $|X|$ means the modulus of X. This is the absolute value of X ignoring the sign, for example the modulus of -5, written $|-5|$ is 5. So
$P(|X| > t) = P(X < -t) + P(X > t)$
$P(|X| < t) = P(-t < X < t)$

A

a $\nu = 10$

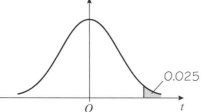

0.025

$t_{10}(0.025) = 2.228$

You are looking for $P(X > t)$ so you can use the tables directly. Look where the $\nu = 10$ row intersects with the 0.025 column.

b

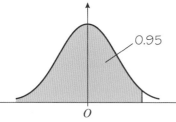

0.95

If $P(X < t) = 0.95$ then $P(X > t) = 1 - 0.95 = 0.05$
From the table $t_{10}(0.05) = 1.812$

The whole area under the curve = 1.
So $P(X > t) = 1 - P(X < t)$. Look where the $\nu = 10$ row intersects with the 0.05 column.

c

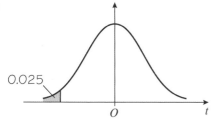

0.025

From **a**, $P(X > t) = 0.025$ when $t = 2.228$
so $P(X < t) = 0.025$ if $t = -2.228$

Because the distribution is symmetrical $P(X < -t) = P(X > t)$. You know from part **a** that $P(X > t) = 0.025$ if $t = 2.228$, so $P(X < t) = 0.025$ when $t = -0.228$.

d

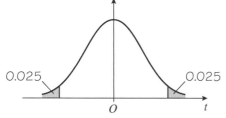

0.025   0.025

From **a** and **c**, $P(|X| > t) = 0.05$ if $X < -2.228$ and $X > 2.228$
There are therefore two values for t and they are
-2.228 and 2.228

This is two-tailed with probability of 0.025 at each tail.

e

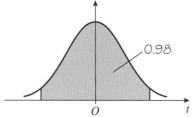

0.98

$P(|X| > t) = 0.01$ if $t = 2.764$ and $-2.764 < X < 2.764$
Again there are two values of t and they are -2.764 and 2.764

Again, a two-tailed problem. From the diagram you can see that you are looking for tails each with probability 0.01.

Example ②

Ⓐ The random variable Y has a t_4-distribution.

Determine:

a $P(Y > 3.747)$ **b** $P(Y < -2.132)$

---

a $\nu = 4$
$P(Y > 3.747) = 0.01$

From the $\nu = 4$ row of the table you can see that 3.747 is in the 0.01 probability column.

b $P(Y > 2.132) = 0.05$
By symmetry $P(Y < -2.132) = 0.05$

From the $\nu = 4$ row of the table you can see that 2.132 is in the 0.05 probability column.

You can use the *t*-distribution to find a confidence interval for the mean of a normal distribution when the variance is unknown.

For a sample taken from a normal population with unknown variance,

$$t = \frac{\overline{X} - \mu}{\dfrac{S}{\sqrt{n}}} \text{ has a } t_{n-1}\text{-distribution.}$$

If you want to find a 95% confidence interval for the population mean, μ, then you start by finding a value of t such that

$$P\left(\frac{\overline{X} - \mu}{\dfrac{S}{\sqrt{n}}} > t\right) = 0.025 \quad \text{and} \quad P\left(\frac{\overline{X} - \mu}{\dfrac{S}{\sqrt{n}}} < -t\right) = 0.025$$

This value of t is called the t_{n-1} value for a probability of 0.025. You can find it using tables or your calculator.

Notation This value is written as $t_{n-1}(0.025)$.

$$\text{Thus } P\left(-t_{n-1}(0.025) < \frac{\overline{X} - \mu}{\dfrac{S}{\sqrt{n}}} < t_{n-1}(0.025)\right) = (1 - 0.025) - 0.025 = 0.95$$

Look at the inequality inside the bracket:

$$-t_{n-1}(0.025) \times \frac{S}{\sqrt{n}} < \overline{X} - \mu < t_{n-1}(0.025) \times \frac{S}{\sqrt{n}}$$

$$-t_{n-1}(0.025) \times \frac{S}{\sqrt{n}} - \overline{X} < -\mu < t_{n-1}(0.025) \times \frac{S}{\sqrt{n}} - \overline{X}$$

$$\overline{X} - t_{n-1}(0.025) \times \frac{S}{\sqrt{n}} < \mu < \overline{X} + t_{n-1}(0.025) \times \frac{S}{\sqrt{n}}$$

You are interested in μ, so here you try to isolate it by

1 multiplying by $\dfrac{S}{\sqrt{n}}$

2 subtracting $\overline{X}$

3 multiplying by -1 and altering the inequality.

For a particular sample with mean $\overline{x}$ and variance s^2, this becomes:

$$P\left(\overline{x} - t_{n-1}(0.025) \times \frac{S}{\sqrt{n}} < \mu < \overline{x} + t_{n-1}(0.025) \times \frac{S}{\sqrt{n}}\right) = 0.95$$

So for a small sample of size n from a normal distribution $N(\mu, \sigma^2)$ with unknown mean and variance:

- the 95% confidence limits for μ are given by $\overline{x} \pm t_{n-1}(0.025) \times \dfrac{S}{\sqrt{n}}$

- the 95% confidence interval for μ is given by

$$\left(\overline{x} - t_{n-1}(0.025) \times \frac{S}{\sqrt{n}}, \overline{x} + t_{n-1}(0.025) \times \frac{S}{\sqrt{n}}\right)$$

A In the same way,

- the 90% confidence limits for μ are given by $\bar{x} \pm t_{n-1}(0.05) \times \dfrac{s}{\sqrt{n}}$

- the 90% confidence interval for μ is given by

$$\left(\bar{x} - t_{n-1}(0.05) \times \frac{s}{\sqrt{n}}, \ \bar{x} + t_{n-1}(0.05) \times \frac{s}{\sqrt{n}}\right)$$

- **In general, for a small sample of size n from a normal distribution $N(\mu, \sigma^2)$ with unknown mean and variance:**

  - **the $100(1 - \alpha)\%$ confidence limits for the population mean are**

    $$\bar{x} \pm t_{n-1}\left(\frac{\alpha}{2}\right) \times \frac{s}{\sqrt{n}}$$

  - **the $100(1 - \alpha)\%$ confidence interval for the population mean is**

    $$\left(\bar{x} - t_{n-1}\left(\frac{\alpha}{2}\right) \times \frac{s}{\sqrt{n}}, \ \bar{x} + t_{n-1}\left(\frac{\alpha}{2}\right) \times \frac{s}{\sqrt{n}}\right)$$

Example 3

A sample of 6 trout taken from a fish farm were caught and their lengths in centimetres were measured. The lengths of the fish were as follows:

26.8	26.0	25.8	25.5	24.3	24.6

Assuming that the lengths of trout are normally distributed, find a 90% confidence interval for the mean length of trout in the fish farm.

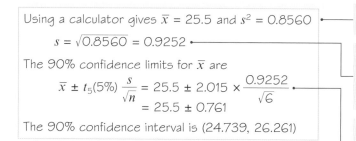

Using a calculator gives $\bar{x} = 25.5$ and $s^2 = 0.8560$ — First find the sample mean and variance.

$s = \sqrt{0.8560} = 0.9252$ — The standard deviation of the sample can be found by taking the square root of the variance.

The 90% confidence limits for $\bar{x}$ are

$\bar{x} \pm t_5(5\%) \dfrac{s}{\sqrt{n}} = 25.5 \pm 2.015 \times \dfrac{0.9252}{\sqrt{6}}$

$= 25.5 \pm 0.761$

The 90% confidence interval is $(24.739, 26.261)$ — Put your values for $\bar{x}$ and s into the formula, and work out the confidence interval.

Example 4

The percentage starch content of potatoes is normally distributed with mean μ. In order to assess the mean value of the starch content, a random sample of 12 potatoes is selected and their starch content measured. The percentages of starch contents obtained were as follows:

23.2	20.3	18.6	20.0	20.8	21.6	19.4	18.7	22.1	19.5	21.3	22.6

Find a 95% confidence interval for the mean.

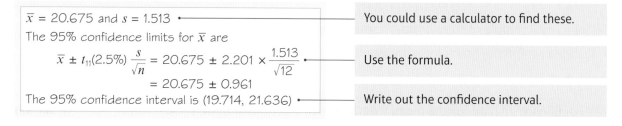

$\bar{x} = 20.675$ and $s = 1.513$ — You could use a calculator to find these.

The 95% confidence limits for $\bar{x}$ are

$\bar{x} \pm t_{11}(2.5\%) \dfrac{s}{\sqrt{n}} = 20.675 \pm 2.201 \times \dfrac{1.513}{\sqrt{12}}$ — Use the formula.

$= 20.675 \pm 0.961$

The 95% confidence interval is $(19.714, 21.636)$ — Write out the confidence interval.

Exercise 7A

A

1 Given that the random variable X has a t_{12}-distribution, find values of t such that:

   a $P(X < t) = 0.025$ **b** $P(X > t) = 0.05$ **c** $P(|X| > t) = 0.95$

2 Find:

   a $t_{26}(0.01)$ **b** $t_{26}(0.05)$

3 The random variable Y has a t_n-distribution. Find a value (or values) of t that satisfies each of the following.

   a $n = 10$, $P(Y < t) = 0.95$ **b** $n = 32$, $P(Y < t) = 0.005$ **c** $n = 5$, $P(Y < t) = 0.025$

   d $n = 16$, $P(|Y| < t) = 0.98$ **e** $n = 18$, $P(|Y| > t) = 0.10$

4 A test on the life (in hours) of a certain make of torch batteries gave the following results.

        20.3   17.3   25.0   18.4   16.3   24.8   24.3   21.2

   Assuming that the lifetime of batteries is normally distributed, find a 90% confidence interval for the mean.

5 A sample of size 16 taken from a normal population with unknown variance gave the following sample values.

        $\bar{x} = 12.4$ $s^2 = 21.0$

   Find a 95% confidence interval for the population mean.

6 The mean heights (measured in centimetres) of six male students at a college were as follows:

        182   178   183   180   169   184

   Calculate:

   a a 90% confidence interval   **b** a 95% confidence interval

   for the mean height of male students at the college.

   You may assume that the heights are normally distributed.

E   7 The masses (in grams) of 10 nails selected at random from a bin of 90 mm long nails were:

        9.7   10.2   11.2   9.4   11.0   11.2   9.8   9.8   10.0   11.3

   a Calculate a 98% confidence interval for the mean mass of the nails in the bin.   **(6 marks)**

   b State one assumption you have made in your calculation.   **(1 mark)**

E   8 A random sample of the feet of 8 adult males gave the following summary statistics of length x (in cm):

        $\Sigma x = 224.1$ $\Sigma x^2 = 6337.39$

   Assuming that the length of men's feet is normally distributed, calculate a 99% confidence interval for the mean length of men's feet based upon these results.   **(6 marks)**

A **9** A random sample of 26 students from the sixth form of a school sat an intelligence test that
E measured their IQs. The results are summarised below.

$$\bar{x} = 122 \qquad s^2 = 225$$

Assuming that IQ is normally distributed, calculate a 95% confidence interval for the mean
IQ of the students. **(6 marks)**

E/P **10** Add ticks to this table to show the distribution you would use when finding a confidence
interval.

	Normal	χ^2	t
For the population mean, using a sample of size 50 from a population of unknown variance			
For the population mean, using a sample of size 6 from a population of known variance			
For the population variance, using a sample of size 20			

(3 marks)

E/P **11** A company manufactures light bulbs which they state have an average lifespan of 500 days.
The manager is concerned that the production process is faulty and that the light bulbs do not
last as long as stated.
He tests a random sample of 15 bulbs and finds their lifespan, x days. The data is summarised as

$$\Sigma x = 7338 \qquad \Sigma x^2 = 3\,618\,260$$

a Explain why you need to use the t-distribution to find a confidence interval for the
population mean. **(1 mark)**

b Find a 90% confidence interval for the mean lifespan, μ days, of the light bulbs. **(6 marks)**

c State one assumption you have made in finding your answer to part **b**. **(1 mark)**

d Find a 95% confidence interval for the population variance. **(4 marks)**

> **Hint** Confidence intervals for the population
> variance are covered in Chapter 6. ← **Section 6.1**

7.2 Hypothesis test for the mean of a normal distribution with unknown variance

Apart from using the t-distribution rather than the normal distribution for finding the critical region,
testing the mean of a normal distribution with unknown variance follows the same steps as you used
when testing the mean of a normal distribution with known variance.

The following steps might help you in answering questions about hypothesis testing of the mean of a
normal distribution with unknown variance.

1 Write down H_0.

2 Write down H_1.

3 Specify the significance level, α.

4 Write down the number of degrees of freedom, ν.

A 5 Write down the critical region.

6 Calculate $\bar{x}$, s^2 and t using

$$\bar{x} = \frac{\sum x}{n}, \quad s^2 = \frac{\sum(x^2 - \bar{x})^2}{n-1} \quad \left(\text{or } s^2 = \frac{\sum x^2 - n\bar{x}^2}{n-1}\right) \text{ and } t = \frac{\bar{x} - \mu}{\frac{s}{\sqrt{n}}}$$

7 Conclusions

The following points should be addressed:

  i Is the result significant or not?

  ii What are the implications in terms of the original problem?

Example 5

A shopkeeper sells jars of jam. The weights of the jars of jam are normally distributed with a mean of 150 g. A customer complains that the mean weight of 8 jars she had bought was only 147 g. An estimate for the standard deviation of the weights of the 8 jars of jam calculated from the 8 observations was 2 g.

a Test, at the 5% significance level, whether 147 g is significantly less than the quoted mean.

b Discuss whether the customer has cause for complaint.

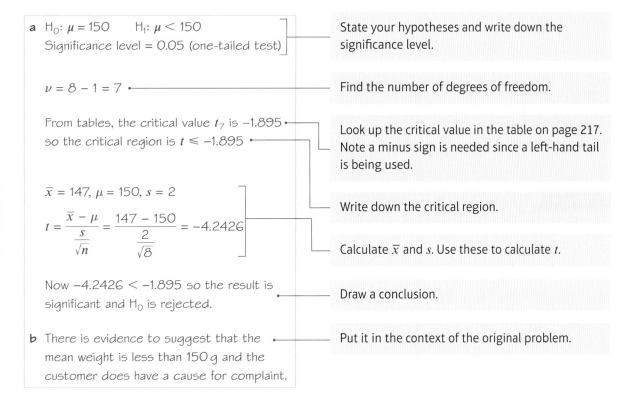

a H_0: $\mu = 150$ H_1: $\mu < 150$
Significance level = 0.05 (one-tailed test)

> State your hypotheses and write down the significance level.

$\nu = 8 - 1 = 7$

> Find the number of degrees of freedom.

From tables, the critical value t_7 is −1.895 so the critical region is $t \leqslant -1.895$

> Look up the critical value in the table on page 217. Note a minus sign is needed since a left-hand tail is being used.

$\bar{x} = 147$, $\mu = 150$, $s = 2$

$$t = \frac{\bar{x} - \mu}{\frac{s}{\sqrt{n}}} = \frac{147 - 150}{\frac{2}{\sqrt{8}}} = -4.2426$$

> Write down the critical region.

> Calculate $\bar{x}$ and s. Use these to calculate t.

Now −4.2426 < −1.895 so the result is significant and H_0 is rejected.

> Draw a conclusion.

b There is evidence to suggest that the mean weight is less than 150 g and the customer does have a cause for complaint.

> Put it in the context of the original problem.

Example 6

A The temperature (°C) was measured at noon on 10 days during the month of March in West Cumbria. The readings were:

12.8   11.4   12.9   15.1   15.4   13.5   14.9   15.0   16.0   15.8

Using a 5% significance level, test whether or not this is an increase over the previous year when the average noon temperature was 13.5 °C.

$H_0: \mu = 13.5$ $H_1: \mu > 13.5$

Significance level 5%

$\nu = 9$

From tables, the critical value is $t_9 = 1.833$

so the critical region is $t \geqslant 1.833$

$\bar{x} = \dfrac{12.8 + 11.4 + 12.9 + 15.1 + 15.4 + 13.5 + 14.9 + 15.0 + 16.0 + 15.8}{10}$

$= 14.28$

$s^2 = \dfrac{\sum x^2 - n\bar{x}^2}{n-1} = \dfrac{2060.28 - 10 \times 14.28^2}{10-1} = 2.344$

$s = 1.531$

$t = \dfrac{\bar{x} - \mu}{\dfrac{s}{\sqrt{n}}} = \dfrac{14.28 - 13.5}{\dfrac{1.531}{\sqrt{10}}} = 1.611$

$1.611 < 1.833$, so the result is not significant.

There is not enough evidence to suggest that the average temperature has increased.

State your hypotheses and write down the significance level.

Write down the critical region.

Calculate $\bar{x}$ and s. Note both of these are easily found using a calculator.

Calculate t.

Draw a conclusion.

Example 7

A concrete manufacturer tests cubes of its concrete at regular intervals, and their compressive strengths in N mm^{-2} are determined. The mean value of the strengths is required to be 0.47 N mm^{-2}. A new supplier of cement offers to supply the firm at a cheaper rate than the present supplier, and a trial bag of cement is used to make 12 concrete cubes. Upon testing, these cubes are found to have strengths (x) such that $\sum x = 5.52$ and $\sum x^2 = 2.542$. Assume that the strengths are normally distributed.

a Stating your hypotheses clearly, test, at the 5% level of significance, whether or not the use of the new cement has altered the mean strength of the concrete.

b In the light of your conclusion to the test in part **a**, what would you recommend the manufacturer to do?

a H_0: $\mu = 0.47$ H_1: $\mu \neq 0.47$

$\nu = 12 - 1 = 11$

Probability in each tail = 0.025

From tables the critical value is 2.201

The critical region is $|t| \geqslant 2.201$

$$\bar{x} = \frac{\sum x}{n} = \frac{5.52}{12} = 0.46$$

$$s^2 = \frac{\sum x^2 - n\bar{x}^2}{n - 1} = \frac{2.542 - 12 \times 0.46^2}{11} = 0.0002545$$

$s = 0.016$

$$t = \frac{\bar{x} - \mu}{\frac{s}{\sqrt{n}}} = \frac{0.46 - 0.47}{\frac{0.016}{\sqrt{12}}} = -2.165$$

Now $|-2.165| < |-2.201|$

The result is not significant. There is not enough evidence to suggest that the mean strength has altered.

b Since the mean strength has not altered, the manufacturer should accept the new supplier because they are cheaper. The two values -2.165 and -2.201 are quite close, however, and a one-tailed test of whether or not the strength had decreased should be done, or failing this a further sample could be taken.

You are looking to see if the strength has altered up or down so use $\neq$ in H_1.

This is a two-tailed test so halve the significance level to find the probability in each tail.

Since you are given $\sum x$ and $\sum x^2$ in the question, use these formulae.

Since $\bar{x} < \mu$, t is negative.

t lies between -2.201 and 2.201.

Draw the conclusion in context.

Base your recommendation on your conclusion.

Exercise 7B

1 Given that the observations 9, 11, 11, 12, 14, have been drawn from a normal distribution, test H_0: $\mu = 11$ against H_1: $\mu > 11$. Use a 5% significance level.

2 A random sample of size 28 taken from a normally distributed variable gave the sample values $\bar{x} = 17.1$ and $s^2 = 4$. Test H_0: $\mu = 19$ against H_1: $\mu < 19$. Use a 1% level of significance.

3 A random sample of size 13 taken from a normally distributed variable gave the sample values $\bar{x} = 3.26$ and $s^2 = 0.64$. Test H_0: $\mu = 3$ against H_1: $\mu \neq 3$. Use a 5% significance level.

 4 A certain brand of blanched hazelnuts for use in cooking is sold in packets. The weights of the packets of hazelnuts follow a normal distribution with mean, μ. The manufacturer claims that $\mu = 100$ g. A sample of 15 packets was taken and the weight, x, of each was measured. The results are summarised by the following statistics: $\sum x = 1473$, $\sum x^2 = 148\,119$.

  a Explain why it is not suitable to use a normal approximation for $\dfrac{\bar{X} - \mu}{\frac{S}{\sqrt{n}}}$ in this instance.   **(1 mark)**

  b Test, at the 5% significance level, whether or not there is evidence to justify the manufacturer's claim.   **(7 marks)**

A **5** A manufacturer claims that the lifetimes of its 100-watt bulbs are normally distributed with a
E mean of 1000 hours. A laboratory tests 8 bulbs and finds their lifetimes to be 985, 920, 1110, 1040, 945, 1165, 1170 and 1055 hours.

Stating your hypotheses clearly, examine whether or not the bulbs have a longer mean lifetime than that claimed. Use a 5% level of significance. **(7 marks)**

E **6** A fertiliser manufacturer claims that by using brand F fertiliser the yield of fruit bushes will be increased. A random sample of 14 fruit bushes was fertilised with brand F and the resulting yields, x, were summarised by $\sum x = 90.8$, $\sum x^2 = 600$. The yield of bushes fertilised by the usual fertiliser was normally distributed with a mean of 6 kg of fruit per bush.
Test, at the 2.5% significance level, the manufacturer's claim. **(7 marks)**

E **7** A nuclear reprocessing company claims that the amount of radiation within a reprocessing building in which there had been an accident had been reduced to an acceptable level by their clean-up team. The amounts of radiation, x, at 20 sites within the building in suitable units are summarised by $\sum x = 21.7$, $\sum x^2 = 28.4$. In the same units, the acceptable level of radiation is given as 1.00.

a By carrying out a suitable test for the population mean, test whether the building falls within acceptable radiation levels. **(7 marks)**

b State one assumption made in carrying out your test. **(1 mark)**

E/P **8** Scores in an aptitude test are assumed to be normally distributed with a population mean of 100. A company claims to be able to train people to improve their scores in the test. A random sample of 20 people is taken and they are trained before taking the test. The sample standard deviation is found to be 15 and the mean of the scores of the 20 people is found to be 110.

a Test, at the 5% level of significance, whether there is evidence of the training improving the scores of participants. State your hypotheses clearly. **(5 marks)**

b Test, at the 10% level of significance, the hypothesis that the population standard deviation is different from 12. State your hypotheses clearly. **(5 marks)**

7.3 The paired t-test

There are many occasions when you might want to compare results before and after some treatment, or the effectiveness of two different types of treatment. You could, for example, be investigating the effect of alcohol on people's reactions, or the difference in intelligence levels of identical twins who were separated at birth and who have been brought up in different family circumstances.

In both cases you need to have a common link between the two sets of results, for instance by taking the same person's result before and after drinking alcohol, or by the twins being identical. It is necessary to have this link so that differences caused by other factors are eliminated as much as possible. It would, for example, be of little use if you tested one person's reactions without drinking alcohol and a different person's reactions after drinking alcohol because any difference could be due to normal variations between their reactions. In the same way you would have to use identical twins

A in the intelligence experiment, otherwise any difference in intelligence might be due to the normal variability of intelligence between different people. In these cases, each result in one of the samples is paired with a result in the other sample; the results are therefore referred to as **paired**.

In paired experiments such as these you are not really interested in the individual results as such, but in the difference, D, between the results. In these circumstances you can treat the differences between pairs of matched subjects as if they were a random sample from a $N(\mu, \sigma^2)$ distribution. You can then proceed as you did for a single sample.

Although you do not need to assume the two populations are normal, you need to assume that the **differences** are normally distributed. Given that you are unlikely to know σ^2 and that n is likely to be small, then

$$t = \frac{\overline{D} - \mu_D}{\frac{S}{\sqrt{n}}} \sim t_{n-1}$$

Taking H_0: $\mu_D = 0$ as your null hypothesis, this reduces to

$$t = \frac{\overline{D} - 0}{\frac{S}{\sqrt{n}}} \sim t_{n-1}$$

> **Note** This is the null hypothesis that on average there is no difference between the two populations.

- **In a paired experiment with a mean of the differences between the samples of $\overline{D}$,**

$$\frac{\overline{D} - \mu_D}{\frac{S}{\sqrt{n}}} \sim t_{n-1}$$

The paired *t*-test proceeds in almost the same way as the *t*-test itself. The steps are given below.

1 Write down the null hypothesis H_0.

2 Write down the alternative hypothesis H_1.

3 Specify α.

4 Write down the degrees of freedom (remembering that $\nu = n - 1$).

5 Write down the critical region.

6 Calculate the differences d.

  Calculate $\overline{d}$ and s^2.

  Calculate the value of the test statistic $t = \dfrac{\overline{d} - \mu_D}{\frac{s}{\sqrt{n}}}$

7 Complete the test and state your conclusions. As before, the following points should be addressed:
  i Is the result significant or not?
  ii What are the implications in terms of the original problem?

Example 8

A In an experiment to test the effects of alcohol on the reaction times of people, a group of 10 students took part in the experiment. The students were asked to react to a light going on by pushing a switch that would switch it off again. Their reaction times were automatically recorded. After the students had each drunk one pint of beer the experiment was repeated. The results are shown below.

Student	A	B	C	D	E	F	G	H	I	J
Reaction time before (seconds)	0.8	0.2	0.4	0.6	0.4	0.6	0.4	0.8	1.0	0.9
Reaction time after (seconds)	0.7	0.5	0.6	0.8	0.8	0.6	0.7	0.9	1.0	0.7
Difference	−0.1	0.3	0.2	0.2	0.4	0	0.3	0.1	0	−0.2

Test, at the 5% significance level, whether or not the consumption of a pint of beer increased the students' reaction times.

$H_0: \mu_d = 0$ $\qquad\qquad$ $H_1: \mu_d > 0$	State your hypotheses.
Significance level = 0.05 (one-tailed test)	Write down the significance level.
$\nu = 10 - 1 = 9$	Find the number of degrees of freedom.
Critical value $t_9(5\%) = 1.833$	Look up the critical value in the table.
The critical region is $t \geqslant 1.833$	Write down the critical region.
$\dfrac{\sum d}{n} = \dfrac{1.2}{10} = 0.12$ $s^2 = \dfrac{\sum d^2 - n\bar{d}^2}{n-1}$ $\quad = \dfrac{0.48 - 10(0.12)^2}{9}$ $\quad = 0.037333$	Calculate $\bar{d}$ and s^2.
$t = \dfrac{0.12 - 0}{\dfrac{\sqrt{0.037333}}{\sqrt{10}}} = 1.9640$	Calculate the value of the test statistic $t = \dfrac{\bar{d} - \mu_D}{\dfrac{s}{\sqrt{n}}}$
1.9640 > 1.833. The result is significant: reject H_0. There is evidence that consuming a pint of beer increased the students' reaction times.	Always state whether you accept or reject H_0 and draw a conclusion (in context if possible).

Example 9

A In order to compare two methods of measuring the hardness of metals, readings of Brinell hardness were taken using each method for 8 different metal specimens. The resulting Brinell hardness readings are given in the table below.

Material	Reading method A	Reading method B
Aluminium	29	31
Magnesium alloy	64	63
Wrought iron	104	105
Duralumin	116	119
Mild steel	138	140
70/30 brass	156	156
Cast iron	199	200
Nickel chrome steel	385	386

Use a paired *t*-test, at the 5% level of significance, to test whether or not there is a difference in the readings given by the two methods.

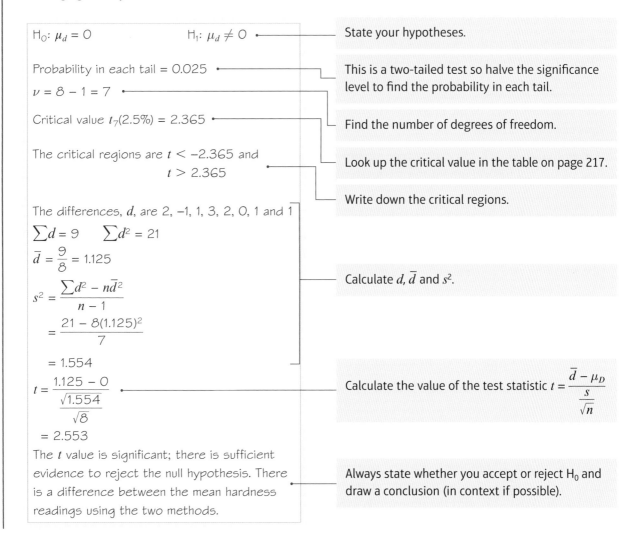

$H_0: \mu_d = 0$ $\qquad H_1: \mu_d \neq 0$ — State your hypotheses.

Probability in each tail = 0.025 — This is a two-tailed test so halve the significance level to find the probability in each tail.

$\nu = 8 - 1 = 7$ — Find the number of degrees of freedom.

Critical value $t_7(2.5\%) = 2.365$ — Look up the critical value in the table on page 217.

The critical regions are $t < -2.365$ and $t > 2.365$ — Write down the critical regions.

The differences, d, are 2, −1, 1, 3, 2, 0, 1 and 1

$\sum d = 9 \qquad \sum d^2 = 21$

$\bar{d} = \dfrac{9}{8} = 1.125$

$s^2 = \dfrac{\sum d^2 - n\bar{d}^2}{n - 1}$

$\quad = \dfrac{21 - 8(1.125)^2}{7}$

$\quad = 1.554$

Calculate d, $\bar{d}$ and s^2.

$t = \dfrac{1.125 - 0}{\dfrac{\sqrt{1.554}}{\sqrt{8}}}$

$\quad = 2.553$

Calculate the value of the test statistic $t = \dfrac{\bar{d} - \mu_D}{\dfrac{s}{\sqrt{n}}}$

The *t* value is significant; there is sufficient evidence to reject the null hypothesis. There is a difference between the mean hardness readings using the two methods.

Always state whether you accept or reject H_0 and draw a conclusion (in context if possible).

Exercise 7C

A
E/P

1 It is claimed that completion of a shorthand course has increased the shorthand speeds of the students.

   a If the suggestion that the mean speed of the students has not altered is to be tested, write down suitable hypotheses for which
     i a two-tailed test would be appropriate
     ii a one-tailed test would be appropriate. **(2 marks)**

The table below gives the shorthand speeds of students before and after the course.

Student	A	B	C	D	E	F
Speed before in words/minute	35	40	28	45	30	32
Speed after	42	45	28	45	40	40

   b Carry out a paired t-test, at the 5% significance level, to determine whether or not there has been an increase in shorthand speeds. **(7 marks)**

E 2 A large number of students took two General Studies papers that were supposed to be of equal difficulty. The results for 10 students chosen at random are shown below.

Candidate	A	B	C	D	E	F	G	H	I	J
Paper 1	18	25	40	10	38	20	25	35	18	43
Paper 2	20	27	39	12	40	23	20	35	20	41

The teacher looked at the marks of the random sample of 10 students, and decided that Paper 2 was easier than Paper 1.

Given that the marks on each paper are normally distributed, carry out an appropriate test of the teacher's claim, at the 1% level of significance. **(7 marks)**

E 3 It is claimed by the manufacturer that by chewing a special flavoured chewing gum, smokers are able to reduce their craving for cigarettes, and thus cut down on the number of cigarettes smoked per day. In a trial of the gum on a random selection of 10 people, the no-gum smoking rate and the smoking rate when chewing the gum were investigated, with the following results.

Person	A	B	C	D	E	F	G	H	I	J
Without-gum smoking rate (cigs./day)	20	35	40	32	45	15	22	30	34	40
With-gum smoking rate (cigs./day)	15	25	35	30	45	15	14	25	28	34

   a Use a paired t-test at the 5% significance level to test the manufacturer's claim. **(7 marks)**
   b State any assumptions you have had to make. **(1 mark)**

E 4 A town council is going to put a new traffic management scheme into operation in the hope that it will make travel to work in the mornings quicker for most people. Before the scheme is put into operation, 10 randomly selected workers are asked to record the time it takes them to come into work on a Wednesday morning. After the scheme is put in place, the same 10 workers are again asked to record the time it takes them to come into work on a particular Wednesday morning.

A The times in minutes are shown in the table below.

Worker	A	B	C	D	E	F	G	H	I	J
Before	23	37	53	42	39	60	54	85	46	38
After	18	35	49	42	34	48	52	79	37	37

Test, at the 5% significance level, whether or not the journey time to work has decreased.

(7 marks)

E **5** A teacher wants to test the idea that students' results in mock examinations are good predictors for their results in actual examinations. He selects 8 students at random from those doing a mock Statistics examination and records their marks out of 100. Later he collects the same students' marks in the actual examination. The resulting marks are as follows:

Student	A	B	C	D	E	F	G	H
Mock examination mark	35	86	70	91	45	64	78	38
Actual examination	45	77	81	86	53	71	68	46

  a Use a paired *t*-test to investigate whether or not the mock examination is a good predictor. (Use a 10% significance level.) **(7 marks)**

  b State any assumptions you have made. **(1 mark)**

E/P **6** The manager of a dress-making company took a random sample of 10 of his employees and recorded the number of dresses made by each. He discovered that the number of dresses made between 3.00 and 5.00 p.m. was fewer than the same employees achieved between 9.00 and 11.00 a.m. He wondered whether a tea break from 2.45 to 3.00 p.m. would increase productivity during these last two hours of the day.

The numbers of dresses made by these workers in the last two hours of the day before and after the introduction of the tea break were as shown below.

Worker	A	B	C	D	E	F	G	H	I	J
Before	75	73	75	81	74	73	77	75	75	72
After	80	84	79	84	85	84	78	78	80	83

  a Why was the comparison made for the same 10 workers? **(1 mark)**

  b Conduct, at the 5% level of significance, a paired *t*-test to see whether the introduction of a tea break has increased productivity between 3.00 and 5.00 p.m. **(7 marks)**

E **7** A drug administered in tablet form to help people sleep and a placebo were given for two weeks to a random sample of eight patients in a clinic. The drug and the placebo were given in random order for one week each. The average numbers of hours sleep that each patient had per night with the drug and with the placebo are given in the table below.

Patient	1	2	3	4	5	6	7	8
Hours of sleep with drug	10.5	6.7	8.9	6.7	9.2	10.9	11.9	7.6
Hours of sleep with placebo	10.3	6.5	9.0	5.3	8.7	7.5	9.3	7.2

Test, at the 1% level of significance, whether or not the drug increases the mean number of hours sleep per night. State your hypotheses clearly. **(7 marks)**

7.4 Difference between means of two independent normal distributions

A You need to be able to find a confidence interval for the difference between two means from independent normal distributions with **equal but unknown variances**.

To do this, you need to find a **pooled estimate of variance**.

Suppose that you take random samples from random variables X and Y that have a common

> **Watch out** In Section 5.3, you carried out hypothesis tests for the difference between the means of normal distributions with **known** variances. In that case the population distributions could have different variances. The techniques of this section and the next section only apply to normal distributions with unknown but **equal** variances.

variance, σ^2. You will have two estimates of σ^2, namely s_x^2 and s_y^2. A better estimate of σ^2 than either s_x^2 or s_y^2 can be obtained by pooling the two estimates. You will recall that, for a single sample, an unbiased estimate of the population variance was given by

$$s^2 = \frac{\sum(x - \bar{x})^2}{n - 1}$$

A similar idea works for two pooled estimates. You have

$$s_x^2 = \frac{\sum(x - \bar{x})^2}{n_x - 1} \quad \text{and} \quad s_y^2 = \frac{\sum(y - \bar{y})^2}{n_y - 1}$$

so that $(n_x - 1)s_x^2 = \sum(x - \bar{x})^2$ and $(n_y - 1)s_y^2 = \sum(y - \bar{y})^2$.

These are the sums of the squares of the differences of each sample value from the sample mean. You can add them together to get a total sum of squares of differences:

$$\sum(x - \bar{x})^2 + \sum(y - \bar{y})^2 = (n_x - 1)s_x^2 + (n_y - 1)s_y^2$$

You can use this sum to calculate a pooled estimate, s_p^2, of σ^2:

$$s_p^2 = \frac{(n_x - 1)\,s_x^2 + (n_y - 1)\,s_y^2}{(n_x - 1) + (n_y - 1)} = \frac{(n_x - 1)\,s_x^2 + (n_y - 1)\,s_y^2}{n_x + n_y - 2}$$

- **If a random sample of n_x observations is taken from a normal distribution with unknown variance σ^2, and an independent sample of n_y observations is taken from a normal distribution that also has unknown variance σ^2, then a pooled estimate for σ^2 is**

$$s_p^2 = \frac{(n_x - 1)s_x^2 + (n_y - 1)s_y^2}{n_x + n_y - 2}$$

where $s_x^2 = \dfrac{\sum x^2 - n_x \bar{x}^2}{n_x - 1}$ **and** $s_y^2 = \dfrac{\sum y^2 - n_y \bar{y}^2}{n_y - 1}$

> **Note**
> Notice that if $n_x = n_y = n$, this reduces to $s_p^2 = \dfrac{(n - 1)(s_x^2 + s_x^2)}{2(n - 1)} = \dfrac{s_x^2 + s_y^2}{2}$ which is the mean of the two variances. The pooled estimate of variance is really a weighted mean of two variances with the two weights being $(n_x - 1)$ and $(n_y - 1)$.

Example 10

A A random sample of 15 observations is taken from a population and gives an unbiased estimate for the population variance of 9.47. A second random sample of 12 observations is taken from a different population that has the same population variance as the first population, and gives an unbiased estimate for the variance as 13.84. Calculate an unbiased estimate of the population variance σ^2 using both samples.

$$s_p^2 = \frac{(14 \times 9.47) + (11 \times 13.84)}{14 + 11}$$

$$= 11.3928$$

Use $s_p^2 = \dfrac{(n_x - 1)s_x^2 + (n_y - 1)s_y^2}{(n_x - 1) + (n_y - 1)}$

In Section 5.4, you saw that if the sample sizes are large then

$$\frac{(\overline{X} - \overline{Y}) - (\mu_x - \mu_y)}{\sqrt{\dfrac{s_x^2}{n_x} + \dfrac{s_y^2}{n_y}}}$$ is approximately normal with distribution $N(0, 1^2)$.

When the sample sizes are small you need to make three assumptions:

1 that the populations are normal

2 that the samples are independent

3 that the variances of the two populations are equal.

> **Links** In many cases, it is reasonable to assume that the variances of the populations are equal. If you are unsure, you can use the *F*-distribution to test for equal variance. **← Section 6.4**

The third assumption enables you to pool the two sample variances to find an estimator for the common variance:

$$S_p^2 = \frac{(n_x - 1)\,S_x^2 + (n_y - 1)\,S_y^2}{(n_x - 1) + (n_y - 1)}$$

Substituting S_p^2 for S_x^2 and S_y^2 gives

$$\frac{(\overline{X} - \overline{Y}) - (\mu_x - \mu_y)}{\sqrt{\dfrac{S_p^2}{n_x} + \dfrac{S_p^2}{n_y}}} = \frac{(\overline{X} - \overline{Y}) - (\mu_x - \mu_y)}{S_p\sqrt{\dfrac{1}{n_x} + \dfrac{1}{n_y}}}$$

Now, because the sample sizes are small, this will not as before follow a $N(0, 1^2)$ distribution. You have already seen that in the single-sample case

$$\frac{\overline{X} - \mu_x}{\dfrac{S}{\sqrt{n_x}}}$$

follows a *t*-distribution, so you will not be surprised to find that

$$\frac{(\overline{X} - \overline{Y}) - (\mu_x - \mu_y)}{S_p\sqrt{\dfrac{1}{n_x} + \dfrac{1}{n_y}}}$$

also follows a *t*-distribution.

A There are $(n_x + n_y)$ observations in the total sample and two calculated restrictions (namely the means $\overline{X}$ and $\overline{Y}$), so the number of degrees of freedom will be $n_x + n_y - 2$.

- **If a random sample of n_x observations is taken from a normal distribution that has unknown variance σ^2, and an independent sample of n_y observations is taken from a normal distribution with equal variance, then**

$$\frac{(\overline{X} - \overline{Y}) - (\mu_x - \mu_y)}{S_p\sqrt{\dfrac{1}{n_x} + \dfrac{1}{n_y}}} \sim t_{n_x + n_y - 2} \text{ where } S_p^2 = \frac{(n_x - 1)S_x^2 + (n_y - 1)S_y^2}{n_x + n_y - 2}$$

You can now use tables of values for the t-distribution to find a confidence interval for $\mu_x - \mu_y$. For example, for a 95% confidence interval you would start by finding the value t_c that is exceeded with probability 0.025. This would give you:

$$P(-t_c < t_{n_x + n_y - 2} < t_c) = 0.95$$

$$P\left(-t_c < \frac{(\overline{x} - \overline{y}) - (\mu_x - \mu_y)}{s_p\sqrt{\dfrac{1}{n_x} + \dfrac{1}{n_y}}} < t_c\right) = 0.95$$

$$P\left(-t_c s_p\sqrt{\dfrac{1}{n_x} + \dfrac{1}{n_y}} < (\overline{x} - \overline{y}) - (\mu_x - \mu_y) < t_c s_p\sqrt{\dfrac{1}{n_x} + \dfrac{1}{n_y}}\right) = 0.95$$

The confidence limits for $(\mu_x - \mu_y)$ are therefore given by

$$(\overline{x} - \overline{y}) \pm t_c s_p\sqrt{\frac{1}{n_x} + \frac{1}{n_y}}$$

and the confidence interval is

$$\left((\overline{x} - \overline{y}) - t_c s_p\sqrt{\frac{1}{n_x} + \frac{1}{n_y}}, \quad (\overline{x} - \overline{y}) + t_c s_p\sqrt{\frac{1}{n_x} + \frac{1}{n_y}}\right)$$

- **The confidence limits for the difference between two means from independent normal distributions, X and Y, when the variances are equal but unknown are given by**

$$(\overline{x} - \overline{y}) \pm t_c s_p\sqrt{\frac{1}{n_x} + \frac{1}{n_y}}$$

where s_p is the pooled estimate of the population variance, and t_c is the relevant value taken from the t-distribution tables.

- **The confidence interval is given by**

$$\left((\overline{x} - \overline{y}) - t_c s_p\sqrt{\frac{1}{n_x} + \frac{1}{n_y}}, \quad (\overline{x} - \overline{y}) + t_c s_p\sqrt{\frac{1}{n_x} + \frac{1}{n_y}}\right)$$

Example 11

A

In a survey on the petrol consumption of cars, a random sample of 12 cars with 2-litre engines was compared with a random sample of 15 cars with 1.6-litre engines. The following results show the consumption, in suitable units, of the cars:

2-litre cars: 34.4, 32.1, 30.1, 32.8, 31.5, 35.8, 28.2, 26.6, 28.8, 28.5, 33.6, 28.8
1.6-litre cars: 35.3, 34.0, 36.7, 40.9, 34.4, 39.8, 33.6, 36.7, 34.0, 39.2, 39.8, 38.7, 40.8, 35.0, 36.7

Calculate a 95% confidence interval for the difference between the two mean petrol consumption figures. You may assume that the variables are normally distributed and that they have the same variance.

For the 2-litre engine, $n_y = 12$, $\bar{y} = 30.933$, $s_y^2 = 8.177$
For the 1.6-litre engine, $n_x = 15$, $\bar{x} = 37.04$, $s_x^2 = 6.894$

$$s_p^2 = \frac{(14 \times 6.894) + (11 \times 8.177)}{25}$$

$\dfrac{(n_x - 1)s_x^2 + (n_y - 1)s_y^2}{n_x + n_y - 2}$

$$= 7.459$$
$$s_p = \sqrt{7.459} = 2.731$$

$t_c = t_{25}(2.5\%) = 2.060$ $\nu = 12 + 15 - 2 = 25$

The confidence limits are

$(37.04 - 30.933) \pm 2.060 \times 2.731\sqrt{\frac{1}{15} + \frac{1}{12}} = 6.107 \pm 2.179$

Use $(\bar{x} - \bar{y}) \pm t_c s_p \sqrt{\frac{1}{n_x} + \frac{1}{n_y}}$

$$= 8.286 \text{ and } 3.928$$

The 95% confidence interval is (3.928, 8.286)
or (3.93, 8.29) to 3 s.f.

Exercise 7D

E **1** A random sample of 10 toothed winkles was taken from a sheltered shore, and a sample of 15 was taken from a non-sheltered shore. The maximum basal width, x mm, of the shells was measured and the results are summarised below.

Sheltered shore: $\bar{x} = 25$, $s^2 = 4$
Non-sheltered shore: $\bar{x} = 22$, $s^2 = 5.3$

a Find a 95% confidence interval for the difference between the means. **(6 marks)**

b State an assumption that you have made when calculating this interval. **(1 mark)**

E/P **2** A packet of plant seeds was sown and, when the seeds had germinated and begun to grow, 8 were transferred into pots containing a soil-less compost and 10 were grown on in a soil-based compost. After 6 weeks of growth, the heights, x, in cm of the plants were measured with the following results.

Soil-less compost: 9.3, 8.7, 7.8, 10.0, 9.2, 9.5, 7.9, 8.9
Soil-based compost: 12.8, 13.1, 11.2, 10.1, 13.1, 12.0, 12.5, 11.7, 11.9, 12.0

a Assuming that the populations are normally distributed, and that there is a difference between the two means, calculate a 90% confidence interval for this difference. **(6 marks)**

b State an additional assumption you have used when calculating this interval and discuss whether this assumption is reasonable in the context given. **(2 marks)**

3 Forty children were randomly selected from all 12-year-old children in a large city to compare two methods of teaching the spelling of 50 words which were likely to be unfamiliar to the children. Twenty children were randomly allocated to each method. Six weeks later the children were tested to see how many of the words they could spell correctly. The summary statistics for the two methods are given in the table below, where $\bar{x}$ is the mean number of words spelled correctly, s^2 is an unbiased estimate of the variance of the number of words spelled correctly and n is the number of children taught using each method.

	$\bar{x}$	s^2	n
Method A	32.7	6.12	20
Method B	38.2	5.22	20

   a Calculate a 99% confidence interval for the difference between the mean numbers of words spelled correctly by children who used Method B and Method A. **(6 marks)**

   b State two assumptions you have made in carrying out part **a**. **(2 marks)**

   c Interpret your result. **(1 mark)**

4 The table below shows summary statistics for the mean daily consumption of cigarettes by a random sample of 10 smokers before and after their attendance at an anti-smoking workshop with $\bar{x}$ representing the mean and s^2 representing the unbiased estimate of population variance in each case.

	$\bar{x}$	s^2	n
Mean daily consumption before the workshop	18.6	32.488	10
Mean daily consumption after the workshop	14.3	33.344	10

Stating clearly any assumption you make, calculate a 90% confidence interval for the difference in the mean daily consumption of cigarettes before and after the workshop. **(7 marks)**

5 Two farmers add different protein supplements to the feed of cows to increase the yield of milk. A sample of 8 cows is taken from the first farmer, who uses supplement A, and the yield of milk is measured. A second sample, of size 7, is taken from the cows of the second farmer, who uses supplement B. The table shows the mean daily yield, in litres, and unbiased estimates for the population variance in each case.

	$\bar{x}$	s^2	n
Supplement A	24.5	1.2	8
Supplement B	26.8	1.6	7

   a Stating your hypotheses clearly, test, at the 10% level of significance, the hypothesis that there is a difference in the variability of the yields. State any assumptions you make. **(5 marks)**

The farmers wish to find a confidence interval for the difference in the average milk yield for the two supplements.

   b Explain how the result from part **a** can be used to justify the use of a t-distribution to find the confidence interval. **(1 mark)**

   c Find, correct to 3 significant figures, a 95% confidence interval for the difference in the average milk yield. **(5 marks)**

7.5 Hypothesis test for the difference between means

A
Apart from using the *t*-distribution rather than the normal distribution for finding the critical values, testing the difference between means of two independent normal distributions with unknown variances follows similar steps to those used for testing the difference of means when the variances are known.

The following steps might help you in answering questions on the difference of means of normal distributions when the variances are unknown.

1 Write down H_0.

2 Write down H_1.

3 Specify the significance level, α.

4 Write down the number of degrees of freedom, ν.

5 Write down the critical region.

6 Calculate the sample means and variances, $\bar{x}$, $\bar{y}$, s_x^2 and s_y^2.

7 Calculate a pooled estimate of the variance:

$$s_p^2 = \frac{(n_x - 1)\,s_x^2 + (n_y - 1)\,s_y^2}{n_x + n_y - 2}$$

8 Calculate the value of *t*:

$$t = \frac{(\bar{x} - \bar{y}) - (\mu_x - \mu_y)}{s_p\sqrt{\dfrac{1}{n_x} + \dfrac{1}{n_y}}}$$

9 Complete the test and state your conclusions. The following points should be addressed:
  i Is the result significant?
  ii What are the implications in terms of the original problem?

Example 12

Two groups of students, *X* and *Y*, were taught by different teachers. At the end of their course, a random sample of students from each class was selected and given a test. The test results out of 50 were as follows:

Group *X*:   40 37 45 34 30 41 42 43 36

Group *Y*:   38 43 36 45 35 44 41

The headteacher wishes to find out if there is a significant difference between the results for these two groups.

a Write down any assumptions that need to be made in order to conduct a difference of means test on this data.

b Assuming that these assumptions apply, test at the 10% level of significance whether or not there is a significant difference between the means.

A

a The assumptions that need to be made are that the two samples come from normal distributions, are independent and that the populations from which they are taken have the same variances.

> State your hypotheses.

b $H_0: \mu_x = \mu_y \qquad H_1: \mu_x \neq \mu_y$

Probability in each tail = 0.05

$\nu = 9 + 7 - 2 = 14$

Critical value $t_{14}(0.05)$ is 1.761

> This is a two-tailed test. Halve the significance level to find the probability in each tail.

> Find the number of degrees of freedom ($n_1 + n_2 - 2$ in this case).

> Look up the critical value in the table.

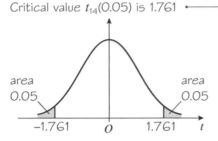

area 0.05   area 0.05

$-1.761 \quad 0 \quad 1.761 \quad t$

The critical regions are $t \leqslant -1.761$ and $t \geqslant 1.761$

> Write down the critical region. There are two regions as it is a two-tailed test.

Using a calculator gives

$n_x = 9, \bar{x} = 38.667, s_x^2 = 23.0$

$n_y = 7, \bar{y} = 40.286, s_y^2 = 15.9$

> Calculate $\bar{x}, \bar{y}, s_x^2$ and s_y^2.

$s_p^2 = \dfrac{(8 \times 23) + (6 \times 15.9)}{9 + 7 - 2}$

$= 19.957$

So $s_p = 4.467$

> Calculate a pooled estimate of the variance using $\dfrac{(n_x - 1)s_x^2 + (n_y - 1)s_y^2}{n_x + n_y - 2}$

$t = \dfrac{38.667 - 40.286}{4.467\sqrt{\dfrac{1}{9} + \dfrac{1}{7}}}$

$= -0.719$

> Calculate t using $\dfrac{(\bar{x} - \bar{y}) - (\mu_x - \mu_y)}{s_p\sqrt{\frac{1}{n_x} + \frac{1}{n_y}}}$
> $\mu_x - \mu_y = 0$ from hypothesis.

$-1.761 < -0.719 < 1.761$ so the result is not significant. Accept H_0. On the evidence given by the two samples there is no difference between the means of the two groups.

> Always state whether you accept or reject H_0 and draw a conclusion (in context if possible).

Example 13

A random sample of the heights, in cm, of sixth form boys and girls was taken with the following results:

    Boys' heights: 152, 148, 147, 157, 158, 140, 141, 144
    Girls' heights: 142, 146, 132, 125, 138, 131, 143

a Carry out a two-sample t-test at the 5% significance level on these data to see whether the mean height of boys exceeds the mean height of girls by more than 4 cm.

b State any assumptions that you have made.

A

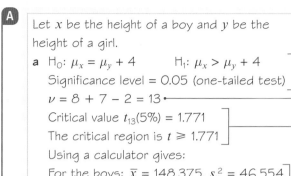

Let x be the height of a boy and y be the height of a girl.

a H_0: $\mu_x = \mu_y + 4$ H_1: $\mu_x > \mu_y + 4$

Significance level = 0.05 (one-tailed test)

$\nu = 8 + 7 - 2 = 13$

Critical value $t_{13}(5\%) = 1.771$

The critical region is $t \geqslant 1.771$

Using a calculator gives:

For the boys: $\bar{x} = 148.375$, $s_x^2 = 46.554$, $n_x = 8$

For the girls: $\bar{y} = 136.714$, $s_y^2 = 57.905$, $n_y = 7$

$s_p^2 = \dfrac{(7 \times 46.554) + (6 \times 57.905)}{8 + 7 - 2}$

$= 51.793$

So $s_p = 7.197$

$t = \dfrac{(148.375 - 136.714) - 4}{7.197\sqrt{\frac{1}{8} + \frac{1}{7}}}$

$= 2.057$

2.057 is in the critical region and so there is sufficient evidence to reject the null hypothesis. The mean height of boys exceeds the mean height of girls by more than 4 cm.

b The assumptions made are that the two samples are independent, that the variances of both populations are equal and that the populations are normally distributed.

State your hypotheses and write down the significance level.

Find the number of degrees of freedom.

Look up the critical value in the table and write down the critical region.

Calculate $\bar{x}$, $\bar{y}$, s_x^2 and s_y^2.

Calculate a pooled estimate of the variance using $\dfrac{(n_x - 1)s_x^2 + (n_y - 1)s_y^2}{n_x + n_y - 2}$

Calculate t using $\dfrac{(\bar{x} - \bar{y}) - (\mu_x - \mu_y)}{s_p\sqrt{\frac{1}{n_x} + \frac{1}{n_y}}}$

$\mu_x - \mu_y = 4$ from hypothesis.

Always state whether you accept or reject H_0 and draw a conclusion (in context if possible).

Exercise 7E

E **1** A random sample of size 20 from a normal population gave $\bar{x} = 16$, $s^2 = 12$.

A second random sample of size 11 from a normal population gave $\bar{x} = 14$, $s^2 = 12$.

a Assuming that both populations have the same variance, write down an unbiased estimate for that variance. **(1 mark)**

b Test, at the 5% level of significance, the suggestion that the two populations have the same mean. **(5 marks)**

E **2** Salmon reared in Scottish fish farms are generally larger than wild salmon. A fisherman measured the length of the first 6 wild salmon caught from a river. Their lengths in centimetres were:

42.8   40.0   38.2   37.5   37.0   36.5

Chefs prefer wild salmon to fish-farmed salmon because of their better flavour. A chef was offered 4 salmon that were claimed to be wild. Their lengths in centimetres were:

42.0   43.0   41.5   40.0

Use the information given above and a suitable *t*-test at the 5% level of significance to help the chef to decide if the claim is likely to be correct. You may assume that the populations are normally distributed. **(8 marks)**

3 In order to check the effectiveness of three drugs against the *E. coli* bacillus, 15 cultures of the bacillus (five for each of three different antibiotics) had discs soaked in the antibiotics placed in their centre. The 15 cultures were left for a time and the area in cm² per microgram of drug where the *E. coli* was killed was measured. The results for the three different drugs are given below.

Streptomycin: 0.210   0.252   0.251   0.210   0.256   0.253
Tetracycline: 0.123   0.090   0.123   0.141   0.142   0.092
Erythromycin: 0.134   0.120   0.123   0.210   0.134   0.134

a It was thought that tetracycline and erythromycin seemed equally effective. Assuming that the populations are normally distributed, test this at the 5% significance level. **(8 marks)**

b Streptomycin was thought to be more effective than either of the other two drugs. Treating the other two as being a single sample of 12, test this assertion at the same level of significance. **(7 marks)**

4 To test whether a new version of a computer programming language enabled faster task completion, the same task was performed by 16 programmers, divided at random into two groups. The first group used the new version of the language, and the time for task completion, in hours, for each programmer was as follows:

4.9   6.3   9.6   5.2   4.1   7.2   4.0

The second group used the old version, and their times were summarised as follows:

$n = 9, \quad \sum x = 71.2, \quad \sum x^2 = 604.92$

a State the null and alternative hypotheses. **(1 mark)**

b Perform an appropriate test at the 5% level of significance. **(7 marks)**

In order to compare like with like, experiments such as this are often performed using the same individuals in the first and the second groups.

c Give a reason why this strategy would not be appropriate in this case. **(1 mark)**

5 A company undertakes investigations to compare the fuel consumption, *x*, in miles per gallon, of two different cars, the Volcera and the Spintono, with a view to purchasing a number as company cars.

For a random sample of 12 Volceras the fuel consumption is summarised by

$\sum v = 384 \quad \sum v^2 = 12480$

A statistician incorrectly combines the figures for the sample of 12 Volceras with those of a random sample of 15 Spintonos, then carries out calculations as if they are all one larger sample and obtains the results $\bar{y} = 34$ and $s^2 = 23$.

a Show that, for the sample of 15 Spintonos, $\sum x = 534$ and $\sum x^2 = 19\,330$. **(2 marks)**

b Given that the variance of the fuel consumption for each make of car is σ^2, obtain an unbiased estimate for σ^2. **(3 marks)**

c Test, at the 5% level of significance, whether there is a difference between the mean fuel consumption of the two models of car. State your hypotheses and conclusion clearly. **(7 marks)**

d State any further assumption you made in order to be able to carry out your test in part **c**. **(1 mark)**

e Give two precautions which could be taken when undertaking an investigation into the fuel consumption of two models of car to ensure that a fair comparison is made. **(2 marks)**

6 A group of scientists is experimenting with different fertilisers. Fertiliser A is given to a crop of potatoes and a sample of 11 plants is taken. The weight of potatoes, in kg, from each plant is measured. Fertiliser B is given to a second crop of potatoes and a sample of 13 plants is taken. The weight of potatoes, in kg, from each plant is also measured. The table shows the mean weight of potatoes and unbiased estimates for the population variance in each case.

	$\bar{x}$	s^2	n
Fertiliser A	42.1	2.1	11
Fertiliser B	46.3	3.3	13

a Stating your hypotheses clearly, test, at the 10% level of significance, the hypothesis that there is a difference in the variability of the weights. State any assumptions you make. **(5 marks)**

The scientists wish to test if there is a difference in the average weight of potatoes for each fertiliser.

b Explain how the result in part **a** can be used to justify the scientists using a two-sample t-test to test their hypothesis. **(1 mark)**

c Stating your hypotheses clearly, test, at the 5% level of significance, whether there is a difference in the average weight of potatoes. **(5 marks)**

Challenge

For samples of sizes n_x and n_y from populations X and Y with equal but unknown variance σ^2, show that the pooled sample variance

$$S_p^2 = \frac{(n_x - 1) S_x^2 + (n_y - 1) S_y^2}{n_x + n_y - 2}$$

is an unbiased estimator of σ^2. You may assume that the sample variances S_x and S_y are each unbiased estimators for σ^2.

Mixed exercise **7**

1 A random sample of 14 observations is taken from a normal distribution. The sample has a mean $\bar{x} = 30.4$ and a sample variance $s^2 = 36$.

It is suggested that the population mean is 28. Test this hypothesis at the 5% level of significance.

2 A random sample of 8 observations is taken from a random variable X that is normally distributed. The sample gave the following summary statistics:

$$\sum x^2 = 970.25 \quad \sum x = 85$$

The population mean is thought to be 10. Test this hypothesis against the alternative hypothesis that the mean is greater than 10. Use the 5% level of significance.

3 Six eggs selected at random from the daily output of a brood of hens had the following weights in grams:

55 50 53 53 52 54

Calculate 95% confidence intervals for:

> **Hint** For part **b**, use the chi-squared distribution. ← **Section 6.1**

a the mean **(5 marks)**

b the variance of the population from which these eggs were taken. **(5 marks)**

c What assumption have you made about the distribution of the weights of eggs? **(1 mark)**

A **4** A sample of size 18 was taken from a random variable X which was normally distributed, producing the following summary statistics:

$$\bar{x} = 9.8 \quad s^2 = 0.49$$

Calculate 95% confidence intervals for:

a the mean

b the variance of the population.

E **5** A manufacturer claims that the lifetime of its batteries is normally distributed with mean 21.5 hours. A laboratory tests 8 batteries and finds the lifetimes of these batteries to be as follows:

19.7   18.4   22.2   20.8   16.9   25.3   23.2   21.1

Stating clearly your hypotheses, examine whether or not these lifetimes indicate that the batteries have a shorter mean lifetime than that claimed by the manufacturer.
Use a 5% level of significance. **(6 marks)**

E **6** A diabetic patient monitors his blood glucose in mmol/l at random times of the day over several days. The following is a random sample of the results for this patient.

5.1   5.8   6.1   6.8   6.2   5.1   6.3   6.6   6.1   7.9   5.8   6.5

Assuming the data to be normally distributed, calculate a 95% confidence interval for:

a the mean of the population of blood glucose readings **(6 marks)**

b the standard deviation of the population of blood glucose readings. **(6 marks)**

The level of blood glucose varies throughout the day according to the consumption of food and the amount of exercise taken during the day.

c Comment on the suitability of the patient's method of data collection. **(1 mark)**

E **7** In order to discover the possible error in using a stopwatch, a student started the watch and stopped it again as quickly as she could. The times taken in centiseconds for 6 such attempts are recorded below:

10   13   14   10   13   9

Assuming that the times are normally distributed, find 95% confidence limits for:

a the mean **(6 marks)**

b the variance. **(6 marks)**

E **8** A manufacturer claims that the car batteries which it produces have a mean lifetime of 24 months, with a standard deviation of 4 months. A garage selling the batteries doubts this claim and suggests that both values are in fact higher.

The garage monitors the lifetimes of 10 randomly selected batteries and finds that they have a mean lifetime of 27.2 months and a standard deviation of 5.2 months.

Stating clearly your hypotheses and using a 5% level of significance, test the claim made by the manufacturer for:

a the standard deviation **(6 marks)**

b the mean. **(6 marks)**

c State an assumption which has to be made when carrying out these tests. **(1 mark)**

9 The distance to takeoff from a standing start of an aircraft was measured on twenty occasions. The results are summarised in the table.

Assuming that distance to takeoff is normally distributed, find 95% confidence intervals for:

Distance (m)	Frequency
700–	3
710–	5
720–	9
730–	2
740–750	1

  a the mean **(6 marks)**

  b the standard deviation. **(6 marks)**

It has been hypothesised that the mean distance to takeoff is 725 m.

  c Comment on this hypothesis in the light of your interval from part **a**. **(1 mark)**

10 A company knows from previous experience that the mean time taken by maintenance engineers to repair a particular electrical fault on a complex piece of electrical equipment is 3.5 hours, with a standard deviation of 0.5 hours.

A new method of repair has been devised, but before converting to this new method the company took a random sample of 10 of its engineers and each engineer carried out a repair using the new method. The time, x hours, it took each of them to carry out the repair was recorded and the data are summarised below:

$$\sum x = 34.2 \quad \sum x^2 = 121.6$$

Assume that the data can be regarded as a random sample from a normal population.

  a For the new repair method, calculate an unbiased estimate of the variance. **(2 marks)**

  b Use your estimate from part **a** to calculate, for the new repair method, a 95% confidence interval for:

    i the mean

    ii the standard deviation. **(10 marks)**

  c Use your calculations and the given data to compare the two repair methods in order to advise the company as to which method to use. **(2 marks)**

  d Suggest an alternative way of comparing the two methods of repair using the 10 randomly chosen engineers. **(1 mark)**

 11 A random sample of 60 female raccoons is taken and their heights recorded. The sample mean is found to be 24 cm and an unbiased estimate for the population variance is found to be 2.1 cm².

  a Given that the underlying population is normally distributed, find a 90% confidence interval for the mean height of female raccoons. State clearly the approximating distribution you have used to determine this confidence interval. **(5 marks)**

A second random sample of 6 male raccoons is taken and their heights recorded. The sample mean is found to be 27 cm and an unbiased estimate for the population variance is found to be 2.7 cm².

A hypothesis test is to be carried out to test if the mean height of male raccoons is greater than 25 cm.

  b Explain why the approximating distribution used in part **a** is no longer valid when carrying out this test. **(1 mark)**

  c Test, at the 5% level of significance, the hypothesis that male raccoons have an average height greater than 25 cm. State your hypotheses clearly. **(5 marks)**

A **12** A chemist has developed a fuel additive and claims that it reduces the fuel consumption of cars.
E To test this claim, 8 randomly selected cars were each filled with 20 litres of fuel and driven
around a race circuit. Each car was tested twice, once with the additive and once without.
The distances, in miles, that each car travelled before running out of fuel are given in the table
below.

Car	1	2	3	4	5	6	7	8
Distance without additive	163	172	195	170	183	185	161	176
Distance with additive	168	185	187	172	180	189	172	175

Assuming that the distances travelled follow a normal distribution and stating your hypotheses
clearly, test, at the 10% level of significance, whether or not there is evidence to support the
chemist's claim. **(7 marks)**

E/P **13** A farmer set up a trial to assess the effect of two different diets on the increase in the weight of
his lambs. He randomly selected 20 lambs. Ten of the lambs were given diet *A* and the other
10 lambs were given diet *B*. The gain in weight, in kg, of each lamb over the period of the trial
was recorded.

a State why a paired *t*-test is not suitable for use with these data. **(1 mark)**

b Suggest an alternative method for selecting the sample which would make the use of a paired
t-test valid. **(1 mark)**

c Suggest two other factors that the farmer might consider when selecting the
sample. **(2 marks)**

The following paired data were collected.

Diet *A*	5	6	7	4.6	6.1	5.7	6.2	7.4	5	3
Diet *B*	7	7.2	8	6.4	5.1	7.9	8.2	6.2	6.1	5.8

d Using a paired *t*-test at the 5% significance level, test whether or not there is evidence of
a difference in the weight gained by the lambs using diet *A* compared with those using
diet *B*. **(7 marks)**

e State, giving a reason, which diet you would recommend the farmer to use for his
lambs. **(1 mark)**

E **14** A medical student is investigating two methods of taking a person's blood pressure. He takes a
random sample of 10 people and measures their blood pressure using an arm cuff and a finger
monitor. The table below shows the blood pressure for each person, measured by each method.

Person	*A*	*B*	*C*	*D*	*E*	*F*	*G*	*H*	*I*	*J*
Arm cuff	140	110	138	127	142	112	122	128	132	160
Finger monitor	154	112	156	152	142	104	126	132	144	180

a Use a paired *t*-test to determine, at the 10% level of significance, whether or not there
is a difference in the mean blood pressure measured using the two methods. State your
hypotheses clearly. **(7 marks)**

b State an assumption about the underlying distribution of measured blood pressure
required for this test. **(1 mark)**

15 The weights, in grams, of mice are normally distributed. A biologist takes a random sample of 10 mice. She weighs each mouse and records its weight.

The 10 mice are then fed on a special diet. They are weighed again after two weeks.

Their weights in grams are as follows:

Mouse	A	B	C	D	E	F	G	H	I	J
Weight before diet	50.0	48.3	47.5	54.0	38.9	42.7	50.1	46.8	40.3	41.2
Weight after diet	52.1	47.6	50.1	52.3	42.2	44.3	51.8	48.0	41.9	43.6

Stating your hypotheses clearly, and using a 1% level of significance, test whether or not the diet causes an increase in the mean weight of the mice. **(7 marks)**

16 A hospital department installed a new, more sophisticated, piece of equipment to replace an ageing one in order to speed up the treatment of patients. The treatment times of random samples of patients during the last week of operation of the old equipment and during the first week of operation of the new equipment were recorded. The summary results, in minutes, are shown in the table.

a Show that the values of s^2 for the old and new equipment are 8.2 and 14.5 respectively. **(2 marks)**

	n	$\sum x$	$\sum x^2$
Old equipment	10	225	5136.3
New equipment	9	234	6200.0

Stating clearly your hypotheses, test:

b whether the variance of the times using the new equipment is greater than the variance of the times using the old equipment, using a 5% significance level **(6 marks)**

c whether there is a difference between the mean times for treatment using the new equipment and the old equipment, using a 2% significance level. **(6 marks)**

d Find 95% confidence limits for the mean difference in treatment times between the new and old equipment. **(5 marks)**

Even if the new equipment would eventually lead to a reduction in treatment times, it might be that to begin with treatment times using the new equipment would be higher than those using the old equipment.

e Give one reason why this might be so. **(1 mark)**

f Suggest how the comparison between the old and new equipment could be improved. **(1 mark)**

17 Two different drugs designed to increase the red blood cell count are administered to two groups of patients. A sample of 25 patients who took the first drug, A, is taken and the red blood cell count, in million cells per microlitre, is recorded. A sample, of size 19, is then taken from the patients who took drug B. The table shows the mean red blood cell count and unbiased estimates for the population variance in each case.

	$\bar{x}$	s^2	n
Drug A	5.9	2.6	25
Drug B	4.8	1.7	19

a Stating your hypotheses clearly, test, at the 10% level of significance, the hypothesis that there is a difference in the variability of the red blood cell counts. State any assumptions you make. **(5 marks)**

Doctors wish to find a confidence interval for the difference in the mean red blood cell counts for the two drugs.

b With reference to your answer to part **a**, comment on the suitability of using a t-distribution to find this confidence interval. **(1 mark)**

c Find, correct to 3 significant figures, a 95% confidence interval for the difference in the mean red blood cell count. **(5 marks)**

Challenge

Three independent random samples of sizes n_x, n_y and n_z (where n_x, n_y, $n_z > 1$) are taken from three populations X, Y and Z respectively, where X, Y and Z have equal but unknown variance σ^2.

a Given that the sample variances S_x^2, S_y^2 and S_z^2 are unbiased estimators for σ^2, find a pooled estimator S_p^2 for σ^2 based on all three samples, giving your answer in terms of n_x, n_y, n_z, S_x^2, S_y^2 and S_z^2.

b Show that the estimator found in part **a** is unbiased.

Summary of key points

1 If a random sample $X_1, X_2, \ldots, X_n$ is selected from a normal distribution with mean μ and unknown variance σ^2 then

$$t = \frac{\overline{X} - \mu}{\dfrac{S}{\sqrt{n}}}$$

has a t_{n-1}-distribution where S^2 is an unbiased estimator of σ^2.

2 In general, for a small sample of size n from a normal distribution $N(\mu, \sigma^2)$ with unknown mean and variance:

- the $100(1 - \alpha)\%$ confidence limits for the population mean are

$$\overline{x} \pm t_{n-1}\left(\frac{\alpha}{2}\right) \times \frac{S}{\sqrt{n}}$$

- the $100(1 - \alpha)\%$ confidence interval for the population mean is

$$\left(\overline{x} - t_{n-1}\left(\frac{\alpha}{2}\right) \times \frac{S}{\sqrt{n}}, \quad \overline{x} + t_{n-1}\left(\frac{\alpha}{2}\right) \times \frac{S}{\sqrt{n}}\right)$$

3 In a paired experiment with a mean of the differences between the samples of $\overline{D}$,

$$\frac{\overline{D} - \mu_D}{\dfrac{S}{\sqrt{n}}} \sim t_{n-1}$$

4 If a random sample of n_x observations is taken from a normal distribution with unknown variance σ^2, and an independent sample of n_y observations is taken from a normal distribution that also has unknown variance σ^2, then a pooled estimate for σ^2 is

$$s_p^2 = \frac{(n_x - 1)\, s_x^2 + (n_y - 1)\, s_y^2}{n_x + n_y - 2}$$

where $s_x^2 = \dfrac{\sum x^2 - n_x \bar{x}^2}{n_x - 1}$ and $s_y^2 = \dfrac{\sum y^2 - n_y \bar{y}^2}{n_y - 1}$

5 If a random sample of n_x observations is taken from a normal distribution that has unknown variance σ^2, and an independent sample of n_y observations is taken from a normal distribution with equal variance, then

$$\frac{(\bar{X} - \bar{Y}) - (\mu_x - \mu_y)}{S_p \sqrt{\dfrac{1}{n_x} + \dfrac{1}{n_y}}} \sim t_{n_x + n_y - 2}$$

where $S_p^2 = \dfrac{(n_x - 1)S_x^2 + (n_y - 1)S_y^2}{n_x + n_y - 2}$

6 The confidence limits for the difference between two means from independent normal distributions, X and Y, when the variances are equal but unknown are given by

$$(\bar{x} - \bar{y}) \pm t_c s_p \sqrt{\frac{1}{n_x} + \frac{1}{n_y}}$$

where s_p is the pooled estimate of the population variance, and t_c is the relevant value taken from the t-distribution tables.

The confidence interval is given by

$$\left((\bar{x} - \bar{y}) - t_c s_p \sqrt{\frac{1}{n_x} + \frac{1}{n_y}}, \quad (\bar{x} - \bar{y}) + t_c s_p \sqrt{\frac{1}{n_x} + \frac{1}{n_y}} \right)$$

2 Review exercise

A **E** **1** The time, in minutes, it takes Robert to complete the puzzle in his morning newspaper each day is normally distributed with mean 18 and standard deviation 3. After taking a holiday, Robert records the times taken to complete a random sample of 15 puzzles and he finds that the mean time is 16.5 minutes. You may assume that the holiday has not changed the standard deviation of times taken to complete the puzzle.

Stating your hypotheses clearly, test, at the 5% level of significance, whether or not there has been a reduction in the mean time Robert takes to complete the puzzle. **(6)**

← SM2, Chapter 3

E/P **2** In a trial of diet A a random sample of 80 participants were asked to record their weight loss, x kg, after their first week of using the diet. The results are summarised as follows:

$$\sum x = 361.6 \qquad \sum x^2 = 1753.95$$

a Find unbiased estimates for the mean and variance of weight lost after the first week of using diet A. **(3)**

The designers of diet A believe it can achieve a greater mean weight loss after the first week than an existing diet B. A random sample of 60 people used diet B. After the first week they had achieved a mean weight loss of 4.06 kg, with an unbiased estimate of variance of weight loss of 2.50 kg^2.

b Test, at the 5% level of significance, whether or not the mean weight loss

after the first week using diet A is greater than that using diet B. State your hypotheses clearly. **(6)**

c Explain the significance of the central limit theorem to the test in part **b**. **(1)**

d State an assumption you have made in carrying out the test in part **b**. **(1)**

← Sections 5.1, 5.4

E/P **3** A random sample of the daily sales (in £s) of a small company is taken and, using tables of the normal distribution, a 99% confidence interval for the mean daily sales is found to be (123.5, 154.7).

Find a 95% confidence interval for the mean daily sales of the company. **(6)**

← Section 5.2

E **4** A machine produces metal containers. The masses of the containers are normally distributed. A random sample of 10 containers was taken and the mass of each container was recorded to the nearest 0.1 kg. The results were as follows:

| 49.7 | 50.3 | 51.0 | 49.5 | 49.9 |
| 50.1 | 50.2 | 50.0 | 49.6 | 49.7 |

a Find unbiased estimates of the mean and variance of the masses of the population of metal containers. **(3)**

The machine is set to produce metal containers whose masses have a population standard deviation of 0.5 kg.

b Find:

i a 95% confidence interval

ii a 99% confidence interval

for the population mean. **(5)**

← Sections 5.1, 5.2

196

5 The drying times of paint can be assumed to be normally distributed. A paint manufacturer paints 10 test areas with a new paint. The following drying times, to the nearest minute, were recorded:

82	98	140	110	90
125	150	130	70	110

a Calculate unbiased estimates for the mean and the variance of the population of drying times of this paint. **(3)**

Given that the population standard deviation is 25,

b find a 95% confidence interval for the mean drying time of this paint. **(5)**

Fifteen similar sets of tests are done and the 95% confidence interval is determined for each set.

c Find the probability that all 15 of these confidence intervals contain the population mean. **(2)**

← Sections 5.1, 5.2

6 Some biologists were studying a large group of wading birds. A random sample of 36 were measured and the wing length, x mm, of each wading bird was recorded. The results are summarised as follows:

$$\sum x = 6046 \qquad \sum x^2 = 1\,016\,338$$

a Calculate unbiased estimates for the mean and the variance of the wing lengths of these birds. **(3)**

Given that the standard deviation of the wing lengths of this particular type of bird is actually 5.1 mm,

b find a 99% confidence interval for the mean wing length of the birds from this group. **(3)**

← Sections 5.1, 5.2

7 A sociologist is studying how much junk food teenagers eat. A random sample of 100 female teenagers and an independent random sample of 200 male teenagers were asked to estimate their weekly expenditure, x, in pounds, on junk food. The results are summarised below.

	n	$\bar{x}$	s
Female teenagers	100	5.48	3.62
Male teenagers	200	6.86	4.51

a Using a 5% significance level, test whether or not there is a difference in the mean amounts spent on junk food by male teenagers and female teenagers. State your hypotheses clearly. **(7)**

b Explain briefly the importance of the central limit theorem in this problem. **(1)**

← Section 5.4

8 A computer company repairs large numbers of PCs and wants to estimate the mean time taken to repair a particular fault. Five repairs are chosen at random from the company's records and the times taken, in seconds, are as follows:

205	310	405	195	320

a Calculate unbiased estimates of the mean and the variance of the population of repair times from which this sample has been taken. **(3)**

It is known from previous results that the standard deviation of the repair time for this fault is 100 seconds. The company manager wants to ensure that there is a probability of at least 0.95 that the estimate of the population mean lies within 20 seconds of its true value.

b Find the minimum sample size required. **(5)**

← Sections 5.1, 5.2

9 A random sample of 15 tomatoes is taken and the mass, x grams, of each tomato is found. The results are summarised as follows:

A

$\sum x = 208 \qquad \sum x^2 = 2962$

a Assuming that the masses of the tomatoes are normally distributed, calculate the 90% confidence interval for the variance σ^2 of the masses of the tomatoes. **(6)**

b State with a reason whether or not the confidence interval supports the assertion $\sigma = 3$. **(1)**

← **Section 6.1**

E **10** A machine is filling bottles of milk. A random sample of 16 bottles was taken and the volume of milk in each bottle was measured and recorded. The volume of milk in a bottle is normally distributed and the unbiased estimate of the variance, s^2, of the volume of milk in a bottle is 0.003.

a Find a 95% confidence interval for the variance of the population of volumes of milk from which the sample was taken. **(6)**

The machine should fill bottles so that the standard deviation of the volumes is equal to 0.07.

b Comment on this with reference to your 95% confidence interval. **(1)**

← **Section 6.1**

E/P **11** A mechanic is required to change car tyres. An inspector timed a random sample of 20 tyre changes and calculated the unbiased estimate of the population variance to be 6.25 minutes^2.

a Test, at the 5% significance level, whether or not the standard deviation of the population of times taken by the mechanic is greater than 2 minutes. State your hypotheses clearly. **(7)**

b State one assumption you have made in carrying out this test. **(1)**

← **Section 6.2**

E **12** The random variable X has an F-distribution with 10 and 12 degrees of

A freedom. Find a and b such that $P(a < X < b) = 0.90$. **(3)**

← **Section 6.3**

E **13** The random variable X has an F-distribution with 8 and 12 degrees of freedom.

Find $P\left(\dfrac{1}{5.67} < X < 2.85\right)$. **(3)**

← **Section 6.3**

E **14** A beach is divided into two areas, A and B. A random sample of pebbles is taken from each of the two areas and the length of each pebble is measured. A sample of size 26 is taken from area A and the unbiased estimate for the population variance is $s_A^2 = 0.495\,\text{mm}^2$. A sample of size 25 is taken from area B and the unbiased estimate for the population variance is $s_B^2 = 1.04\,\text{mm}^2$.

a Stating your hypotheses clearly, test, at the 10% significance level, whether or not there is a difference in variability of pebble length between area A and area B. **(7)**

b State the assumption you have made about the populations of pebble lengths in order to carry out the test. **(1)**

← **Section 6.4**

E **15** The masses, in grams, of apples are assumed to follow a normal distribution.

The masses of apples sold by a supermarket have variance σ_S^2. A random sample of 4 apples from the supermarket had the following masses:

    114   110   119   123

a Find a 95% confidence interval for σ_S^2. **(4)**

The masses of apples sold on a market stall have variance σ_M^2. A second random sample of 7 apples was taken from the market stall. The sample variance s_M^2 of the apples was 318.8.

A **b** Stating your hypotheses clearly, test, at the 1% level of significance, whether or not there is evidence that $\sigma_M^2 > \sigma_S^2$. **(7)**

← Sections 6.1, 6.4

E/P **16** A nutritionist studied the levels of cholesterol, X mg/cm³, of male students at a large college. She assumed that $X \sim N(\mu, \sigma^2)$ and examined a random sample of 25 male students. Using this sample, she obtained the following unbiased estimates for μ and σ^2:

$$\hat{\mu} = 1.68 \qquad \hat{\sigma}^2 = 1.79$$

a Using the t-distribution, find a 95% confidence interval for μ. **(6)**

b Obtain a 95% confidence interval for σ^2. **(6)**

A cholesterol reading of more than 2.5 mg/cm³ is regarded as high.

c Use appropriate confidence limits from parts **a** and **b** to find the lowest estimate of the proportion of male students in the college with high cholesterol. **(2)**

← Sections 6.1, 7.1

E/P **17** A doctor wishes to study the level of blood glucose in males. The level of blood glucose is normally distributed. The doctor measured the blood glucose of 10 randomly selected male students from a school. The results, in mmol/litre, are given below.

| 4.7 | 3.6 | 3.8 | 4.7 | 4.1 |
| 2.2 | 3.6 | 4.0 | 4.4 | 5.0 |

a Calculate a 95% confidence interval for the mean. **(6)**

b Calculate a 95% confidence interval for the variance. **(6)**

A blood glucose reading of more than 7 mmol/litre is counted as high.

c Use appropriate confidence limits from parts **a** and **b** to find the highest estimate of the proportion of male students in the school with a high blood glucose level. **(2)**

← Sections 6.1, 7.1

A **E/P** **18** A supervisor wishes to check the typing speed of a new typist. On 10 randomly selected occasions, the supervisor records the time taken for the new typist to type 100 words. The results, in seconds, are given below.

| 110 | 125 | 130 | 126 | 128 |
| 127 | 118 | 120 | 122 | 125 |

The supervisor assumes that the time taken to type 100 words is normally distributed.

a Calculate a 95% confidence interval for
   i the mean
   ii the variance
   of the population of times taken by this typist to type 100 words. **(12)**

The supervisor requires the average time needed to type 100 words to be no more than 130 seconds and the standard deviation to be no more than 4 seconds.

b Comment on whether or not the supervisor should be concerned about the speed of the new typist. **(2)**

← Sections 6.1, 7.1

E/P **19** The length, X mm, of a spring made by a machine is normally distributed with $X \sim N(\mu, \sigma^2)$. A random sample of 20 springs is selected and their lengths measured in mm. Using this sample, the unbiased estimates of μ and σ^2 are:

$$\bar{x} = 100.6 \qquad s^2 = 1.5$$

Stating your hypotheses clearly, test, at the 10% level of significance,

a whether or not the variance of the lengths of springs is different from 0.9 **(7)**

b whether or not the mean length of the springs is greater than 100 mm. You should use the t-distribution to carry out your test. **(6)**

← Sections 6.2, 7.2

E/P **20** A grocer receives deliveries of cauliflowers from two different growers, A and B. The grocer takes random samples

A of cauliflowers from those supplied by each grower. He measures the weight, x, in grams, of each cauliflower. The results are summarised in the table below.

	Sample size	Σx	Σx^2
A	11	6600	3 960 540
B	13	9815	7 410 579

a Show, at the 10% significance level, that the variances of the populations from which the samples are drawn can be assumed to be equal by testing the hypothesis $H_0: \sigma_A^2 = \sigma_B^2$ against hypothesis $H_1: \sigma_A^2 \neq \sigma_B^2$.

(You may assume that the two samples come from normal populations.) **(7)**

The grocer believes that the mean weight of cauliflowers provided by B is at least 150 g more than the mean weight of cauliflowers provided by A.

b Use a 5% significance level to test the grocer's belief. **(6)**

c Justify your choice of test. **(1)**

← Sections 6.3, 7.5

E/P 21 An educational researcher is testing the effectiveness of a new method of teaching a topic in mathematics. A random sample of 10 children were taught by the new method and a second random sample of 9 children, of similar age and ability, were taught by the conventional method. At the end of the teaching, the same test was given to both groups of children.

The marks obtained by the two groups are summarised in the table below.

	New method	Conventional method
Mean ($\bar{x}$)	82.3	78.2
Standard deviation (s)	3.5	5.7
Number of students (n)	10	9

A a Stating your hypotheses clearly and using a 5% level of significance, investigate whether or not

   i the variance of the marks of children taught by the conventional method is greater than that of children taught by the new method

   ii the mean score of children taught by the conventional method is lower than the mean score of those taught by the new method.

   [In each case you should give full details of the calculation of the test statistics.] **(12)**

b State any assumptions you made in order to carry out these tests. **(2)**

c Find a 95% confidence interval for the common variance of the marks of the two groups. **(5)**

← Sections 6.3, 7.5

E/P 22 A large number of students are split into two groups, A and B. The students sit the same test but under different conditions. Group A has music playing in the room during the test, and group B has no music playing during the test. Small samples are then taken from each group and their marks recorded. The marks are normally distributed.

The marks are as follows:

Sample from group A
42   40   35   37   34   43   42   44   49

Sample from group B
40   44   38   47   38   37   33

a Stating your hypotheses clearly, and using a 10% level of significance, test whether or not there is evidence of a difference between the variances of the marks of the two groups. **(6)**

b State clearly an assumption you have made to enable you to carry out the test in part **a**. **(1)**

A

c Use a two tailed test, with a 5% level of significance, to determine whether playing music during the test made any difference to the mean marks of the two groups. State your hypotheses clearly. **(6)**

d Write down what you can conclude about the effect of music on a student's performance during the test. **(1)**

← Sections 6.3, 7.5

E/P **23** A company undertakes investigations to compare fuel consumption x, in miles per gallon, of two different cars, the *Relaxant* and the *Elegane*, with a view to purchasing a number of cars. A random sample of 13 *Relaxants* and an independent random sample of 7 *Eleganes* were taken and the following statistics calculated.

Car	Sample size n	Sample mean $\bar{x}$	Sample variance s^2
Relaxant	13	32.31	14.48
Elegane	7	28.43	35.79

The company assumes that fuel consumption for each make of car follows a normal distribution.

a Stating your hypotheses clearly test, at the 10% level of significance, whether or not the two distributions have the same variance. **(6)**

b Stating your hypotheses clearly test, at the 5% level of significance, whether or not there is a difference in mean fuel consumption between the two types of car. **(6)**

c Explain the importance of the conclusion to the test in part **a** in justifying the use of the test in part **b**. **(2)**

d State two factors which might be considered when undertaking an investigation into fuel consumption of two models of car to ensure that a fair comparison is made. **(2)**

← Sections 6.3, 7.5

A **24** A town council is concerned that the mean price of renting two-bedroom flats in the town has exceeded £650 per month. A random sample of eight two-bedroom flats had the following rent prices, x, in pounds per month.

E

705 640 560 680
800 620 580 760

[You may assume
$\Sigma x = 5345$, $\Sigma x^2 = 3\,621\,025$.]

a Find a 90% confidence interval for the mean price of renting a two-bedroom flat. **(6)**

b State an assumption that is required for the validity of your interval in part **a**. **(1)**

c Comment on whether or not the town council is justified in being concerned. Give a reason for your answer. **(2)**

← Section 7.1

E **25** Historical records from a large colony of squirrels show that the weight of squirrels is normally distributed with a mean of 1012 g. Following a change in the diet of squirrels, a biologist is interested in whether or not the mean weight has changed.

A random sample of 14 squirrels is weighed and their weights, x, in grams, recorded. The results are summarised as follows:

$\Sigma x = 13\,700$ $\Sigma x^2 = 13\,448\,750$

Stating your hypotheses clearly, and using a suitable t-distribution, test, at the 5% level of significance, whether or not there has been a change in the mean weight of the squirrels. **(7)**

← Section 7.2

26 A machine is set to fill bags with flour such that the mean weight is 1010 grams.

E

To check that the machine is working properly, a random sample of 8 bags is

A

selected. The weight of flour, in grams, in each bag is as follows:

> 1010   1015   1005   1000
> 998   1008   1012   1007

Carry out a suitable test, at the 5% significance level, to determine whether or not the mean weight of flour in the bags is less than 1010 grams. (You may assume that the weight of flour delivered by the machine is normally distributed.) **(7)**

← **Section 7.2**

E **27** A doctor believes that the span of a person's dominant hand is greater than that of the weaker hand. To test this theory, the doctor measures the spans of the dominant and weaker hands of a random sample of 8 people. The spans, in mm, are summarised in the table below.

Person	A	B	C	D	E	F	G	H
Dominant hand	202	251	215	235	210	195	191	230
Weaker hand	195	249	218	234	211	197	181	225

Carry out a paired t-test, at the 5% significance level, to determine whether the doctor's belief is correct. **(7)**

← **Section 7.3**

E **28** A group of 10 technology students is assessed by coursework and a written examination. The marks, given as percentages, are shown in the table below.

Student	Coursework	Written exam
1	65	61
2	73	76
3	62	65
4	81	77
5	78	72
6	74	71
7	68	72
8	59	42
9	76	69
10	70	63

A

a Use a suitable t-test to determine whether or not the coursework marks are significantly higher than the written examination marks. Use a 5% level of significance. **(7)**

b State an assumption about the distribution of marks that is needed to make the above test valid. **(1)**

← **Section 7.3**

E **29** An engineer decided to investigate whether or not the strength of rope was affected by water. A random sample of 9 pieces of rope was taken and each piece was cut in half. One half of each piece was soaked in water over night, and then each piece of rope was tested to find its strength. The results, in coded units, are given in the table below.

Rope number	1	2	3	4	5	6	7	8	9
Dry rope	9.7	8.5	6.3	8.3	7.2	5.4	6.8	8.1	5.9
Wet rope	9.1	9.5	8.2	9.7	8.5	4.9	8.4	8.7	7.7

Assuming that the strength of rope follows a normal distribution, test whether or not there is any difference between the mean strengths of dry and wet rope. State your hypotheses clearly and use a 1% level of significance. **(7)**

← **Section 7.3**

E **30** As part of an investigation into the effectiveness of solar heating, a pair of houses was identified where the mean weekly fuel consumption was the same. One of the houses was then fitted with solar heating and the other was not. Following the fitting of the solar heating, a random sample of 9 weeks was taken. The table below shows the weekly fuel consumption for each house.

Week	1	2	3	4	5	6	7	8	9
Without solar heating	19	19	18	14	6	7	5	31	43
With solar heating	13	22	11	16	14	1	0	20	38

A

a Stating your hypotheses clearly, test, at the 5% level of significance, whether or not there is evidence that the solar heating reduces the mean weekly fuel consumption. **(7)**

b State an assumption about weekly fuel consumption that is required to carry out this test. **(1)**

← Section 7.3

E/P **31** Two methods of extracting juice from an orange are to be compared. Eight oranges are halved. One half of each orange is chosen at random and allocated to Method A and the other half is allocated to Method B. The amounts of juice extracted, in ml, are given in the table.

	Orange							
	1	**2**	**3**	**4**	**5**	**6**	**7**	**8**
Method A	29	30	26	25	26	22	23	28
Method B	27	25	28	24	23	26	22	25

One statistician suggests performing a two-sample t-test to investigate whether or not there is a difference between the mean amounts of juice extracted by the two methods.

a Stating your hypotheses clearly and using a 5% significance level, carry out this test.

(You may assume $\bar{x}_A = 26.125$, $s_A^2 = 7.84$, $\bar{x}_B = 25$, $s_B^2 = 4$ and $\sigma_A^2 = \sigma_B^2$) **(7)**

Another statistician suggests analysing these data using a paired t-test.

b Using a 5% significance level, carry out this test. **(7)**

c State which of these two tests you consider to be more appropriate. Give a reason for your choice. **(2)**

← Sections 7.3, 7.5

E/P **32** Brickland and Goodbrick are two manufacturers of bricks. The lengths of the bricks produced by each manufacturer can be assumed to be normally distributed. A random sample of 20 bricks is taken from Brickland

A

and the length, x mm, of each brick is recorded. The mean of this sample is 207.1 mm and the variance is 3.2 mm^2.

a Calculate the 98% confidence interval for the mean length of brick from Brickland. **(6)**

A random sample of 10 bricks is selected from those manufactured by Goodbrick. The length of each brick, y mm, is recorded. The results are summarised as follows:

$$\sum y = 2046.2 \qquad \sum y^2 = 418\,785.4$$

The variances of the length of brick for each manufacturer are assumed to be the same.

b Find a 90% confidence interval for the value by which the mean length of brick made by Brickland exceeds the mean length of brick made by Goodbrick. **(7)**

← Sections 7.1, 7.5

Challenge

1 A random sample of three independent variables X_1, X_2 and X_3 is taken from a distribution with mean μ and variance σ^2.

a Show that $\frac{2}{3}X_1 - \frac{1}{2}X_2 + \frac{5}{6}X_3$ is an unbiased estimator for μ.

An unbiased estimator for μ is given by $\hat{\mu} = aX_1 + bX_2$ where a and b are constants.

b Show that $\text{Var}(\hat{\mu}) = (2a^2 - 2a + 1)\sigma^2$.

c Hence determine the value of a and the value of b for which $\hat{\mu}$ has minimum variance.

← Section 5.1

2 The random variable X has the continuous uniform distribution U[0, 1]. A random sample of three independent observations X_1, X_2 and X_3 is taken from X. The random variable M is defined as the **median** of these three observations.

a Show that the probability density function of M is given by:

$$h(x) = \begin{cases} 6x(1-x) & 0 \leqslant x \leqslant 1 \\ 0 & \text{otherwise} \end{cases}$$

b Hence show that M is an unbiased estimator for the median of X.

c Find the standard error of this estimator.

← Sections 3.1, 5.1

Exam-style practice
Further Mathematics
AS Level
Further Statistics 2

Time: 50 minutes
You must have: Mathematical Formulae and Statistical Tables, Calculator

1 A continuous random variable X has a probability density function given by

$$f(x) = \begin{cases} 2 - kx^2 & 1 \leqslant x \leqslant 2 \\ 0 & \text{otherwise} \end{cases}$$

where k is a positive constant.

a Sketch $f(x)$. **(2)**

b State the mode of X. **(1)**

c Show that $k = \frac{3}{7}$ **(2)**

d Use algebraic integration to find $E(X)$. **(3)**

e Define fully the cumulative distribution function $F(x)$. **(4)**

f Show that the median, m, of X satisfies the equation $2m^3 - 28m + 33 = 0$. **(2)**

Given that $m = 1.357$,

g Comment on the skewness of the distribution of X. **(1)**

2 The time taken, in seconds, for the lift to arrive at Gladys' floor of her hotel is modelled by the continuous random variable X, which is uniformly distributed over $0 \leqslant x \leqslant 40$.

a Write down the probability distribution function $f(x)$. **(1)**

Find:

b $E(X)$ **(1)**

c $Var(X)$ **(2)**

d $P(15 \leqslant X \leqslant 30)$ **(2)**

e Given that Gladys has already spent 15 seconds waiting for the lift, find the probability that it will arrive in the next ten seconds. **(2)**

3 A group of 10 languages students sat tests in French and Spanish. Their results were as follows, where x represents their French score and y their Spanish score.

x	21	5	13	16	12	15	18	28	17	19
y	40	31	23	27	26	38	21	47	33	30

a Calculate Spearman's rank correlation coefficient for these data. **(4)**

b Stating your hypotheses clearly, test, at the 5% level of significance, whether or not there is an association between French and Spanish scores. **(4)**

The product-moment correlation coefficient for these results is 0.568.

c Stating your hypotheses clearly, test, at the 5% level of significance, whether or not the correlation coefficient is greater than zero. **(3)**

d State a reason why the conclusions of the two tests seem to conflict. **(1)**

4 The owner of a riverside café is investigating the relationship between the daily mean temperature and the sales of ice-cream and coffee. He takes a random sample of 5 days in August and records the temperature, t (°C), and the sales of coffee, c (£100s). The data is shown in the table below.

t	12	15	19	20	22
c	7	6.8	6.5	6.1	5.9

Summary statistics are calculated and found to be

$$S_{tt} = 65.2 \qquad S_{cc} = 0.852 \qquad \sum tc = 561.3 \qquad \sum t = 88 \qquad \sum c = 32.3$$

He also collects data on sales of ice-cream on the same five days and calculates that the residual sum of squares (RSS) is 0.0524.

Explain, with clear reasons, whether the data for the sales of ice-cream or the data for the sales of coffee is more likely to fit a linear model. **(5)**

Exam-style practice
Further Mathematics
A Level
Further Statistics 2

Time: 1 hour 30 minutes
You must have: Mathematical Formulae and Statistical Tables, Calculator

1 The weights of male warthogs are normally distributed with a mean of 90 kg and a standard deviation of 10 kg.

The weights of female warthogs are normally distributed with a mean of 60 kg and a standard deviation of 5 kg.

Given that the weights of male and female warthogs are independent, find the probability that:

a 5 randomly chosen males and 2 randomly chosen females will weigh more than 560 kg in total, **(5)**

b a randomly chosen male will weigh less than 1.4 times a randomly chosen female. **(6)**

2 A forestry worker is testing the effect of using a fertiliser on willow saplings. Two independent random samples of saplings are selected and their height gained over a 20-day period is recorded. One sample of 10 saplings is given the fertiliser while the other sample of 13 saplings is placed in an identical environment but without fertiliser. The heights gained (x cm) by both groups of saplings are summarised by the statistics in the table below.

	Sample size	Mean $\bar{x}$	Standard deviation s
With fertiliser	10	23.36	5.29
Without fertiliser	13	19.96	6.84

a Use a two-tailed test to show that, at the 10% level of significance, the variances of the heights gained by the saplings with and without fertiliser can be assumed to be equal. State your hypotheses clearly. **(4)**

b Stating your hypotheses clearly test, at the 5% level of significance, whether or not there is a difference in the mean height gained by the two groups of saplings. **(7)**

c State the importance of the test in **a** to your test in part **b**. **(1)**

3 A doctor believes that the span of an adult male's hand, in mm, is normally distributed with a mean of μ mm and a standard deviation of σ mm. A random sample of 6 men's hands were measured. Using this sample, she obtained unbiased estimates of μ and σ^2 as $\hat{\mu}$ and $\hat{\sigma}^2$.

A 95% confidence interval for μ was found to be (206.2, 223.5).

a Show that $\hat{\sigma}^2 = 67.9$ (correct to 3 significant figures). **(4)**

b Obtain a 95% confidence interval for σ. **(3)**

c Use appropriate confidence limits to find, to 2 decimal places, the highest estimate of the proportion of adult males with a hand span greater than 230 mm. **(4)**

4 A researcher thinks that there is a link between a person's confidence and their height. She devises a test to measure the confidence, c, of nine people and their height, h cm. The data are shown in the table below.

h	179	169	187	166	162	193	161	177	168
c	569	561	579	561	540	598	542	565	573

$\sum h = 1562$, $\sum c = 5088$, $\sum hc = 884\,484$, $\sum h^2 = 272\,094$, $\sum c^2 = 2\,878\,966$

  a Find the values of S_{hh}, S_{cc} and S_{hc}.   **(3)**

  b Calculate the equation of the regression line of c on h, giving your answer in the form

     $c = a + bh$

     where a and b are given to 3 significant figures.   **(2)**

  c Give an interpretation of the value of b.   **(1)**

  d Explain why it would not make sense to provide an interpretation of a.   **(1)**

The researcher decides to use this regression model to predict a person's confidence but is told that one of the values for confidence is incorrectly recorded.

  e **i** Calculate the residual values.

     ii Hence identify the incorrect value, giving a reason for your answer.   **(3)**

  f Ignoring the incorrect value, produce a new model.   **(2)**

  g Use the new model to predict the confidence of a person who is 172 cm tall.   **(1)**

5 The table shows the qualifying lap-times of eight drivers in an amateur motorbike race.

Driver	Amy	Carl	Dhruv	David	Ali	Paula	Sarah	Jake
Qualifying time (mm:ss)	3:45	2:52	3:07	2:49	3:49	2:50	2:57	3:11

In the actual race, the drivers finished in the following order, with the quickest driver first:

    Carl, Paula, Sarah, David, Dhruv, Amy, Jake, Ali

  a Calculate Spearman's rank correlation coefficient for these results.   **(4)**

  b Stating your hypotheses clearly, test, at the 5% significance level, whether or not there is a positive correlation between qualifying lap-times and actual race results.   **(4)**

  c Justify the use of Spearman's rank correlation coefficient for this test.   **(1)**

In another race, qualifying was abandoned due to poor weather, resulting in four drivers being allocated a qualifying time of 3:00.

  d Briefly explain how you could calculate Spearman's rank correlation coefficient in this situation.   **(2)**

6 A continuous random variable X has a cumulative distribution function given by

$$F(x) = \begin{cases} 0 & x < 0 \\ \frac{1}{2}x^3(x + 1) & 0 \leqslant x \leqslant 1 \\ 1 & x > 1 \end{cases}$$

  a Use algebraic integration to find $E(X)$ and $Var(X)$.   **(5)**

  b Find the mode of X, giving a reason for your answer.   **(2)**

  c Describe the skewness of the distribution of X. Give a reason for your answer.   **(1)**

k is a constant such that $0 < k \leqslant \frac{1}{3}$.

  d Show that $P(k < X < 3k) = k^3(40k + 13)$.   **(3)**

7 A wooden pole AB is 6 m long. The pole is sawn into two pieces at a randomly chosen point P. The random variable $X \sim U[0, 6]$ represents the length in metres of the section AP.

The two sections of the pole are used to form a framework for the cross section of a tent, with angle $APB = 90°$, as shown in the diagram below:

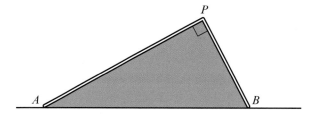

Find the expected area enclosed by the framework and the ground, shown shaded on the diagram. **(6)**

BINOMIAL CUMULATIVE DISTRIBUTION FUNCTION

The tabulated value is $P(X \le x)$, where X has a binomial distribution with index n and parameter p.

$p =$	0.05	0.10	0.15	0.20	0.25	0.30	0.35	0.40	0.45	0.50
$n = 5, x = 0$	0.7738	0.5905	0.4437	0.3277	0.2373	0.1681	0.1160	0.0778	0.0503	0.0312
1	0.9774	0.9185	0.8352	0.7373	0.6328	0.5282	0.4284	0.3370	0.2562	0.1875
2	0.9988	0.9914	0.9734	0.9421	0.8965	0.8369	0.7648	0.6826	0.5931	0.5000
3	1.0000	0.9995	0.9978	0.9933	0.9844	0.9692	0.9460	0.9130	0.8688	0.8125
4	1.0000	1.0000	0.9999	0.9997	0.9990	0.9976	0.9947	0.9898	0.9815	0.9688
$n = 6, x = 0$	0.7351	0.5314	0.3771	0.2621	0.1780	0.1176	0.0754	0.0467	0.0277	0.0156
1	0.9672	0.8857	0.7765	0.6554	0.5339	0.4202	0.3191	0.2333	0.1636	0.1094
2	0.9978	0.9842	0.9527	0.9011	0.8306	0.7443	0.6471	0.5443	0.4415	0.3438
3	0.9999	0.9987	0.9941	0.9830	0.9624	0.9295	0.8826	0.8208	0.7447	0.6563
4	1.0000	0.9999	0.9996	0.9984	0.9954	0.9891	0.9777	0.9590	0.9308	0.8906
5	1.0000	1.0000	1.0000	0.9999	0.9998	0.9993	0.9982	0.9959	0.9917	0.9844
$n = 7, x = 0$	0.6983	0.4783	0.3206	0.2097	0.1335	0.0824	0.0490	0.0280	0.0152	0.0078
1	0.9556	0.8503	0.7166	0.5767	0.4449	0.3294	0.2338	0.1586	0.1024	0.0625
2	0.9962	0.9743	0.9262	0.8520	0.7564	0.6471	0.5323	0.4199	0.3164	0.2266
3	0.9998	0.9973	0.9879	0.9667	0.9294	0.8740	0.8002	0.7102	0.6083	0.5000
4	1.0000	0.9998	0.9988	0.9953	0.9871	0.9712	0.9444	0.9037	0.8471	0.7734
5	1.0000	1.0000	0.9999	0.9996	0.9987	0.9962	0.9910	0.9812	0.9643	0.9375
6	1.0000	1.0000	1.0000	1.0000	0.9999	0.9998	0.9994	0.9984	0.9963	0.9922
$n = 8, x = 0$	0.6634	0.4305	0.2725	0.1678	0.1001	0.0576	0.0319	0.0168	0.0084	0.0039
1	0.9428	0.8131	0.6572	0.5033	0.3671	0.2553	0.1691	0.1064	0.0632	0.0352
2	0.9942	0.9619	0.8948	0.7969	0.6785	0.5518	0.4278	0.3154	0.2201	0.1445
3	0.9996	0.9950	0.9786	0.9437	0.8862	0.8059	0.7064	0.5941	0.4770	0.3633
4	1.0000	0.9996	0.9971	0.9896	0.9727	0.9420	0.8939	0.8263	0.7396	0.6367
5	1.0000	1.0000	0.9998	0.9988	0.9958	0.9887	0.9747	0.9502	0.9115	0.8555
6	1.0000	1.0000	1.0000	0.9999	0.9996	0.9987	0.9964	0.9915	0.9819	0.9648
7	1.0000	1.0000	1.0000	1.0000	1.0000	0.9999	0.9998	0.9993	0.9983	0.9961
$n = 9, x = 0$	0.6302	0.3874	0.2316	0.1342	0.0751	0.0404	0.0207	0.0101	0.0046	0.0020
1	0.9288	0.7748	0.5995	0.4362	0.3003	0.1960	0.1211	0.0705	0.0385	0.0195
2	0.9916	0.9470	0.8591	0.7382	0.6007	0.4628	0.3373	0.2318	0.1495	0.0898
3	0.9994	0.9917	0.9661	0.9144	0.8343	0.7297	0.6089	0.4826	0.3614	0.2539
4	1.0000	0.9991	0.9944	0.9804	0.9511	0.9012	0.8283	0.7334	0.6214	0.5000
5	1.0000	0.9999	0.9994	0.9969	0.9900	0.9747	0.9464	0.9006	0.8342	0.7461
6	1.0000	1.0000	1.0000	0.9997	0.9987	0.9957	0.9888	0.9750	0.9502	0.9102
7	1.0000	1.0000	1.0000	1.0000	0.9999	0.9996	0.9986	0.9962	0.9909	0.9805
8	1.0000	1.0000	1.0000	1.0000	1.0000	1.0000	0.9999	0.9997	0.9992	0.9980
$n = 10, x = 0$	0.5987	0.3487	0.1969	0.1074	0.0563	0.0282	0.0135	0.0060	0.0025	0.0010
1	0.9139	0.7361	0.5443	0.3758	0.2440	0.1493	0.0860	0.0464	0.0233	0.0107
2	0.9885	0.9298	0.8202	0.6778	0.5256	0.3828	0.2616	0.1673	0.0996	0.0547
3	0.9990	0.9872	0.9500	0.8791	0.7759	0.6496	0.5138	0.3823	0.2660	0.1719
4	0.9999	0.9984	0.9901	0.9672	0.9219	0.8497	0.7515	0.6331	0.5044	0.3770
5	1.0000	0.9999	0.9986	0.9936	0.9803	0.9527	0.9051	0.8338	0.7384	0.6230
6	1.0000	1.0000	0.9999	0.9991	0.9965	0.9894	0.9740	0.9452	0.8980	0.8281
7	1.0000	1.0000	1.0000	0.9999	0.9996	0.9984	0.9952	0.9877	0.9726	0.9453
8	1.0000	1.0000	1.0000	1.0000	1.0000	0.9999	0.9995	0.9983	0.9955	0.9893
9	1.0000	1.0000	1.0000	1.0000	1.0000	1.0000	1.0000	0.9999	0.9997	0.9990

$p =$	0.05	0.10	0.15	0.20	0.25	0.30	0.35	0.40	0.45	0.50
$n = 12, x = 0$	0.5404	0.2824	0.1422	0.0687	0.0317	0.0138	0.0057	0.0022	0.0008	0.0002
1	0.8816	0.6590	0.4435	0.2749	0.1584	0.0850	0.0424	0.0196	0.0083	0.0032
2	0.9804	0.8891	0.7358	0.5583	0.3907	0.2528	0.1513	0.0834	0.0421	0.0193
3	0.9978	0.9744	0.9078	0.7946	0.6488	0.4925	0.3467	0.2253	0.1345	0.0730
4	0.9998	0.9957	0.9761	0.9274	0.8424	0.7237	0.5833	0.4382	0.3044	0.1938
5	1.0000	0.9995	0.9954	0.9806	0.9456	0.8822	0.7873	0.6652	0.5269	0.3872
6	1.0000	0.9999	0.9993	0.9961	0.9857	0.9614	0.9154	0.8418	0.7393	0.6128
7	1.0000	1.0000	0.9999	0.9994	0.9972	0.9905	0.9745	0.9427	0.8883	0.8062
8	1.0000	1.0000	1.0000	0.9999	0.9996	0.9983	0.9944	0.9847	0.9644	0.9270
9	1.0000	1.0000	1.0000	1.0000	1.0000	0.9998	0.9992	0.9972	0.9921	0.9807
10	1.0000	1.0000	1.0000	1.0000	1.0000	1.0000	0.9999	0.9997	0.9989	0.9968
11	1.0000	1.0000	1.0000	1.0000	1.0000	1.0000	1.0000	1.0000	0.9999	0.9998
$n = 15, x = 0$	0.4633	0.2059	0.0874	0.0352	0.0134	0.0047	0.0016	0.0005	0.0001	0.0000
1	0.8290	0.5490	0.3186	0.1671	0.0802	0.0353	0.0142	0.0052	0.0017	0.0005
2	0.9638	0.8159	0.6042	0.3980	0.2361	0.1268	0.0617	0.0271	0.0107	0.0037
3	0.9945	0.9444	0.8227	0.6482	0.4613	0.2969	0.1727	0.0905	0.0424	0.0176
4	0.9994	0.9873	0.9383	0.8358	0.6865	0.5155	0.3519	0.2173	0.1204	0.0592
5	0.9999	0.9978	0.9832	0.9389	0.8516	0.7216	0.5643	0.4032	0.2608	0.1509
6	1.0000	0.9997	0.9964	0.9819	0.9434	0.8689	0.7548	0.6098	0.4522	0.3036
7	1.0000	1.0000	0.9994	0.9958	0.9827	0.9500	0.8868	0.7869	0.6535	0.5000
8	1.0000	1.0000	0.9999	0.9992	0.9958	0.9848	0.9578	0.9050	0.8182	0.6964
9	1.0000	1.0000	1.0000	0.9999	0.9992	0.9963	0.9876	0.9662	0.9231	0.8491
10	1.0000	1.0000	1.0000	1.0000	0.9999	0.9993	0.9972	0.9907	0.9745	0.9408
11	1.0000	1.0000	1.0000	1.0000	1.0000	0.9999	0.9995	0.9981	0.9937	0.9824
12	1.0000	1.0000	1.0000	1.0000	1.0000	1.0000	0.9999	0.9997	0.9989	0.9963
13	1.0000	1.0000	1.0000	1.0000	1.0000	1.0000	1.0000	1.0000	0.9999	0.9995
14	1.0000	1.0000	1.0000	1.0000	1.0000	1.0000	1.0000	1.0000	1.0000	1.0000
$n = 20, x = 0$	0.3585	0.1216	0.0388	0.0115	0.0032	0.0008	0.0002	0.0000	0.0000	0.0000
1	0.7358	0.3917	0.1756	0.0692	0.0243	0.0076	0.0021	0.0005	0.0001	0.0000
2	0.9245	0.6769	0.4049	0.2061	0.0913	0.0355	0.0121	0.0036	0.0009	0.0002
3	0.9841	0.8670	0.6477	0.4114	0.2252	0.1071	0.0444	0.0160	0.0049	0.0013
4	0.9974	0.9568	0.8298	0.6296	0.4148	0.2375	0.1182	0.0510	0.0189	0.0059
5	0.9997	0.9887	0.9327	0.8042	0.6172	0.4164	0.2454	0.1256	0.0553	0.0207
6	1.0000	0.9976	0.9781	0.9133	0.7858	0.6080	0.4166	0.2500	0.1299	0.0577
7	1.0000	0.9996	0.9941	0.9679	0.8982	0.7723	0.6010	0.4159	0.2520	0.1316
8	1.0000	0.9999	0.9987	0.9900	0.9591	0.8867	0.7624	0.5956	0.4143	0.2517
9	1.0000	1.0000	0.9998	0.9974	0.9861	0.9520	0.8782	0.7553	0.5914	0.4119
10	1.0000	1.0000	1.0000	0.9994	0.9961	0.9829	0.9468	0.8725	0.7507	0.5881
11	1.0000	1.0000	1.0000	0.9999	0.9991	0.9949	0.9804	0.9435	0.8692	0.7483
12	1.0000	1.0000	1.0000	1.0000	0.9998	0.9987	0.9940	0.9790	0.9420	0.8684
13	1.0000	1.0000	1.0000	1.0000	1.0000	0.9997	0.9985	0.9935	0.9786	0.9423
14	1.0000	1.0000	1.0000	1.0000	1.0000	1.0000	0.9997	0.9984	0.9936	0.9793
15	1.0000	1.0000	1.0000	1.0000	1.0000	1.0000	1.0000	0.9997	0.9985	0.9941
16	1.0000	1.0000	1.0000	1.0000	1.0000	1.0000	1.0000	1.0000	0.9997	0.9987
17	1.0000	1.0000	1.0000	1.0000	1.0000	1.0000	1.0000	1.0000	1.0000	0.9998
18	1.0000	1.0000	1.0000	1.0000	1.0000	1.0000	1.0000	1.0000	1.0000	1.0000

$p =$	0.05	0.10	0.15	0.20	0.25	0.30	0.35	0.40	0.45	0.50
$n = 25, x = 0$	0.2774	0.0718	0.0172	0.0038	0.0008	0.0001	0.0000	0.0000	0.0000	0.0000
1	0.6424	0.2712	0.0931	0.0274	0.0070	0.0016	0.0003	0.0001	0.0000	0.0000
2	0.8729	0.5371	0.2537	0.0982	0.0321	0.0090	0.0021	0.0004	0.0001	0.0000
3	0.9659	0.7636	0.4711	0.2340	0.0962	0.0332	0.0097	0.0024	0.0005	0.0001
4	0.9928	0.9020	0.6821	0.4207	0.2137	0.0905	0.0320	0.0095	0.0023	0.0005
5	0.9988	0.9666	0.8385	0.6167	0.3783	0.1935	0.0826	0.0294	0.0086	0.0020
6	0.9998	0.9905	0.9305	0.7800	0.5611	0.3407	0.1734	0.0736	0.0258	0.0073
7	1.0000	0.9977	0.9745	0.8909	0.7265	0.5118	0.3061	0.1536	0.0639	0.0216
8	1.0000	0.9995	0.9920	0.9532	0.8506	0.6769	0.4668	0.2735	0.1340	0.0539
9	1.0000	0.9999	0.9979	0.9827	0.9287	0.8106	0.6303	0.4246	0.2424	0.1148
10	1.0000	1.0000	0.9995	0.9944	0.9703	0.9022	0.7712	0.5858	0.3843	0.2122
11	1.0000	1.0000	0.9999	0.9985	0.9893	0.9558	0.8746	0.7323	0.5426	0.3450
12	1.0000	1.0000	1.0000	0.9996	0.9966	0.9825	0.9396	0.8462	0.6937	0.5000
13	1.0000	1.0000	1.0000	0.9999	0.9991	0.9940	0.9745	0.9222	0.8173	0.6550
14	1.0000	1.0000	1.0000	1.0000	0.9998	0.9982	0.9907	0.9656	0.9040	0.7878
15	1.0000	1.0000	1.0000	1.0000	1.0000	0.9995	0.9971	0.9868	0.9560	0.8852
16	1.0000	1.0000	1.0000	1.0000	1.0000	0.9999	0.9992	0.9957	0.9826	0.9461
17	1.0000	1.0000	1.0000	1.0000	1.0000	1.0000	0.9998	0.9988	0.9942	0.9784
18	1.0000	1.0000	1.0000	1.0000	1.0000	1.0000	1.0000	0.9997	0.9984	0.9927
19	1.0000	1.0000	1.0000	1.0000	1.0000	1.0000	1.0000	0.9999	0.9996	0.9980
20	1.0000	1.0000	1.0000	1.0000	1.0000	1.0000	1.0000	1.0000	0.9999	0.9995
21	1.0000	1.0000	1.0000	1.0000	1.0000	1.0000	1.0000	1.0000	1.0000	0.9999
22	1.0000	1.0000	1.0000	1.0000	1.0000	1.0000	1.0000	1.0000	1.0000	1.0000
$n = 30, x = 0$	0.2146	0.0424	0.0076	0.0012	0.0002	0.0000	0.0000	0.0000	0.0000	0.0000
1	0.5535	0.1837	0.0480	0.0105	0.0020	0.0003	0.0000	0.0000	0.0000	0.0000
2	0.8122	0.4114	0.1514	0.0442	0.0106	0.0021	0.0003	0.0000	0.0000	0.0000
3	0.9392	0.6474	0.3217	0.1227	0.0374	0.0093	0.0019	0.0003	0.0000	0.0000
4	0.9844	0.8245	0.5245	0.2552	0.0979	0.0302	0.0075	0.0015	0.0002	0.0000
5	0.9967	0.9268	0.7106	0.4275	0.2026	0.0766	0.0233	0.0057	0.0011	0.0002
6	0.9994	0.9742	0.8474	0.6070	0.3481	0.1595	0.0586	0.0172	0.0040	0.0007
7	0.9999	0.9922	0.9302	0.7608	0.5143	0.2814	0.1238	0.0435	0.0121	0.0026
8	1.0000	0.9980	0.9722	0.8713	0.6736	0.4315	0.2247	0.0940	0.0312	0.0081
9	1.0000	0.9995	0.9903	0.9389	0.8034	0.5888	0.3575	0.1763	0.0694	0.0214
10	1.0000	0.9999	0.9971	0.9744	0.8943	0.7304	0.5078	0.2915	0.1350	0.0494
11	1.0000	1.0000	0.9992	0.9905	0.9493	0.8407	0.6548	0.4311	0.2327	0.1002
12	1.0000	1.0000	0.9998	0.9969	0.9784	0.9155	0.7802	0.5785	0.3592	0.1808
13	1.0000	1.0000	1.0000	0.9991	0.9918	0.9599	0.8737	0.7145	0.5025	0.2923
14	1.0000	1.0000	1.0000	0.9998	0.9973	0.9831	0.9348	0.8246	0.6448	0.4278
15	1.0000	1.0000	1.0000	0.9999	0.9992	0.9936	0.9699	0.9029	0.7691	0.5722
16	1.0000	1.0000	1.0000	1.0000	0.9998	0.9979	0.9876	0.9519	0.8644	0.7077
17	1.0000	1.0000	1.0000	1.0000	0.9999	0.9994	0.9955	0.9788	0.9286	0.8192
18	1.0000	1.0000	1.0000	1.0000	1.0000	0.9998	0.9986	0.9917	0.9666	0.8998
19	1.0000	1.0000	1.0000	1.0000	1.0000	1.0000	0.9996	0.9971	0.9862	0.9506
20	1.0000	1.0000	1.0000	1.0000	1.0000	1.0000	0.9999	0.9991	0.9950	0.9786
21	1.0000	1.0000	1.0000	1.0000	1.0000	1.0000	1.0000	0.9998	0.9984	0.9919
22	1.0000	1.0000	1.0000	1.0000	1.0000	1.0000	1.0000	1.0000	0.9996	0.9974
23	1.0000	1.0000	1.0000	1.0000	1.0000	1.0000	1.0000	1.0000	0.9999	0.9993
24	1.0000	1.0000	1.0000	1.0000	1.0000	1.0000	1.0000	1.0000	1.0000	0.9998
25	1.0000	1.0000	1.0000	1.0000	1.0000	1.0000	1.0000	1.0000	1.0000	1.0000

$p =$	0.05	0.10	0.15	0.20	0.25	0.30	0.35	0.40	0.45	0.50
$n = 40, x = 0$	0.1285	0.0148	0.0015	0.0001	0.0000	0.0000	0.0000	0.0000	0.0000	0.0000
1	0.3991	0.0805	0.0121	0.0015	0.0001	0.0000	0.0000	0.0000	0.0000	0.0000
2	0.6767	0.2228	0.0486	0.0079	0.0010	0.0001	0.0000	0.0000	0.0000	0.0000
3	0.8619	0.4231	0.1302	0.0285	0.0047	0.0006	0.0001	0.0000	0.0000	0.0000
4	0.9520	0.6290	0.2633	0.0759	0.0160	0.0026	0.0003	0.0000	0.0000	0.0000
5	0.9861	0.7937	0.4325	0.1613	0.0433	0.0086	0.0013	0.0001	0.0000	0.0000
6	0.9966	0.9005	0.6067	0.2859	0.0962	0.0238	0.0044	0.0006	0.0001	0.0000
7	0.9993	0.9581	0.7559	0.4371	0.1820	0.0553	0.0124	0.0021	0.0002	0.0000
8	0.9999	0.9845	0.8646	0.5931	0.2998	0.1110	0.0303	0.0061	0.0009	0.0001
9	1.0000	0.9949	0.9328	0.7318	0.4395	0.1959	0.0644	0.0156	0.0027	0.0003
10	1.0000	0.9985	0.9701	0.8392	0.5839	0.3087	0.1215	0.0352	0.0074	0.0011
11	1.0000	0.9996	0.9880	0.9125	0.7151	0.4406	0.2053	0.0709	0.0179	0.0032
12	1.0000	0.9999	0.9957	0.9568	0.8209	0.5772	0.3143	0.1285	0.0386	0.0083
13	1.0000	1.0000	0.9986	0.9806	0.8968	0.7032	0.4408	0.2112	0.0751	0.0192
14	1.0000	1.0000	0.9996	0.9921	0.9456	0.8074	0.5721	0.3174	0.1326	0.0403
15	1.0000	1.0000	0.9999	0.9971	0.9738	0.8849	0.6946	0.4402	0.2142	0.0769
16	1.0000	1.0000	1.0000	0.9990	0.9884	0.9367	0.7978	0.5681	0.3185	0.1341
17	1.0000	1.0000	1.0000	0.9997	0.9953	0.9680	0.8761	0.6885	0.4391	0.2148
18	1.0000	1.0000	1.0000	0.9999	0.9983	0.9852	0.9301	0.7911	0.5651	0.3179
19	1.0000	1.0000	1.0000	1.0000	0.9994	0.9937	0.9637	0.8702	0.6844	0.4373
20	1.0000	1.0000	1.0000	1.0000	0.9998	0.9976	0.9827	0.9256	0.7870	0.5627
21	1.0000	1.0000	1.0000	1.0000	1.0000	0.9991	0.9925	0.9608	0.8669	0.6821
22	1.0000	1.0000	1.0000	1.0000	1.0000	0.9997	0.9970	0.9811	0.9233	0.7852
23	1.0000	1.0000	1.0000	1.0000	1.0000	0.9999	0.9989	0.9917	0.9595	0.8659
24	1.0000	1.0000	1.0000	1.0000	1.0000	1.0000	0.9996	0.9966	0.9804	0.9231
25	1.0000	1.0000	1.0000	1.0000	1.0000	1.0000	0.9999	0.9988	0.9914	0.9597
26	1.0000	1.0000	1.0000	1.0000	1.0000	1.0000	1.0000	0.9996	0.9966	0.9808
27	1.0000	1.0000	1.0000	1.0000	1.0000	1.0000	1.0000	0.9999	0.9988	0.9917
28	1.0000	1.0000	1.0000	1.0000	1.0000	1.0000	1.0000	1.0000	0.9996	0.9968
29	1.0000	1.0000	1.0000	1.0000	1.0000	1.0000	1.0000	1.0000	0.9999	0.9989
30	1.0000	1.0000	1.0000	1.0000	1.0000	1.0000	1.0000	1.0000	1.0000	0.9997
31	1.0000	1.0000	1.0000	1.0000	1.0000	1.0000	1.0000	1.0000	1.0000	0.9999
32	1.0000	1.0000	1.0000	1.0000	1.0000	1.0000	1.0000	1.0000	1.0000	1.0000

$p =$	0.05	0.10	0.15	0.20	0.25	0.30	0.35	0.40	0.45	0.50
$n = 50, x = 0$	0.0769	0.0052	0.0003	0.0000	0.0000	0.0000	0.0000	0.0000	0.0000	0.0000
1	0.2794	0.0338	0.0029	0.0002	0.0000	0.0000	0.0000	0.0000	0.0000	0.0000
2	0.5405	0.1117	0.0142	0.0013	0.0001	0.0000	0.0000	0.0000	0.0000	0.0000
3	0.7604	0.2503	0.0460	0.0057	0.0005	0.0000	0.0000	0.0000	0.0000	0.0000
4	0.8964	0.4312	0.1121	0.0185	0.0021	0.0002	0.0000	0.0000	0.0000	0.0000
5	0.9622	0.6161	0.2194	0.0480	0.0070	0.0007	0.0001	0.0000	0.0000	0.0000
6	0.9882	0.7702	0.3613	0.1034	0.0194	0.0025	0.0002	0.0000	0.0000	0.0000
7	0.9968	0.8779	0.5188	0.1904	0.0453	0.0073	0.0008	0.0001	0.0000	0.0000
8	0.9992	0.9421	0.6681	0.3073	0.0916	0.0183	0.0025	0.0002	0.0000	0.0000
9	0.9998	0.9755	0.7911	0.4437	0.1637	0.0402	0.0067	0.0008	0.0001	0.0000
10	1.0000	0.9906	0.8801	0.5836	0.2622	0.0789	0.0160	0.0022	0.0002	0.0000
11	1.0000	0.9968	0.9372	0.7107	0.3816	0.1390	0.0342	0.0057	0.0006	0.0000
12	1.0000	0.9990	0.9699	0.8139	0.5110	0.2229	0.0661	0.0133	0.0018	0.0002
13	1.0000	0.9997	0.9868	0.8894	0.6370	0.3279	0.1163	0.0280	0.0045	0.0005
14	1.0000	0.9999	0.9947	0.9393	0.7481	0.4468	0.1878	0.0540	0.0104	0.0013
15	1.0000	1.0000	0.9981	0.9692	0.8369	0.5692	0.2801	0.0955	0.0220	0.0033
16	1.0000	1.0000	0.9993	0.9856	0.9017	0.6839	0.3889	0.1561	0.0427	0.0077
17	1.0000	1.0000	0.9998	0.9937	0.9449	0.7822	0.5060	0.2369	0.0765	0.0164
18	1.0000	1.0000	0.9999	0.9975	0.9713	0.8594	0.6216	0.3356	0.1273	0.0325
19	1.0000	1.0000	1.0000	0.9991	0.9861	0.9152	0.7264	0.4465	0.1974	0.0595
20	1.0000	1.0000	1.0000	0.9997	0.9937	0.9522	0.8139	0.5610	0.2862	0.1013
21	1.0000	1.0000	1.0000	0.9999	0.9974	0.9749	0.8813	0.6701	0.3900	0.1611
22	1.0000	1.0000	1.0000	1.0000	0.9990	0.9877	0.9290	0.7660	0.5019	0.2399
23	1.0000	1.0000	1.0000	1.0000	0.9996	0.9944	0.9604	0.8438	0.6134	0.3359
24	1.0000	1.0000	1.0000	1.0000	0.9999	0.9976	0.9793	0.9022	0.7160	0.4439
25	1.0000	1.0000	1.0000	1.0000	1.0000	0.9991	0.9900	0.9427	0.8034	0.5561
26	1.0000	1.0000	1.0000	1.0000	1.0000	0.9997	0.9955	0.9686	0.8721	0.6641
27	1.0000	1.0000	1.0000	1.0000	1.0000	0.9999	0.9981	0.9840	0.9220	0.7601
28	1.0000	1.0000	1.0000	1.0000	1.0000	1.0000	0.9993	0.9924	0.9556	0.8389
29	1.0000	1.0000	1.0000	1.0000	1.0000	1.0000	0.9997	0.9966	0.9765	0.8987
30	1.0000	1.0000	1.0000	1.0000	1.0000	1.0000	0.9999	0.9986	0.9884	0.9405
31	1.0000	1.0000	1.0000	1.0000	1.0000	1.0000	1.0000	0.9995	0.9947	0.9675
32	1.0000	1.0000	1.0000	1.0000	1.0000	1.0000	1.0000	0.9998	0.9978	0.9836
33	1.0000	1.0000	1.0000	1.0000	1.0000	1.0000	1.0000	0.9999	0.9991	0.9923
34	1.0000	1.0000	1.0000	1.0000	1.0000	1.0000	1.0000	1.0000	0.9997	0.9967
35	1.0000	1.0000	1.0000	1.0000	1.0000	1.0000	1.0000	1.0000	0.9999	0.9987
36	1.0000	1.0000	1.0000	1.0000	1.0000	1.0000	1.0000	1.0000	1.0000	0.9995
37	1.0000	1.0000	1.0000	1.0000	1.0000	1.0000	1.0000	1.0000	1.0000	0.9998
38	1.0000	1.0000	1.0000	1.0000	1.0000	1.0000	1.0000	1.0000	1.0000	1.0000

PERCENTAGE POINTS OF THE NORMAL DISTRIBUTION

The values z in the table are those which a random variable $Z \sim N(0, 1)$ exceeds with probability p; that is, $P(Z > z) = 1 - \Phi(z) = p$.

p	z	p	z
0.5000	0.0000	0.0500	1.6449
0.4000	0.2533	0.0250	1.9600
0.3000	0.5244	0.0100	2.3263
0.2000	0.8416	0.0050	2.5758
0.1500	1.0364	0.0010	3.0902
0.1000	1.2816	0.0005	3.2905

PERCENTAGE POINTS OF THE χ^2 DISTRIBUTION

The values in the table are those which a random variable with the χ^2 distribution on ν degrees of freedom exceeds with the probability shown.

ν	0.995	0.990	0.975	0.950	0.900	0.100	0.050	0.025	0.010	0.005
1	0.000	0.000	0.001	0.004	0.016	2.705	3.841	5.024	6.635	7.879
2	0.010	0.020	0.051	0.103	0.211	4.605	5.991	7.378	9.210	10.597
3	0.072	0.115	0.216	0.352	0.584	6.251	7.815	9.348	11.345	12.838
4	0.207	0.297	0.484	0.711	1.064	7.779	9.488	11.143	13.277	14.860
5	0.412	0.554	0.831	1.145	1.610	9.236	11.070	12.832	15.086	16.750
6	0.676	0.872	1.237	1.635	2.204	10.645	12.592	14.449	16.812	18.548
7	0.989	1.239	1.690	2.167	2.833	12.017	14.067	16.013	18.475	20.278
8	1.344	1.646	2.180	2.733	3.490	13.362	15.507	17.535	20.090	21.955
9	1.735	2.088	2.700	3.325	4.168	14.684	16.919	19.023	21.666	23.589
10	2.156	2.558	3.247	3.940	4.865	15.987	18.307	20.483	23.209	25.188
11	2.603	3.053	3.816	4.575	5.580	17.275	19.675	21.920	24.725	26.757
12	3.074	3.571	4.404	5.226	6.304	18.549	21.026	23.337	26.217	28.300
13	3.565	4.107	5.009	5.892	7.042	19.812	22.362	24.736	27.688	29.819
14	4.075	4.660	5.629	6.571	7.790	21.064	23.685	26.119	29.141	31.319
15	4.601	5.229	6.262	7.261	8.547	22.307	24.996	27.488	30.578	32.801
16	5.142	5.812	6.908	7.962	9.312	23.542	26.296	28.845	32.000	34.267
17	5.697	6.408	7.564	8.672	10.085	24.769	27.587	30.191	33.409	35.718
18	6.265	7.015	8.231	9.390	10.865	25.989	28.869	31.526	34.805	37.156
19	6.844	7.633	8.907	10.117	11.651	27.204	30.144	32.852	36.191	38.582
20	7.434	8.260	9.591	10.851	12.443	28.412	31.410	34.170	37.566	39.997
21	8.034	8.897	10.283	11.591	13.240	29.615	32.671	35.479	38.932	41.401
22	8.643	9.542	10.982	12.338	14.042	30.813	33.924	36.781	40.289	42.796
23	9.260	10.196	11.689	13.091	14.848	32.007	35.172	38.076	41.638	44.181
24	9.886	10.856	12.401	13.848	15.659	33.196	36.415	39.364	42.980	45.558
25	10.520	11.524	13.120	14.611	16.473	34.382	37.652	40.646	44.314	46.928
26	11.160	12.198	13.844	15.379	17.292	35.563	38.885	41.923	45.642	48.290
27	11.808	12.879	14.573	16.151	18.114	36.741	40.113	43.194	46.963	49.645
28	12.461	13.565	15.308	16.928	18.939	37.916	41.337	44.461	48.278	50.993
29	13.121	14.256	16.047	17.708	19.768	39.088	42.557	45.722	49.588	52.336
30	13.787	14.953	16.791	18.493	20.599	40.256	43.773	46.979	50.892	53.672

CRITICAL VALUES FOR CORRELATION COEFFICIENTS

These tables concern tests of the hypothesis that a population correlation coefficient ρ is 0. The values in the tables are the minimum values which need to be reached by a sample correlation coefficient in order to be significant at the level shown, on a one-tailed test.

Product Moment Coefficient					Sample	Spearman's Coefficient		
Level					Size, n	Level		
0.10	0.05	0.025	0.01	0.005		0.05	0.025	0.01
0.8000	0.9000	0.9500	0.9800	0.9900	4	1.0000	–	–
0.6870	0.8054	0.8783	0.9343	0.9587	5	0.9000	1.0000	1.0000
0.6084	0.7293	0.8114	0.8822	0.9172	6	0.8286	0.8857	0.9429
0.5509	0.6694	0.7545	0.8329	0.8745	7	0.7143	0.7857	0.8929
0.5067	0.6215	0.7067	0.7887	0.8343	8	0.6429	0.7381	0.8333
0.4716	0.5822	0.6664	0.7498	0.7977	9	0.6000	0.7000	0.7833
0.4428	0.5494	0.6319	0.7155	0.7646	10	0.5636	0.6485	0.7455
0.4187	0.5214	0.6021	0.6851	0.7348	11	0.5364	0.6182	0.7091
0.3981	0.4973	0.5760	0.6581	0.7079	12	0.5035	0.5874	0.6783
0.3802	0.4762	0.5529	0.6339	0.6835	13	0.4835	0.5604	0.6484
0.3646	0.4575	0.5324	0.6120	0.6614	14	0.4637	0.5385	0.6264
0.3507	0.4409	0.5140	0.5923	0.6411	15	0.4464	0.5214	0.6036
0.3383	0.4259	0.4973	0.5742	0.6226	16	0.4294	0.5029	0.5824
0.3271	0.4124	0.4821	0.5577	0.6055	17	0.4142	0.4877	0.5662
0.3170	0.4000	0.4683	0.5425	0.5897	18	0.4014	0.4716	0.5501
0.3077	0.3887	0.4555	0.5285	0.5751	19	0.3912	0.4596	0.5351
0.2992	0.3783	0.4438	0.5155	0.5614	20	0.3805	0.4466	0.5218
0.2914	0.3687	0.4329	0.5034	0.5487	21	0.3701	0.4364	0.5091
0.2841	0.3598	0.4227	0.4921	0.5368	22	0.3608	0.4252	0.4975
0.2774	0.3515	0.4133	0.4815	0.5256	23	0.3528	0.4160	0.4862
0.2711	0.3438	0.4044	0.4716	0.5151	24	0.3443	0.4070	0.4757
0.2653	0.3365	0.3961	0.4622	0.5052	25	0.3369	0.3977	0.4662
0.2598	0.3297	0.3882	0.4534	0.4958	26	0.3306	0.3901	0.4571
0.2546	0.3233	0.3809	0.4451	0.4869	27	0.3242	0.3828	0.4487
0.2497	0.3172	0.3739	0.4372	0.4785	28	0.3180	0.3755	0.4401
0.2451	0.3115	0.3673	0.4297	0.4705	29	0.3118	0.3685	0.4325
0.2407	0.3061	0.3610	0.4226	0.4629	30	0.3063	0.3624	0.4251
0.2070	0.2638	0.3120	0.3665	0.4026	40	0.2640	0.3128	0.3681
0.1843	0.2353	0.2787	0.3281	0.3610	50	0.2353	0.2791	0.3293
0.1678	0.2144	0.2542	0.2997	0.3301	60	0.2144	0.2545	0.3005
0.1550	0.1982	0.2352	0.2776	0.3060	70	0.1982	0.2354	0.2782
0.1448	0.1852	0.2199	0.2597	0.2864	80	0.1852	0.2201	0.2602
0.1364	0.1745	0.2072	0.2449	0.2702	90	0.1745	0.2074	0.2453
0.1292	0.1654	0.1966	0.2324	0.2565	100	0.1654	0.1967	0.2327

PERCENTAGE POINTS OF STUDENT'S t-DISTRIBUTION

The values in the table are those which a random variable with Student's t-distribution on ν degrees of freedom exceeds with the probability shown.

ν	0.10	0.05	0.025	0.01	0.005
1	3.078	6.314	12.706	31.821	63.657
2	1.886	2.920	4.303	6.965	9.925
3	1.638	2.353	3.182	4.541	5.841
4	1.533	2.132	2.776	3.747	4.604
5	1.476	2.015	2.571	3.365	4.032
6	1.440	1.943	2.447	3.143	3.707
7	1.415	1.895	2.365	2.998	3.499
8	1.397	1.860	2.306	2.896	3.355
9	1.383	1.833	2.262	2.821	3.250
10	1.372	1.812	2.228	2.764	3.169
11	1.363	1.796	2.201	2.718	3.106
12	1.356	1.782	2.179	2.681	3.055
13	1.350	1.771	2.160	2.650	3.012
14	1.345	1.761	2.145	2.624	2.977
15	1.341	1.753	2.131	2.602	2.947
16	1.337	1.746	2.120	2.583	2.921
17	1.333	1.740	2.110	2.567	2.898
18	1.330	1.734	2.101	2.552	2.878
19	1.328	1.729	2.093	2.539	2.861
20	1.325	1.725	2.086	2.528	2.845
21	1.323	1.721	2.080	2.518	2.831
22	1.321	1.717	2.074	2.508	2.819
23	1.319	1.714	2.069	2.500	2.807
24	1.318	1.711	2.064	2.492	2.797
25	1.316	1.708	2.060	2.485	2.787
26	1.315	1.706	2.056	2.479	2.779
27	1.314	1.703	2.052	2.473	2.771
28	1.313	1.701	2.048	2.467	2.763
29	1.311	1.699	2.045	2.462	2.756
30	1.310	1.697	2.042	2.457	2.750
32	1.309	1.694	2.037	2.449	2.738
34	1.307	1.691	2.032	2.441	2.728
36	1.306	1.688	2.028	2.435	2.719
38	1.304	1.686	2.024	2.429	2.712
40	1.303	1.684	2.021	2.423	2.704
45	1.301	1.679	2.014	2.412	2.690
50	1.299	1.676	2.009	2.403	2.678
55	1.297	1.673	2.004	2.396	2.668
60	1.296	1.671	2.000	2.390	2.660
70	1.294	1.667	1.994	2.381	2.648
80	1.292	1.664	1.990	2.374	2.639
90	1.291	1.662	1.987	2.369	2.632
100	1.290	1.660	1.984	2.364	2.626
110	1.289	1.659	1.982	2.361	2.621
120	1.289	1.658	1.980	2.358	2.617

PERCENTAGE POINTS OF THE F-DISTRIBUTION

The values in the table are those which a random variable with the F-distribution on ν_1 and ν_2 degrees of freedom exceeds with probability 0.05 or 0.01.

Probability	ν_2 \ ν_1	1	2	3	4	5	6	8	10	12	24	∞
0.05	1	161.4	199.5	215.7	224.6	230.2	234.0	238.9	241.9	243.9	249.1	254.3
	2	18.51	19.00	19.16	19.25	19.30	19.33	19.37	19.40	19.41	19.46	19.50
	3	10.13	9.55	9.28	9.12	9.01	8.94	8.85	8.79	8.74	8.64	8.53
	4	7.71	6.94	6.59	6.39	6.26	6.16	6.04	5.96	5.91	5.77	5.63
	5	6.61	5.79	5.41	5.19	5.05	4.95	4.82	4.74	4.68	4.53	4.37
	6	5.99	5.14	4.76	4.53	4.39	4.28	4.15	4.06	4.00	3.84	3.67
	7	5.59	4.74	4.35	4.12	3.97	3.87	3.73	3.64	3.57	3.41	3.23
	8	5.32	4.46	4.07	3.84	3.69	3.58	3.44	3.35	3.28	3.12	2.93
	9	5.12	4.26	3.86	3.63	3.48	3.37	3.23	3.14	3.07	2.90	2.71
	10	4.96	4.10	3.71	3.48	3.33	3.22	3.07	2.98	2.91	2.74	2.54
	11	4.84	3.98	3.59	3.36	3.20	3.09	2.95	2.85	2.79	2.61	2.40
	12	4.75	3.89	3.49	3.26	3.11	3.00	2.85	2.75	2.69	2.51	2.30
	14	4.60	3.74	3.34	3.11	2.96	2.85	2.70	2.60	2.53	2.35	2.13
	16	4.49	3.63	3.24	3.01	2.85	2.74	2.59	2.49	2.42	2.24	2.01
	18	4.41	3.55	3.16	2.93	2.77	2.66	2.51	2.41	2.34	2.15	1.92
	20	4.35	3.49	3.10	2.87	2.71	2.60	2.45	2.35	2.28	2.08	1.84
	25	4.24	3.39	2.99	2.76	2.60	2.49	2.34	2.24	2.16	1.96	1.71
	30	4.17	3.32	2.92	2.69	2.53	2.42	2.27	2.16	2.09	1.89	1.62
	40	4.08	3.23	2.84	2.61	2.45	2.34	2.18	2.08	2.00	1.79	1.51
	60	4.00	3.15	2.76	2.53	2.37	2.25	2.10	1.99	1.92	1.70	1.39
	120	3.92	3.07	2.68	2.45	2.29	2.18	2.02	1.91	1.83	1.61	1.25
	∞	3.84	3.00	2.60	2.37	2.21	2.10	1.94	1.83	1.75	1.52	1.00
0.01	1	4052.	5000.	5403.	5625.	5764.	5859.	5982.	6056.	6106.	6235.	6366.
	2	98.50	99.00	99.17	99.25	99.30	99.33	99.37	99.40	99.42	99.46	99.50
	3	34.12	30.82	29.46	28.71	28.24	27.91	27.49	27.23	27.05	26.60	26.13
	4	21.20	18.00	16.69	15.98	15.52	15.21	14.80	14.55	14.37	13.93	13.45
	5	16.26	13.27	12.06	11.39	10.97	10.67	10.29	10.05	9.89	9.47	9.02
	6	13.70	10.90	9.78	9.15	8.75	8.47	8.10	7.87	7.72	7.31	6.88
	7	12.20	9.55	8.45	7.85	7.46	7.19	6.84	6.62	6.47	6.07	5.65
	8	11.30	8.65	7.59	7.01	6.63	6.37	6.03	5.81	5.67	5.28	4.86
	9	10.60	8.02	6.99	6.42	6.06	5.80	5.47	5.26	5.11	4.73	4.31
	10	10.00	7.56	6.55	5.99	5.64	5.39	5.06	4.85	4.17	4.33	3.91
	11	9.65	7.21	6.22	5.67	5.32	5.07	4.74	4.54	4.40	4.02	3.60
	12	9.33	6.93	5.95	5.41	5.06	4.82	4.50	4.30	4.16	3.78	3.36
	14	8.86	6.51	5.56	5.04	4.70	4.46	4.14	3.94	3.80	3.43	3.00
	16	8.53	6.23	5.29	4.77	4.44	4.20	3.89	3.69	3.55	3.18	2.75
	18	8.29	6.01	5.09	4.58	4.25	4.01	3.71	3.51	3.37	3.00	2.57
	20	8.10	5.85	4.94	4.43	4.10	3.87	3.56	3.37	3.23	2.86	2.42
	25	7.77	5.57	4.68	4.18	3.86	3.63	3.32	3.13	2.99	2.62	2.17
	30	7.56	5.39	4.51	4.02	3.70	3.47	3.17	2.98	2.84	2.47	2.01
	40	7.31	5.18	4.31	3.83	3.51	3.29	2.99	2.80	2.66	2.29	1.80
	60	7.08	4.98	4.13	3.65	3.34	3.12	2.82	2.63	2.50	2.12	1.60
	120	6.85	4.79	3.95	3.48	3.17	2.96	2.66	2.47	2.34	1.95	1.38
	∞	6.63	4.61	3.78	3.32	3.02	2.80	2.51	2.32	2.18	1.79	1.00

If an *upper* percentage point of the F-distribution on ν_1 and ν_2 degrees of freedom is f, then the corresponding *lower* percentage point of the F-distribution on ν_2 and ν_1 degrees of freedom is $\dfrac{1}{f}$

POISSON CUMULATIVE DISTRIBUTION FUNCTION

The tabulated value is $P(X \leqslant x)$, where X has a Poisson distribution with parameter λ.

$\lambda =$	0.5	1.0	1.5	2.0	2.5	3.0	3.5	4.0	4.5	5.0
$x = 0$	0.6065	0.3679	0.2231	0.1353	0.0821	0.0498	0.0302	0.0183	0.0111	0.0067
1	0.9098	0.7358	0.5578	0.4060	0.2873	0.1991	0.1359	0.0916	0.0611	0.0404
2	0.9856	0.9197	0.8088	0.6767	0.5438	0.4232	0.3208	0.2381	0.1736	0.1247
3	0.9982	0.9810	0.9344	0.8571	0.7576	0.6472	0.5366	0.4335	0.3423	0.2650
4	0.9998	0.9963	0.9814	0.9473	0.8912	0.8153	0.7254	0.6288	0.5321	0.4405
5	1.0000	0.9994	0.9955	0.9834	0.9580	0.9161	0.8576	0.7851	0.7029	0.6160
6	1.0000	0.9999	0.9991	0.9955	0.9858	0.9665	0.9347	0.8893	0.8311	0.7622
7	1.0000	1.0000	0.9998	0.9989	0.9958	0.9881	0.9733	0.9489	0.9134	0.8666
8	1.0000	1.0000	1.0000	0.9998	0.9989	0.9962	0.9901	0.9786	0.9597	0.9319
9	1.0000	1.0000	1.0000	1.0000	0.9997	0.9989	0.9967	0.9919	0.9829	0.9682
10	1.0000	1.0000	1.0000	1.0000	0.9999	0.9997	0.9990	0.9972	0.9933	0.9863
11	1.0000	1.0000	1.0000	1.0000	1.0000	0.9999	0.9997	0.9991	0.9976	0.9945
12	1.0000	1.0000	1.0000	1.0000	1.0000	1.0000	0.9999	0.9997	0.9992	0.9980
13	1.0000	1.0000	1.0000	1.0000	1.0000	1.0000	1.0000	0.9999	0.9997	0.9993
14	1.0000	1.0000	1.0000	1.0000	1.0000	1.0000	1.0000	1.0000	0.9999	0.9998
15	1.0000	1.0000	1.0000	1.0000	1.0000	1.0000	1.0000	1.0000	1.0000	0.9999
16	1.0000	1.0000	1.0000	1.0000	1.0000	1.0000	1.0000	1.0000	1.0000	1.0000
17	1.0000	1.0000	1.0000	1.0000	1.0000	1.0000	1.0000	1.0000	1.0000	1.0000
18	1.0000	1.0000	1.0000	1.0000	1.0000	1.0000	1.0000	1.0000	1.0000	1.0000
19	1.0000	1.0000	1.0000	1.0000	1.0000	1.0000	1.0000	1.0000	1.0000	1.0000

$\lambda =$	5.5	6.0	6.5	7.0	7.5	8.0	8.5	9.0	9.5	10.0
$x = 0$	0.0041	0.0025	0.0015	0.0009	0.0006	0.0003	0.0002	0.0001	0.0001	0.0000
1	0.0266	0.0174	0.0113	0.0073	0.0047	0.0030	0.0019	0.0012	0.0008	0.0005
2	0.0884	0.0620	0.0430	0.0296	0.0203	0.0138	0.0093	0.0062	0.0042	0.0028
3	0.2017	0.1512	0.1118	0.0818	0.0591	0.0424	0.0301	0.0212	0.0149	0.0103
4	0.3575	0.2851	0.2237	0.1730	0.1321	0.0996	0.0744	0.0550	0.0403	0.0293
5	0.5289	0.4457	0.3690	0.3007	0.2414	0.1912	0.1496	0.1157	0.0885	0.0671
6	0.6860	0.6063	0.5265	0.4497	0.3782	0.3134	0.2562	0.2068	0.1649	0.1301
7	0.8095	0.7440	0.6728	0.5987	0.5246	0.4530	0.3856	0.3239	0.2687	0.2202
8	0.8944	0.8472	0.7916	0.7291	0.6620	0.5925	0.5231	0.4557	0.3918	0.3328
9	0.9462	0.9161	0.8774	0.8305	0.7764	0.7166	0.6530	0.5874	0.5218	0.4579
10	0.9747	0.9574	0.9332	0.9015	0.8622	0.8159	0.7634	0.7060	0.6453	0.5830
11	0.9890	0.9799	0.9661	0.9467	0.9208	0.8881	0.8487	0.8030	0.7520	0.6968
12	0.9955	0.9912	0.9840	0.9730	0.9573	0.9362	0.9091	0.8758	0.8364	0.7916
13	0.9983	0.9964	0.9929	0.9872	0.9784	0.9658	0.9486	0.9261	0.8981	0.8645
14	0.9994	0.9986	0.9970	0.9943	0.9897	0.9827	0.9726	0.9585	0.9400	0.9165
15	0.9998	0.9995	0.9988	0.9976	0.9954	0.9918	0.9862	0.9780	0.9665	0.9513
16	0.9999	0.9998	0.9996	0.9990	0.9980	0.9963	0.9934	0.9889	0.9823	0.9730
17	1.0000	0.9999	0.9998	0.9996	0.9992	0.9984	0.9970	0.9947	0.9911	0.9857
18	1.0000	1.0000	0.9999	0.9999	0.9997	0.9993	0.9987	0.9976	0.9957	0.9928
19	1.0000	1.0000	1.0000	1.0000	0.9999	0.9997	0.9995	0.9989	0.9980	0.9965
20	1.0000	1.0000	1.0000	1.0000	1.0000	0.9999	0.9998	0.9996	0.9991	0.9984
21	1.0000	1.0000	1.0000	1.0000	1.0000	1.0000	0.9999	0.9998	0.9996	0.9993
22	1.0000	1.0000	1.0000	1.0000	1.0000	1.0000	1.0000	0.9999	0.9999	0.9997

FORMULAE

PROBABILITY

$P(A') = 1 - P(A)$

$P(A \cup B) = P(A) + P(B) - P(A \cap B)$

$P(A \cap B) = P(A)P(B|A)$

$$P(A|B) = \frac{P(B|A)\,P(A)}{P(B|A)\,P(A) + P(B|A')\,P(A')}$$

For independent events A and B,

$P(B|A) = P(B)$

$P(A|B) = P(A)$

$P(A \cap B) = P(A)P(B)$

STANDARD DEVIATION

Standard deviation $= \sqrt{\text{variance}}$

Interquartile range $= \text{IQR} = Q_3 - Q_1$

For a set of n values $x_1, x_2, \dots x_i, \dots x_n$

$$S_{xx} = \sum(x_i - \bar{x})^2 = \sum x_i^2 - \frac{\left(\sum x_i\right)^2}{n}$$

Standard deviation $= \sqrt{\dfrac{S_{xx}}{n}}$ or $\sqrt{\dfrac{\sum x^2}{n} - \bar{x}^2}$

DISCRETE DISTRIBUTIONS

For a discrete random variable X taking values x_i with probabilities $P(X = x_i)$

Expectation (mean): $E(X) = \mu = \sum x_i P(X = x_i)$

Variance: $\text{Var}(X) = \sigma^2 = \sum(x_i - \mu)^2 P(X = x_i) = \sum x_i^2 P(X = x_i) - \mu^2$

For a function $g(X)$: $E(g(X)) = \sum g(x_i) P(X = x_i)$

The probability generating function of X is $G_X(t) = E(t^X)$ and

$E(X) = G'_X(1)$ and $\text{Var}(X) = G''_X(1) + G'_X(1) - [G'_X(1)]^2$

For $Z = X + Y$, where X and Y are independent: $G_Z(t) = G_X(t) \times G_Y(t)$

Standard discrete distributions

Distribution of X	$P(X = x)$	Mean	Variance	P.G.F.
Binomial B(n, p)	$\binom{n}{x} p^x (1 - p)^{n-x}$	np	$np(1 - p)$	$(1 - p + pt)^n$
Poisson Po(λ)	$e^{-\lambda} \dfrac{\lambda^x}{x!}$	λ	λ	$e^{\lambda(t-1)}$
Geometric Geo(p) on 1, 2, ...	$p(1 - p)^{x-1}$	$\dfrac{1}{p}$	$\dfrac{1 - p}{p^2}$	$\dfrac{pt}{1 - (1 - p)t}$
Negative binomial on $r, r + 1, ...$	$\binom{x - 1}{r - 1} p^r (1 - p)^{x-r}$	$\dfrac{r}{p}$	$\dfrac{r(1 - p)}{p^2}$	$\left(\dfrac{pt}{1 - (1 - p)t} \right)^r$

CONTINUOUS DISTRIBUTIONS

For a continuous random variable X having probability density function f

$\quad\quad$ Expectation (mean): $\quad E(X) = \mu = \int x f(x)\,dx$

$\quad\quad$ Variance: $\quad\quad\quad\quad Var(X) = \sigma^2 = \int (x - \mu)^2 f(x)\,dx = \int x^2 f(x)\,dx - \mu^2$

For a function g(X): $\quad E(g(X)) = \int g(x) f(x)\,dx$

Cumulative distribution function: $F(x_0) = P(X \leqslant x_0) = \displaystyle\int_{-\infty}^{x_0} f(t)\,dt$

Standard continuous distributions

Distribution of X	p.d.f.	Mean	Variance
Normal N(μ, σ^2)	$\dfrac{1}{\sigma \sqrt{2\pi}} e^{-\frac{1}{2}\left(\frac{x-\mu}{\sigma}\right)^2}$	μ	σ^2
Uniform (rectangular) on $[a, b]$	$\dfrac{1}{b - a}$	$\frac{1}{2}(a + b)$	$\frac{1}{12}(b - a)^2$

CORRELATION AND REGRESSION

For a set of n pairs of values (x_i, y_i)

$$S_{xx} = \sum(x_i - \bar{x})^2 = \sum x_i^2 - \frac{(\sum x_i)^2}{n}$$

$$S_{yy} = \sum(y_i - \bar{y})^2 = \sum y_i^2 - \frac{(\sum y_i)^2}{n}$$

$$S_{xy} = \sum(x_i - \bar{x})(y_i - \bar{y}) = \sum x_i y_i - \frac{(\sum x_i)(\sum y_i)}{n}$$

The product moment correlation coefficient is:

$$r = \frac{S_{xy}}{\sqrt{S_{xx} S_{yy}}} = \frac{\sum(x_i - \bar{x})(y_i - \bar{y})}{\sqrt{(\sum(x_i - \bar{x})^2)(\sum(y_i - \bar{y})^2)}} = \frac{\sum x_i y_i - \frac{(\sum x_i)(\sum y_i)}{n}}{\sqrt{\left(\sum x_i^2 - \frac{(\sum x_i)^2}{n}\right)\left(\sum y_i^2 - \frac{(\sum y_i)^2}{n}\right)}}$$

The regression coefficient of y on x is $b = \dfrac{S_{xy}}{S_{xx}} = \dfrac{\sum(x_i - \bar{x})(y_i - \bar{y})}{\sum(x_i - \bar{x})^2}$

Least squares regression line of y on x is $y = a + bx$ where $a = \bar{y} - b\bar{x}$

Residual sum of squares (RSS) $= S_{yy} - \dfrac{(S_{xy})^2}{S_{xx}} = S_{yy}(1 - r^2)$

Spearman's rank correlation coefficient is $r_s = 1 - \dfrac{6\sum d^2}{n(n^2 - 1)}$

EXPECTATION ALGEBRA

For independent random variables X and Y

$$E(XY) = E(X)E(Y)$$

$$Var(aX \pm bY) = a^2 Var(X) + b^2 Var(Y)$$

SAMPLING DISTRIBUTIONS

Tests for mean when σ is known

For a random sample $X_1, X_2, \ldots, X_n$ of n independent observations from a distribution having mean μ and variance σ^2

$\overline{X}$ is an unbiased estimator of μ, with $\mathrm{Var}(\overline{X}) = \dfrac{\sigma^2}{n}$

S^2 is an unbiased estimator of σ^2, where $S^2 = \dfrac{\sum(X_i - \overline{X})^2}{n-1}$

For a random sample of n observations from $\mathrm{N}(\mu, \sigma^2)$

$$\frac{\overline{X} - \mu}{\sigma/\sqrt{n}} \sim \mathrm{N}(0, 1)$$

For a random sample of n_x observations from $\mathrm{N}(\mu_x, \sigma_x^2)$ and, independently, a random sample of n_y observations from $\mathrm{N}(\mu_y, \sigma_y^2)$,

$$\frac{(\overline{X} - \overline{Y}) - (\mu_x - \mu_y)}{\sqrt{\dfrac{\sigma_x^2}{n_x} + \dfrac{\sigma_y^2}{n_y}}} \sim \mathrm{N}(0, 1)$$

Tests for variance and mean when σ is not known

For a random sample of n observations from $\mathrm{N}(\mu, \sigma^2)$

$$\frac{(n-1)S^2}{\sigma^2} \sim \chi_{n-1}^2$$

$$\frac{\overline{X} - \mu}{S/\sqrt{n}} \sim t_{n-1} \text{ (also valid in matched-pairs situations)}$$

For a random sample of n_x observations from $\mathrm{N}(\mu_x, \sigma_x^2)$ and, independently, a random sample of n_y observations from $\mathrm{N}(\mu_y, \sigma_y^2)$,

$$\frac{S_x^2/\sigma_x^2}{S_y^2/\sigma_y^2} \sim F_{n_x-1,\, n_y-1}$$

If $\sigma_x^2 = \sigma_y^2 = \sigma^2$ (unknown) then

$$\frac{(\overline{X} - \overline{Y}) - (\mu_x - \mu_y)}{\sqrt{S_p^2\left(\dfrac{1}{n_x} + \dfrac{1}{n_y}\right)}} \sim t_{n_x+n_y-2} \text{ where } S_p^2 = \frac{(n_x-1)S_x^2 + (n_y-1)S_y^2}{n_x + n_y - 2}$$

NON-PARAMETRIC TESTS

Goodness-of-fit tests and contingency tables: $\sum \dfrac{(O_i - E_i)^2}{E_i} \sim \chi_\nu^2$

Answers

CHAPTER 1
Prior knowledge check

1 a Every minute the reaction produces 3.85 g more of a product.
  b i is reliable as 5 lies inside the range of the data.
    ii is unreliable as 10 lies outside the range of the data.
    iii is unreliable as the regresson equation should only be used to predict a value of p given t.

Exercise 1A

1 $a = -3, b = 6$
2 $y = -14 + 5.5x$
3 $y = 2x$
4 a $\bar{x} = 2.5, \bar{y} = 12, S_{xx} = 5, S_{xy} = 20$
  b $y = 2 + 4x$
5 a $S_{xx} = 40.8, S_{xy} = 69.6$ b $y = -0.294 + 1.71x$
6 a $y = -59 + 57(6) = 283$
  b For each dexterity point, productivity increases by 57.
  c i No, because this is extrapolation as it is outside the range of data.
    ii No, because this is extrapolation as it is outside the range of data.
7 $g = 1.50 + 1.44h$
8 a $p = 65.4 - 1.38w$
  b $w = 47.4 - 0.72p$
  c The gradient of the second regression line is calculated using different summary statistics rather than just the reciprocal of the summary statistics used for the first regression line.
  d i The first one.   ii The second one.
9 a $y = 78.0 - 0.294x$
  b
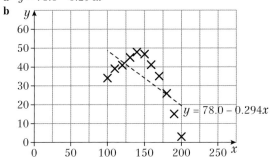
  c Model is not valid since data does not follow a linear pattern.
10 a $S_{nn} = 6486, S_{np} = 6344$
   b $p = 21.0 + 0.978n$
   c £60 100 (3 s.f.)
   d Reliable as 40 000 items lies inside the range of the data.
11 a $S_{nn} = 589.6, S_{np} = 1474$
   b $p = 20 + 2.5n$
   c The increase in cost, in pounds, for every 100 leaflets printed.
   d $t > 8$
12 a $y = -0.07 + 1.45x$
   b Number of years protection per coat of paint.
   c Unreliable as 7 coats lies outside the range of the data.

d 10.08 years
e i 0.4779 + 1.247x
  ii 9.2 years (2 s.f.)
  iii The answer now uses interpolation not extrapolation and the number of data points has increased, which increases accuracy in prediction.

Exercise 1B

1 $y = 6 - x$
2 $s = 88 + p$
3 $y = 32 - 5.33x$
4 $t = 9 + 3s$
5 a $y = 3.5 + 0.5x$ b $d = 35 + 2.5c$
6 a $S_{xy} = 162.2, S_{xx} = 190.8; y = 7.87 + 0.850x$ (3 s.f.)
  b $c = 22.3 + 2.13a$ (3 s.f.)
  c £90.46 or £90.56
7 a $p = 3.03 + 1.49v$ (3 s.f.)   b 10.1 tonnes (3 s.f.)

Exercise 1C

1 Residuals: 0.07, –0.496, 0.471, p – 20.728, –0.094 (3 s.f.) hence $p = 20.8$ (3 s.f.)
2 a –0.99, –1.025, 1.94, 1.595, –0.095, –1.475
  b
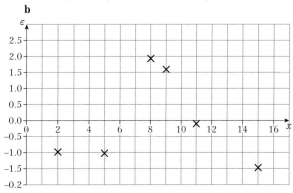
  c No – residuals not randomly scattered about zero.
3 a 0.8892, –0.2956, 0.5196, 0.9888, 0.2732, –5.2576, 1.496, 0.9036, 0.7804, –0.3428
    The outlier is (7.2, 84)
  b Yes: She may have just had a bad day; No: It could have been incorrectly recorded.
  c $p = 51.6 + 5.31t$ (3 s.f.)
  d 93%
4 a –0.176, 0.234, 0.048, –0.038, –0.124, a – 7.216, 0.402, $a = 6.87$
  b Yes: Residuals are randomly distributed about zero.
5 a The RSS measures the reasonableness of linear fit.
  b 0.949 (3 s.f.)
  c Obstacle course – lower RSS
6 a 0.100 (3 s.f.)   b October – lower RSS
7 a $y = 17.55 - 3.117x$
  b The value/price of the car brand new (£17 550)
  c £11 316 (nearest pound)
  d 0.0319, 0.0021, 0.1606, –0.2809, –0.0107, 0.0946
  e Suitable since residuals are close to zero and scattered about zero.
  f 0.1148 (4 d.p.)
  g First sample since RSS is smaller.

Challenge

$p = 13$, $q = 36$

Mixed exercise 1

1 a $t = 1.96 + 0.95s$ **b** 49.5 (3 s.f.)

2 a $S_{xx} = 6.429$, $S_{xy} = 11.68$ **b** $y = 0.554 + 1.82x$

 c 6.01 cm (3 s.f.)

3 a $S_{xx} = 16\,350$, $S_{xx} = 210\,331$

 b $y = 224.5 + 12.86x$

 c 1511

 d 255

 e This answer is unreliable since a Gross National Product of 3500 is a long way outside the range of the data. Also, the regression equation should only be used to estimate values of y given x.

4 a $y = 0.343 + 0.449x$ **b** $t = 2.34 + 0.224m$

 c 4.6 cm (2 s.f.)

5 a $S_{xx} = 90.9$, $S_{xy} = 190$ **b** $y = -1.82 + 2.09x$

 c $p = 25.5 + 2.09r$ **d** 71 (2 s.f.)

 e This answer is reliable since 22 breaths per minute is within the range of the data.

6 a 0.79 kg is the average amount of food consumed in 1 week by 1 hen.

 b 23.9 kg (3 s.f.)   **c** £47.59

7 a & e

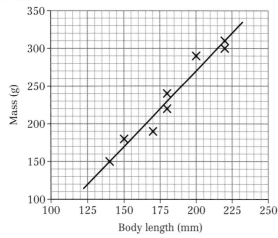

 b There appears to be a linear relationship between body length and body mass.

 c $w = -12.7 + 1.98l$

 d $y = -127 + 1.98x$

 f 290 g (2 s.f.). This is reliable since 210 mm is within the range of the data.

 g Water voles B and C were probably removed from the river since they are both underweight. Water vole A was probably left in the river since it is slightly overweight.

8 a $S_{xy} = 78$, $S_{xx} = 148$ **b** $y = 7.311 + 0.5270x$

 c $w = 816.2 + 210.8n$ **d** 5032 kg

 e 100 items is a long way outside the range of the data.

 f −0.311, 0.108, 1.36, −0.946, −1.69, 0.527, 0.946

 g −124, 43.4, 546, −378, −675, 211, 379

 h They are related by the same code as that used for y in terms of w.

9 a 0.133, 0.422, 0,−0.711, −0.422, 0.156, 0.445

 b No – residuals not randomly scattered about zero.

 c 1.10 (3 s.f.)

 d The second one – lower RSS

10 a $S_{ss} = 39.315$, $S_{sf} = -27.9225$

 b $f = 8.48 - 0.710s$

 c 3.2 mm (2 s.f.)

 d 0.347

 e 0.37

 f No – residuals go negative, positive, negative, so not randomly scattered.

11 a $y = 1.61 + 1.25x$

 b $98.89 - \dfrac{75^2}{60} = 5.14$

 c −0.11

  −0.3611

  0.39

  0.14

  −0.11

  0.64

  0.39

  −1.86

  0.89

 d (9, 11)

 e i Ignore: recording error; Do not ignore: could be a 'runt'.

  ii $y = 1.14 + 1.38x$

  iii 28.7 grams (3 s.f.)

  iv Unreliable as 20 days lies outside the range of the data.

12 a $s = -0.215 + 1.09t$

 b 229

 c −0.117 (3 s.f.)

 d Linear model is suitable since residuals are randomly scattered around zero.

 e 0.0516 (3 s.f.)

 f The UK sample since RSS is smaller.

Challenge

a $\displaystyle\sum \in_i = \sum(y_i - a - bx_i)$

$= \sum(y_i - (\bar{y} - b\bar{x}) - bx_i)$

$= \sum y_i - \sum \bar{y} + \sum b\bar{x} - \sum bx_i$

$= \sum y_i - \bar{y}\sum 1 + b\bar{x}\sum 1 - \sum bx_i$

$= \sum y_i - n\bar{y} + nb\bar{x} - \sum bx_i$

$= \sum y_i - n\dfrac{\sum y_i}{n} + nb\dfrac{\sum x_i}{n} - \sum bx_i$

$= \sum y_i - \sum y_i + b\sum x_i - \sum bx_i$

$= 0$

b The property doesn't take into account if residuals are randomly scattered about 0 and are close to 0. It doesn't take into account outliers and doesn't mean RSS is close to 0.

CHAPTER 2
Prior knowledge check

1 a Strong negative correlation

 b As the age of the car goes up, the value goes down.

2 a 52   **b** 84   **c** 64

Exercise 2A

1 0.985 (3 s.f.)

2 0.202 (3 s.f.)

3 a 9.71 (3 s.f.)

 b 0.968 (3 s.f.)

 c There is positive correlation. The greater the age, the taller the person.

4 a $S_{ll} = 30.3, S_{tt} = 25.1, S_{lt} = 25.35$

b 0.919 (3 s.f.)

c The value of the correlation coefficient is close to 1 and the points lie on an approximate straight line, therefore a linear regression model is suitable.

5 a 0.866 (3 s.f.)

b There is positive correlation. The higher the IQ, the higher the mark in the general knowledge test.

6 0.973

7 a

p	0	5	3	2	1
q	0	17	12	10	6

b 0.974 (3 s.f.)   **c** 0.974 (3 s.f.)

8 a $S_{pp} = 10, S_{tt} = 5.2, S_{pt} = 7$

b 0.971 (3 s.f.)

c 0.971 (3 s.f.)

9 a $S_{xx} = 1601, S_{yy} = 1282, S_{xy} = -899$

b −0.627 (3 s.f.)

c The shopkeeper is wrong. There is negative correlation. Sweet sales actually decrease as newspaper sales increase.

10 a $S_{ff} = \sum f^2 - \dfrac{(\sum f)^2}{n} = \sum (10x)^2 - \dfrac{(\sum 10x)^2}{n}$

$= 100\sum x^2 - \dfrac{100(\sum x)^2}{n} = 100\left(\sum x^2 - \dfrac{(\sum x)^2}{n}\right)$

$= 100 S_{xx} = 100 \times 111.48 = 11\,148$

b 0.934 (3 s.f.)

c The PMCC suggests strong linear correlation but the scatter diagram suggests non-linear fit so a linear regression model is not suitable.

11 a $S_{xx} = \sum x^2 - \dfrac{(\sum x)^2}{n} = 22.02 - \dfrac{12^2}{7} = 1.448...$

$S_{xy} = \sum xy - \dfrac{\sum x \sum y}{n} = 180.37 - \dfrac{12 \times 97.7}{7}$

$= 12.884...$

$S_{yy} = \sum y^2 - \dfrac{(\sum y)^2}{n} = 1491.69 - \dfrac{97.7^2}{7}$

$= 128.077...$

$r = \dfrac{S_{xy}}{\sqrt{S_{xx} S_{yy}}} = \dfrac{12.884...}{\sqrt{1.448... \times 128.077...}} = 0.946$ (3 s.f.)

b −2.29345, 0.22765, 0.8382, 1.16985, 1.39095, 0.61205, −1.94575

c Residuals are not randomly scattered about zero (they 'rise and fall') so this indicates that a linear model is not a good model for this data.

Exercise 2B

1 a The data clearly follows a linear trend.

b Spearman's rank correlation coefficient is easier to calculate.

2 The relationship is clearly non-linear.

3 The number of attempts taken to score a free throw is not normally distributed (it is geometric) so the researcher should use Spearman's rank correlation coefficient.

4 a $\sum d^2 = 10, r_s = 0.714...$ limited evidence of positive correlation between the pairs of ranks

b $\sum d^2 = 18, r_s = 0.8909...$ evidence of positive correlation between the pairs of ranks

c $\sum d^2 = 158, r_s = -0.8809...$ evidence of negative correlation between the pairs of ranks

5 a 1   **b** −1

    c 0.9   **d** 0.5

6 a $\sum d^2 = 48$

b $r_s = 0.832...$ The more goals a team scores, the higher they are likely to be in the league table.

7 $\sum d^2 = 20, r_s = 0.762...$ The trainee vet is doing quite well as there is a fair degree of agreement between the trainee vet and the qualified vet. The trainee still has more to learn as r_s is less than 1.

8 a The marks are discrete rather than continuous values.

The marks are not normally distributed.

b $\sum d^2 = 28$, $r_s = 0.8303...$

This shows a fairly strong positive correlation between the pairs of ranks of the marks awarded by the two judges so it appears they are judging the ice dances using similar criteria and with similar standards.

c Give each of the equal values a rank equal to the mean of the tied ranks.

9 a The emphasis here is on ranks/marks so the data sets are unlikely to be from a bivariate normal distribution.

b 0.580

c Both show positive correlation but the judges agree more on the second dive.

10 a 0.971 (3 s.f.)

b i No change since the rank does not change.

    ii Will increase since $d = 0$ and change in $\sum d^2$ is zero but n increases.

c Use PMCC with tied ranks given mean of ranks.

Exercise 2C

1 a PMCC = −0.975 (3 s.f.)

b Assume data are normally distributed. Critical values are ±0.8745. −0.975 < −0.8745 so reject H_0. There is evidence of correlation.

2 a $r = 0.677...$

b Assume data are jointly normally distributed. $H_0: \rho = 0$; $H_1: \rho > 0$, 5% critical value is 0.5214. Reject H_0. There is evidence to suggest that the taller you are the older you are.

3 a $H_0; \rho_s = 0$; $H_1: \rho_s \neq 0$. Critical region is $r_s < -0.3624$ and $r_s > 0.3624$

b Reject H_0. There is reason to believe that engine size and fuel consumption are related.

4 a $H_0: \rho = 0, H_1: \rho > 0$

Critical value = 0.7887

Since 0.774 is not in the critical region there is insufficient evidence of positive correlation.

b $\sum d^2 = 10$

$r_s = 1 - \dfrac{6 \times 10}{8 \times 63} = 0.881$ (3 s.f.)

c e.g. The data are discrete results in a limited range. They are judgements, not measurements. It is also unlikely that these scores will both be normally distributed.

d $H_0: \rho = 0, H_1: \rho > 0$

Critical value: 0.8333

Since 0.8333 is in the critical region there is evidence of positive correlation.

5 a Ranks are given rather than raw data.

b $\sum d^2 = 46$

$$r_s = 1 - \frac{6 \times 46}{8 \times 63}$$

$r_s = 0.452$

c $H_0; \rho_s = 0; H_1: \rho_s \neq 0$; critical values are ± 0.6249
$0.452 < 0.6249$ or not significant or insufficient
evidence to reject H_0. There is no evidence of
agreement between the two judges.

6 $\sum d^2 = 76, r_s = -0.357$ (3 s.f.) $H_0: \rho_s = 0; H_1: \rho_s < 0$.
Critical value $= -0.8929$. There is no reason to reject
H_0. There is insufficient evidence to show that a team
that scores a lot of goals concedes very few goals.

7 a $\sum d^2 = 2, r_s = 0.943$

b $H_0: \rho_s = 0; H_1: \rho_s > 0$. Critical value $= 0.8286$.
Reject H_0. There is evidence that profits and takings
are positively correlated.

8 a $\sum d^2 = 58, r_s = 0.797...$

b $H_0: \rho_s = 0; H_1: \rho_s \neq 0$. Critical values $= \pm 0.5874$.
Reject H_0: On this evidence it would seem that
students who do well in mathematics are likely to
do well in music.

9 Using Spearman's rank correlation coefficient:
$\sum d^2 = 54, r_s = 0.6727...$
$H_0: \rho_s = 0; H_1: \rho_s > 0$. Critical value 0.5636.
Reject H_0: the child shows some ability in this task.

10 Using Spearman's rank correlation coefficient:
$\sum d^2 = 64, r_s = -0.829$ (3 d.p.)
$H_0: \rho_s = 0; H_1: \rho_s \neq 0$. Critical values $= \pm 0.8857$.
Do not reject H_0: There is insufficient evidence of
correlation between crop yield and wetness.

11 a PMCC since data is likely to be bivariate normal.

b 2.5%   **c** 11

Mixed exercise 2

1 a -0.147 (3 s.f.)   **b** -0.147 (3 s.f.)

c This is a weak negative correlation. There is little
evidence to suggest that science marks are related
to art marks.

2 a $S_{jj} = 4413, S_{pp} = 5145, S_{jp} = 3972$

b 0.834 (3 s.f.)

c There is strong positive correlation, so Nimer is
correct.

3 a $S_{pp} = \sum p^2 - \dfrac{\left(\sum p\right)^2}{n} = \sum(x-10)^2 - \dfrac{\left(\sum(x-10)\right)^2}{n}$

$= \sum(x^2 - 20x + 100) - \dfrac{\left(\left(\sum x\right) - 10n\right)^2}{n}$

$= \sum x^2 - 20\sum x + 100n$
$\quad - \left(\dfrac{\left(\int x\right)^2 - 20n \int x + 100n^2}{n}\right)$

$= \sum x^2 - 20\sum x + 100n$
$\quad - \left(\dfrac{\left(\sum x\right)^2}{n} - 20\sum x + 100n\right)$

$= \sum x^2 - \dfrac{\left(\sum x\right)^2}{n} = S_{xx}$

b -0.964 (3 s.f.)   **c** -0.964 (3 s.f.)

d The PMCC suggets strong (negative) linear
correlation but the scatter diagram suggests
non-linear fit so a linear regression model is not
suitable.

4 a Data is given in ranks rather than raw scores.

b $r_s = 0.648$ (3 s.f.)

c The null hypothesis is only rejected in favour
of the alternative hypothesis if by doing so the
probability of being wrong is less than or equal to
the significance level.

d $H_0: \rho_s = 0; H_1: \rho_s > 0$. Critical value at 5% is 0.5636,
so reject H_0. There is evidence of agreement
between the two judges.

5 a Data given in rank/place order. The populations are
not jointly normal. The relationship between data
sets is non-linear.

b $\sum d^2 = 28, r_s = 0.766...$
$H_0: \rho = 0, H_1: \rho > 0$, 2.5% critical value $= 0.700$
Reject H_0. There is evidence of agreement between
the tutors at the 2.5% significance level. At the 1%
significance level the test statistic and critical value
are very close so it is inconclusive at this level of
significance.

6 a $\sum d^2 = 56, r_s = 0.66$

b $H_0: \rho_s = 0, H_1: \rho_s > 0$, critical value $= 0.5636$
Reject H_0. There is a degree of agreement between
the jumps.

7 Ranking so use Spearman's, $r_s = 0.714$. $H_0: \rho_s = 0$,
$H_1: \rho_s > 0$. Critical value for 0.025 significance level $=$
0.7857. Do not reject H_0. There is insufficient evidence
that the expert can judge relative age accurately.

8 a PMCC $= 0.375$ (3 s.f.)

b $H_0: \rho = 0; H_1: \rho > 0$. Critical value $= 0.3783$.
Do not reject H_0. There is insufficient evidence of
postive correlation between distance and time.

c Both distance and time are normally distributed.

9 a $\sum d^2 = 36, r_s = 0.5714...$

b $H_0: \rho = 0, H_1: \rho \neq 0$, critical values $= \pm 0.7381$
No reason to reject H_0. Students who do well in
geography do not necessarily do well in statistics.

10 a 0.7857 (4 d.p.)

b Critical values $= \pm 0.7381$
Reject H_0. There is evidence to suggest correlation
between life expectancy and literacy.

c Only interested in order OR Cannot assume
normality.

d i No change

ii Would increase since $d = 0$ and n is bigger.

e Use PMCC with tied ranks given mean of those
ranks.

11 a $r_s = 0.314$ (3 s.f.)

b $H_0: \rho_s = 0; H_1: \rho_s > 0$. Critical value $= 0.8286$.
Do not reject H_0. There is insufficient evidence of
agreement between the rankings of the judge and
the vet.

12 a $S_{xx} = 1038.1, S_{yy} = 340.4, S_{xy} = 202.2$,
$r = 0.340$ (3 d.p.)

b One or both given in rank order.
Population is not normal.
Relationship between data sets non-linear.

c $\sum d^2 = 112, r_s = 0.321$ (3 d.p.)

d $H_0: \rho_s = 0, H_1: \rho_s \neq 0$, critical value $= \pm 0.6485$
Do not reject H_0. There is insufficient evidence of
correlation.

13 a $S_{xx} = 6908.1, S_{yy} = 50\,288.1, S_{xy} = 17\,462$,
$r = 0.937$ (3 d.p.)

b $\sum d^2 = 4, r_s = 0.976$ (3 d.p.)

Online Full worked solutions are available in SolutionBank.

c $H_0: \rho_s = 0$, $H_1: \rho_s \neq 0$, critical values $= \pm 0.6485$
Reject H_0. For this machine there is insufficient evidence of correlation between age and maintenance costs.

14 a PMCC $= -0.975$ (3 s.f.)

b $H_0: \rho = 0$, $H_1: \rho < 0$, critical value $= -0.5822$, reject H_0.
The greater the altitude the lower the temperature

c $H_0: \rho_s = 0$ (no association between hours of sunshine and temperature); $H_1: \rho_s \neq 0$, critical value $= \pm 0.7000$
$0.767 > 0.7000$ so reject H_0. There is evidence of a positive association between hours of sunshine and temperature.

15 a You use a rank correlation coefficient if at least one of the sets of data isn't from a normal distribution, or if at least one of the sets of data is already ranked. It is also used if there is a non-linear association between the two data sets.

b $\sum d^2 = 78$, $r_s = 0.527...$

c $H_0: \rho_s = 0$; $H_1: \rho_s > 0$. Critical value $= 0.5636$.
Do not reject H_0. There is insufficient evidence of agreement between the qualified judge and the trainee judge.

16 $\sum d^2 = 120$, $r_s = 0.4285...$
There is only a small degree of positive correlation between league position and home attendance.

17 a $H_0: \rho = 0$; $H_1: \rho > 0$. Critical value 0.5822.
$0.972 > 0.5822$. Evidence to reject H_0. Age and weight are positively associated.

b $\sum d^2 = 26$, $r_s = 1 - \dfrac{6 \times 26}{9 \times 80} = 0.783$ (3 s.f.)

c Critical value $= 0.6000$
$0.783 > 0.600$ is evidence that actual weight and the boy's guesses are associated.

Challenge

a Since there are no ties, both the x's and the y's consist of the integers from 1 to n.
$$\sum x_i^2 = \sum y_i^2 = \sum_{r=1}^{n} r^2 = \frac{n(n+1)(2n+1)}{6}$$

b Since there are no ties, the x_i's and y_i's both consist of the integers from 1 to n. Hence $\sqrt{\sum(x_i - \overline{x})^2 \sum(y_i - \overline{y})^2}$
$$= \sqrt{\sum(x_i - x)^4} = \sum(x_i - \overline{x})^2$$
$$\overline{x} = \frac{\sum x_i}{n} = \frac{\frac{n(n+1)}{2}}{n} = \frac{n+1}{2}$$
$$\sum(x_i - \overline{x})^2 = \sum(x_i^2 - 2x_i\overline{x} + \overline{x}^2)$$
$$= \sum x_i^2 - 2\overline{x}\sum x_i + n\overline{x}^2$$
$$= \frac{n(n+1)(2n+1)}{6} - 2\left(\frac{n(n+1)}{2}\right)\left(\frac{n+1}{2}\right) + n\left(\frac{n+1}{2}\right)^2$$
$$= \frac{n(n+1)(2n+1)}{6} - n\left(\frac{n+1}{2}\right)^2$$
$$= n(n+1)\left(\frac{2n+1}{6} - \frac{n+1}{4}\right)$$
$$= \frac{n(n+1)(4n+2-3n-3)}{12}$$
$$= \frac{n(n+1)(n-1)}{12} = \frac{n(n^2-1)}{12}$$

c $\sum d_i^2 = \sum(y_i - x_i)^2$
$$= \sum x_i^2 - 2\sum x_i y_i + \sum y_i^2$$
$$= 2\sum x_i^2 - 2\sum x_i y_i$$
$$\Rightarrow \sum x_i y_i = \sum x_i^2 - \frac{\sum d_i^2}{2}$$

d $\sum(x_i - \overline{x})(y_i - \overline{y}) = \sum x_i(y_i - \overline{y}) - \sum \overline{x}(y_i - \overline{y})$
$$= \sum x_i y_i - \overline{y}\sum x_i - \overline{x}\sum y_i + n\overline{x}\overline{y}$$
$$= \sum x_i y_i - n\overline{x}\overline{x} - n\overline{x}\overline{x} + n\overline{x}\overline{x}$$
$$= \sum x_i y_i - n\overline{x}\overline{x}$$
$$= \sum x_i^2 - \frac{\sum d_i^2}{2} - n\overline{x}\overline{x}$$
$$= \frac{n(n+1)(2n+1)}{6} - \frac{n(n+1)(n+1)}{4} - \frac{\sum d_i^2}{2}$$
$$= \frac{n(n+1)(4n+2-3n-3)}{12} - \frac{\sum d_i^2}{2}$$
$$= \frac{n(n+1)(n-1)}{12} - \frac{\sum d_i^2}{2}$$
$$= \frac{n(n^2-1)}{12} - \frac{\sum d_i^2}{2}$$

e From parts **b** and **d**
$$\frac{\sum(x_i - \overline{x})(y_i - \overline{y})}{\sqrt{\sum(x_i - \overline{x})^2 \sum(y_i - \overline{y})^2}} = \frac{\frac{n(n^2-1)}{12} - \frac{\sum d_i^2}{2}}{\frac{n(n^2-1)}{12}}$$
$$= \frac{\frac{n(n^2-1)}{12}}{\frac{n(n^2-1)}{12}} - \frac{\frac{\sum d_i^2}{2}}{\frac{n(n^2-1)}{12}}$$
$$= 1 - \frac{6\sum d_i^2}{n(n^2-1)}$$

CHAPTER 3
Prior knowledge check

1 a $k = \frac{1}{20}$ **b** $P(X \geqslant 5) = \frac{7}{10}$ **c** $E(X) = 5$

2 a 4 **b** 3 **c** 48

3 $a = 6$

Exercise 3A

1 a There are negative values for f(x) when $x < 0$ so this is not a probability density function.

b Area of $8\frac{2}{3}$ not equal to 1 therefore it is not a probability density function.

c There are negative values for f(x) so this is not a probability density function.

2 $k = \frac{3}{50}$

3 a

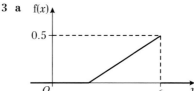

b f(x)

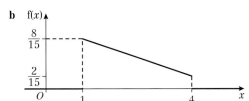

4 a $k = \frac{1}{4}$　　**b** $k = \frac{3}{27} = \frac{1}{9}$　　**c** $k = \frac{1}{6}$

5 a $k = \frac{1}{6}$

b

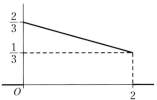

c $\frac{5}{12}$

6 a $\frac{3}{4}$　　**b** $\frac{5}{16}$

7 a $k = \frac{4}{255}$　　**b** $\frac{1}{17}$

8 a $k = \frac{1}{4}$ or 0.25

b

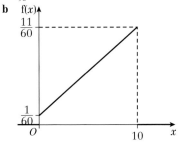

c $\frac{9}{64}$

9 a $\frac{1}{96}$

b f(x)

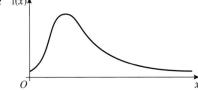

c f(x)

10 a $k = \dfrac{1}{\ln 5}$　　**b** $\dfrac{\ln 2}{\ln 5}$

11 a $k = \dfrac{1}{\ln 6}$　　**b** 0.285

12 a $k = \dfrac{\pi}{2}$

b f(x)

c $\frac{1}{4}$

Challenge

a $k = 2$

b i $\frac{8}{9}$　　**ii** $\frac{1}{400}$

c $p = 2.5$

Exercise 3B

1
$$F(x) = \begin{cases} 0 & x < 0 \\ \dfrac{x^3}{8} & 0 \leqslant x \leqslant 2 \\ 1 & x > 2 \end{cases}$$

2
$$F(x) = \begin{cases} 0 & x < 1 \\ x - \dfrac{x^2}{8} - \dfrac{7}{8} & 1 \leqslant x \leqslant 3 \\ 1 & x > 3 \end{cases}$$

3
$$F(x) = \begin{cases} 0 & x \leqslant 0 \\ \dfrac{x^2}{18} & 0 < x < 3 \\ \dfrac{2x}{3} - \dfrac{x^2}{18} - 1 & 3 \leqslant x \leqslant 6 \\ 1 & x > 6 \end{cases}$$

4 a f(x)

b $k = \frac{1}{9}$

c
$$F(x) = \begin{cases} 0 & x < 0 \\ \dfrac{x}{9} & 0 \leqslant x < 3 \\ \dfrac{x^2}{9} - \dfrac{5x}{9} + 1 & 3 \leqslant x \leqslant 5 \\ 1 & x > 5 \end{cases}$$

5
$$f(x) = \begin{cases} \dfrac{2x}{5} & 2 \leqslant x \leqslant 3 \\ 0 & \text{otherwise} \end{cases}$$

6 a 0.75　　**b** 0.75　　**c** 0.5

7 $\frac{1}{6}2^p + q = 0$　(1)　and　$\frac{1}{6}4^p + q = 1$　(2)

(2) − (1): $\frac{1}{6}4^p - \frac{1}{6}2^p = 1 \Rightarrow 2^{2p} - 2^p = 6$

Let $y = 2^p$, then $y^2 - y - 6 = 0 \Rightarrow (y - 3)(y + 2) = 0$

Taking the positive value, $y = 3 \Rightarrow 2^p = 3$

$p = \dfrac{\ln 3}{\ln 2}$

From (1), $q = -\frac{1}{2}$

8 a
$$f(x) = \begin{cases} \dfrac{3}{2}x^2 - 2x + \dfrac{1}{2} & 1 \leqslant x \leqslant 2 \\ 0 & \text{otherwise} \end{cases}$$

Online Full worked solutions are available in SolutionBank.

b f(x)

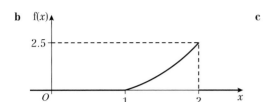

c $\frac{3}{16}$

b 0

c $\mathrm{Var}(X) = E(X^2) - (E(X))^2$

$$= \int_{-1}^{1} \frac{3}{8}x^2(1 + x^2)\,dx - 0^2$$

$$= \frac{3}{8}\left[\frac{x^3}{3} + \frac{x^5}{5}\right]_{-1}^{1}$$

$$= \frac{3}{8}\left(\frac{1}{3} + \frac{1}{5} - \left(-\frac{1}{3}\right) - \left(-\frac{1}{5}\right)\right) = 0.4$$

9 a $\int_{0}^{2} k(4 - x^2)\,dx = 1 = \left[k\left(4x - \frac{x^3}{3}\right)\right]_{0}^{2}$

$\Rightarrow \frac{16k}{3} = 1 \Rightarrow k = \frac{3}{16}$

b

$$F(x) = \begin{cases} 0 & x < 0 \\ \frac{3}{16}\left(4x - \frac{x^3}{3}\right) & 0 \le x \le 2 \\ 1 & x > 2 \end{cases}$$

c 0.007 (1 s.f.)

10 F(x) > 1 for $x > $ e

11 a $k = 49$ **b** 0.25

12 $F(x) = \begin{cases} 0 & x < 1 \\ \dfrac{\ln x}{\ln 7} & 1 \le x \le 7 \\ 1 & x > 7 \end{cases}$

13 $F(x) = \begin{cases} 0 & x < 0 \\ \sin(\pi x) & 0 \le x \le \frac{1}{2} \\ 1 & x > \frac{1}{2} \end{cases}$

14 a $k = \dfrac{1}{2 + \ln 3}$

b

$$f(x) = \begin{cases} \dfrac{1}{2 + \ln 3}\left(1 + \dfrac{1}{x}\right) & 1 \le x \le 3 \\ 0 & \text{otherwise} \end{cases}$$

Challenge

a $F(t) = \begin{cases} 0 & t < 0 \\ 1 - e^{-1.25t} & t \ge 0 \end{cases}$

b 0.2044 (4 d.p.) **c** 0.0235 (4 d.p.)

Exercise 3C

1 a $k = \frac{3}{8}$ **b** $\frac{3}{2}$ **c** $\frac{3}{20}$

2 a 2.25 **b** 0.3375 **c** 0.581 (3 s.f.)

3 a $\frac{8}{3}$ **b** $\frac{8}{9}$ **c** 0.943 (3 s.f.)

d 0.556 (3 s.f.) **e** 8 **f** $\frac{14}{3}$

4 a $k = 2$ **b** $\frac{1}{3}$

c $\mathrm{Var}(X) = E(X^2) - (E(X))^2$

$$= \int_{0}^{1} 2x^2(1 - x)\,dx - \left(\frac{1}{3}\right)^2$$

$$= \left[\frac{2x^3}{3} - \frac{2x^4}{4}\right]_{0}^{1} - \frac{1}{9} = \frac{1}{6} - \frac{1}{9} = \frac{1}{18}$$

d $\frac{4}{9}$

5 a $\frac{5}{16}$ or 0.3125 **b** 0.6 or $\frac{3}{5}$

6 a f(x)

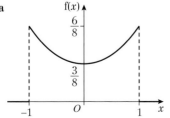

d 0.538 (3 s.f.)

7 a $k = \frac{1}{4}$

b $E(T) = \frac{1}{4}\int_{0}^{2} t^4\,dt = \frac{1}{4}\left[\frac{t^5}{5}\right]_{0}^{2} = \frac{1}{4} \times \frac{32}{5} = 1.6$

c 6.2

d $\frac{8}{75}$ **e** $\frac{32}{75}$ **f** $\frac{1}{16}$

8 a f(x)

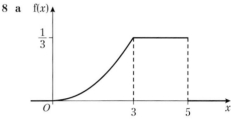

b 3.417 **c** 1.0152 **d** 1.01

9 a f(x)

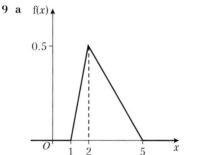

b $\frac{8}{3}$ **c** $\frac{13}{18}$

10 a $\int_{0}^{10} kt^2\,dt = 1 = \left[k\frac{t^3}{3}\right]_{0}^{10} = \frac{1000k}{3}$

$\Rightarrow k = \frac{3}{1000} = 0.003$

b 7.5 **c** 3.75 **d** 0.386

e f(x)

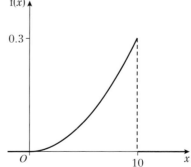

11 a $\frac{3}{4}$ **b** $\frac{4}{5}$ **c** $\frac{19}{80}$

12 a $\frac{50}{9}$

b $E(X^3) = \int_{0}^{10} \frac{x^4}{50}\,dx = \left[\frac{1}{250}x^5\right]_{0}^{10} = 400$

13 a $\dfrac{1}{\ln 3}$ **b** $\dfrac{2}{\ln 3}$ **c** 0.3268

14 a $1 = \int_1^2 \dfrac{c}{x(3-x)}\,dx = \dfrac{c}{3}\int_1^2\left(\dfrac{1}{x}+\dfrac{1}{3-x}\right)dx$

$= \dfrac{c}{3}[\ln x - \ln(3-x)]_1^2 = \dfrac{c}{3}(2\ln 2) = \dfrac{c\ln 4}{3}$

$\Rightarrow c = \dfrac{3}{\ln 4}$

b $E(X) = 1.5$, $Var(X) = 0.0860$

15 $-\dfrac{1}{2}$

Challenge

$Var(X) = \displaystyle\int_{-\infty}^{\infty}(x-\mu)^2 f(x)\,dx$

$= \displaystyle\int_{-\infty}^{\infty}(x^2 - 2\mu x + \mu^2)f(x)\,dx$

$= \displaystyle\int_{-\infty}^{\infty}x^2 f(x)\,dx - 2\mu\int_{-\infty}^{\infty}x f(x)\,dx + \mu^2\int_{-\infty}^{\infty}f(x)\,dx$

$= \displaystyle\int_{-\infty}^{\infty}x^2 f(x)\,dx - 2\mu(\mu) + \mu^2(1)$

$= \displaystyle\int_{-\infty}^{\infty}x^2 f(x)\,dx - \mu^2$

$= E(X^2) - (E(X))^2$

Exercise 3D

1 a f(x)

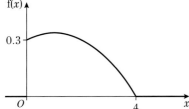

b The mode is 1.

2 a $F(x) = \begin{cases} 0 & x < 0 \\ \dfrac{1}{16}x^2 & 0 \le x \le 4 \\ 1 & x > 4 \end{cases}$

b i 2.83   **ii** 1.26   **iii** 3.58

3 a Median = 1.732 since −1.732 is not in the range.

b $Q_1 = 1.225$, $Q_3 = 2.134$, IQR = 2.134 − 1.225 = 0.909

4 a f(x)

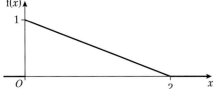

b 0

c $F(x) = \begin{cases} 0 & x < 0 \\ x - \dfrac{1}{4}x^2 & 0 \le x \le 2 \\ 1 & x > 2 \end{cases}$

d Median = $2 - \sqrt{2} = 0.586$ (3 s.f.) as $2 + \sqrt{2}$ is not in range.

e 1

f 0.0506 (3 s.f.)

5 a f(y)

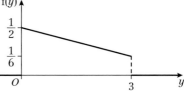

b Positive skew

c 0

d $F(y) = \begin{cases} 0 & y < 0 \\ \dfrac{y}{2} - \dfrac{1}{18}y^2 & 0 \le y \le 3 \\ 1 & y > 3 \end{cases}$

e Median = $\dfrac{9 - 3\sqrt{5}}{2} = 1.15$ (3 s.f.)

f 2.28 (3 s.f.)

6 a f(x)

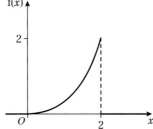

b 2

c $F(x) = \begin{cases} 0 & x < 0 \\ \dfrac{1}{16}x^4 & 0 \le x \le 2 \\ 1 & x > 2 \end{cases}$

d 1.68 (3 s.f.)

7 a

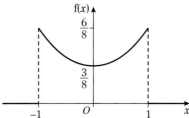

b Bimodal −1 and 1

c Median = 0

d $F(x) = \begin{cases} 0 & x < -1 \\ \dfrac{1}{8}x^3 + \dfrac{3}{8}x + \dfrac{1}{2} & -1 \le x \le 1 \\ 1 & x > 1 \end{cases}$

8 a f(x)

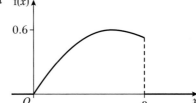

b Negative skew

Online Full worked solutions are available in SolutionBank.

c Mode = 1.5

d
$$F(x) = \begin{cases} 0 & x < 0 \\ \dfrac{9}{20}x^2 - \dfrac{1}{10}x^3 & 0 \leqslant x \leqslant 2 \\ 1 & x > 2 \end{cases}$$

e $F(1.23) = 0.495$ and $F(1.24) = 0.501$. Since 0.5 lies between 0.495 and 0.501 the median lies between 1.23 and 1.24.

9 a
$$f(x) = \begin{cases} \dfrac{1}{4}x & 1 \leqslant x \leqslant 3 \\ 0 & \text{otherwise} \end{cases}$$

b Mode = 3

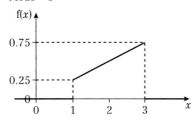

c $\sqrt{5}$

d Mode > median so negative skew.

e $k = 1.9$

10 a
$$f(x) = \begin{cases} 12x^2(1-x) & 0 \leqslant x \leqslant 1 \\ 0 & \text{otherwise} \end{cases}$$

b mode = $\frac{2}{3}$

c 0.2853

11 a
$$F(w) = \begin{cases} 0 & w < 0 \\ \dfrac{w^4}{5^5}(25 - 4w) & 0 \leqslant w \leqslant 5 \\ 1 & w > 5 \end{cases}$$

b $F(3.4) = 0.487\ldots$, $F(3.5) = 0.528\ldots$, so median lies between 3.4 and 3.5.

c The maximum of $f(w)$ is at $w = \frac{15}{4}$, so the mode is $\frac{15}{4}$

d Median < mode so negative skew.

12 a 1.365 (3 d.p.)

b
$$F(x) = \begin{cases} 0 & x < 0 \\ \dfrac{x}{4} & 0 \leqslant x < 1 \\ \dfrac{x^4}{20} + \dfrac{1}{5} & 1 \leqslant x \leqslant 2 \\ 1 & x > 2 \end{cases}$$

c Median = 1.565 (3 d.p.)   IQR = 0.821 (3 d.p.)

d Mean < median so negative skew.

13 a e.g.

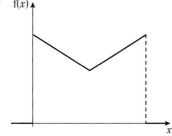

b e.g.

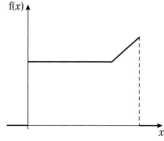

14 e.g. $f(x) = \begin{cases} x & 0 \leqslant x \leqslant 1 \\ \dfrac{1}{2} & 1 < x \leqslant 2 \\ 0 & \text{otherwise} \end{cases}$

15 a Mode = 2

b
$$F(x) = \begin{cases} 0 & x < 2 \\ \dfrac{\ln\left(\dfrac{x}{2}\right)}{\ln 5} & 2 \leqslant x \leqslant 10 \\ 1 & x > 10 \end{cases}$$

c $2\sqrt{5}$

d $Q_1 = 2.991$, $Q_3 = 6.687$, IQR = 3.697

16 a 277 hours

b $Q_1 = 115$ hours, $Q_3 = 555$ hours, IQR = 439 hours

17 a $k = \pi$

b
$$F(x) = \begin{cases} 0 & x < 0 \\ \tan(\pi x) & 0 \leqslant x \leqslant 0.25 \\ 1 & x > 0.25 \end{cases}$$

c 0.1476

18 a $k = \dfrac{5}{\ln 6}$ **b** 3.066 (3 d.p.)   **c** 0.349 (3 d.p.)

d
$$F(x) = \begin{cases} 0 & x < 2 \\ \dfrac{1}{\ln 6}\left(\ln\left(\dfrac{3x}{10-2x}\right)\right) & 2 \leqslant x \leqslant 4 \\ 1 & x > 4 \end{cases}$$

e 3.101 (3 d.p.) **f** 4

g Negative skew: mean < median < mode

Exercise 3E

1 a 0.4   **b** 0.6

2 a $k = 12.6$ **b** 0.39

3 a $k = \frac{1}{8}$ **b** 0.6875   **c** $p = 4$

d $\frac{1}{6}$ **e** $\frac{3}{5}$ **f** $\frac{1}{2}$

4

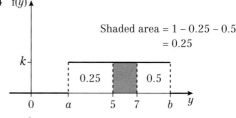

Shaded area = $1 - 0.25 - 0.5$
= 0.25

$k = \frac{1}{8}$ $b = 11$ $a = 3$

5 a $Y \sim U[9, 21]$ **b** $\frac{2}{3}$

6 a Continuous uniform distribution

b $E(Y) = 6$ **c** $\frac{2}{5}$ **d** $\frac{3}{4}$

7 a 1   **b** $\frac{16}{3}$ **c** $\frac{19}{3}$

d

$$F(x) = \begin{cases} 0 & x < -3 \\ \dfrac{x+3}{8} & -3 \leqslant x \leqslant 5 \\ 1 & x > 5 \end{cases}$$

8 a $E(X) = 3$, $Var(X) = \frac{4}{3}$ **b** $E(X) = 2$, $Var(X) = \frac{16}{3}$

9 a 4.5   **b** $\frac{1}{3}$ **c** $20\frac{7}{12} = 20.6$

d

$$F(x) = \begin{cases} 0 & x < 3.5 \\ \dfrac{x}{2} - 1.75 & 3.5 \leqslant x \leqslant 5.5 \\ 1 & x > 5.5 \end{cases}$$

10 $a = -1$ and $b = 3$

11 $E(X) = \dfrac{5 + (-1)}{2} = 2$

$\qquad Var(X) = \dfrac{(5 - (-1))^2}{12} = 3$

$\qquad E(Y) = 4E(X) - 6$
$\qquad\quad\; = 8 - 6 = 2$
$\qquad Var(Y) = 16\,Var(X)$
$\qquad\qquad\quad = 48$

12 a $\frac{4}{7}$ **b** $\frac{2}{5}$

13 a $\alpha = 3$, $\beta = 7$ **b** 0.55

14 a $f(x) = \begin{cases} \dfrac{1}{\beta - \alpha} & \alpha \leqslant x \leqslant \beta \\ 0 & \text{otherwise} \end{cases}$

   b $\alpha = -3$, $\beta = 8$

15 a $f(x) = \begin{cases} \dfrac{1}{9} & -5 \leqslant x \leqslant 4 \\ 0 & \text{otherwise} \end{cases}$

   b

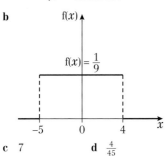

   c 7   **d** $\frac{4}{45}$

16 a $\frac{3}{7}$ **b** $f(x) = \begin{cases} \dfrac{1}{7} & -3 \leqslant x \leqslant 4 \\ 0 & \text{otherwise} \end{cases}$

   c Uniform   **d** Mean = 0.5, Variance = $\frac{49}{12}$

17 a 1.5   **b** $\frac{25}{12}$ **c** $\frac{13}{3}$

   d 0.48   **e** 0.2153

18 a $\alpha = 1.5$, $\beta = 13.5$

   b i $c = 5.5$ **ii** $\frac{7}{24}$

Exercise 3F

1 $E(Y) = E(X^2) = 25\frac{1}{12}$

2 a $f(x) = \begin{cases} \dfrac{1}{6} & 5 \leqslant x \leqslant 11 \\ 0 & \text{otherwise} \end{cases}$

   b 0.5   **c** $67\pi\,cm^2$

3 a 0.2   **b** 0.5   **c** $\frac{1}{12}$

4 a $\frac{3}{8}$ **b** $\frac{27}{512}$ **c** $\frac{3}{5}$

5 a $f(x) = \begin{cases} \dfrac{1}{40} & 175 \leqslant x \leqslant 215 \\ 0 & \text{otherwise} \end{cases}$

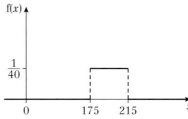

   b i 0.3   **ii** 0
   c 20   **d** 189   **e** 0.1323 (4 d.p.)

6 a $\frac{7}{60}$ **b** $\frac{1}{3}$ **c** 0.2276 (4 d.p.)

7 a $\frac{1}{4}$ **b** 0.2142 (4 d.p.)

8 a 0.6   **b** 0.3222 (4 d.p.)

9 $\frac{200}{3}$

Mixed exercise 3

1 a $\frac{10}{9}, \frac{16}{3}$ **b** $\frac{26}{81}, \frac{26}{9}$ **c** $\frac{5}{12}$

   d $\frac{128}{243}$ **e** 0.5

2 a $\frac{1}{3}$ **b** $\frac{1}{18}$ **c** $\frac{5}{3}, \frac{2}{9}$

   d

$$F(x) = \begin{cases} 0 & x < 0 \\ 2x - x^2 & 0 \leqslant x \leqslant 1 \\ 1 & x > 1 \end{cases}$$

   e Median = 0.293 as 1.71 is not in the range

3 a $F(2) = 1$; $F(y) = k(y^2 - y)$
     $k(4 - 2) = 1 \Rightarrow k = \frac{1}{2}$

   b 0.375

   c Median = 1.62 as -0.618 is not in the range

   d

$$f(y) = \begin{cases} y - \dfrac{1}{2} & 1 \leqslant y \leqslant 2 \\ 0 & \text{otherwise} \end{cases}$$

4 a 0.648

   b Median = 2.55 as -2.55 is not in the range

   c

$$f(x) = \begin{cases} \dfrac{2x}{5} & 2 \leqslant x \leqslant 3 \\ 0 & \text{otherwise} \end{cases}$$

   d $\frac{38}{15}$ **e** Mode = 3

5 a $\displaystyle\int_0^2 kx^2\,dx = 1$; $\left[\dfrac{kx^3}{3}\right]_0^2 = 1$

     $\dfrac{8k}{3} = 1 \Rightarrow k = \dfrac{3}{8}$

   b 1.5

   c

$$F(x) = \begin{cases} 0 & x < 0 \\ \dfrac{x^3}{8} & 0 \leqslant x \leqslant 2 \\ 1 & x > 2 \end{cases}$$

   d 1.59 (3 s.f.)   **e** Mode = 2

6 a $\displaystyle\int_1^3 k(y^2 + 2y + 2)\,dx = 1$

     $\left[k\left(\dfrac{y^3}{3} + y^2 + 2y\right)\right]_1^3 = 1$

     $k\left(\dfrac{3^3}{3} + 3^2 + 6\right) - k\left(\dfrac{1}{3} + 1 + 2\right) = 1$

               $\dfrac{62}{3}k = 1$

                 $k = \dfrac{3}{62}$

Online Full worked solutions are available in SolutionBank.

b
$$F(y) = \begin{cases} 0 & y < 1 \\ \dfrac{y^3}{62} + \dfrac{3y^2}{62} + \dfrac{3y}{31} - \dfrac{5}{31} & 1 \leqslant y \leqslant 3 \\ 1 & y > 3 \end{cases}$$

c $\dfrac{11}{31}$

7 a

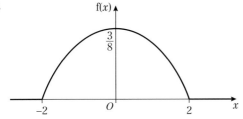

b Mode = 0

c
$$F(x) = \begin{cases} 0 & x < -2 \\ \dfrac{12x}{32} - \dfrac{x^3}{32} + \dfrac{1}{2} & -2 \leqslant x \leqslant 2 \\ 1 & x > 2 \end{cases}$$

d $\dfrac{35}{128}$ or 0.273

8 a $\dfrac{26}{21}$

b
$$F(x) = \begin{cases} 0 & x < 0 \\ \dfrac{x}{3} & 0 \leqslant x < 1 \\ \dfrac{2x^3}{21} + \dfrac{5}{21} & 1 \leqslant x \leqslant 2 \\ 1 & x > 2 \end{cases}$$

c i 1.401   **ii** 0.45

d Mean < median, negative skew

9 $F(1) = 0 \Rightarrow 0.05a - b = 0$
$F(2) = 1 \Rightarrow 0.05a^2 - b = 1$
$0.05(a^2 - a) = 1 \Rightarrow a^2 - a - 20 = 0$
$(a + 4)(a - 5) = 0$
Given that a is positive, $a = 5$ and $b = 0.05a = \dfrac{1}{4}$

10 $F'(x) < 0$ for $8 < x < 10$

11 a $\displaystyle\int_1^3 kx - k\,dx = 1$
$\left[\dfrac{kx^2}{2} - kx\right]_1^3 = 1$
$\left(\dfrac{9k}{2} - 3k\right) - \left(\dfrac{k}{2} - k\right) = 1$
$2k = 1 \Rightarrow k = \dfrac{1}{2}$

b $\dfrac{7}{3}$

c
$$F(x) = \begin{cases} 0 & x < 1 \\ \dfrac{x^2}{4} - \dfrac{x}{2} + \dfrac{1}{4} & 1 \leqslant x \leqslant 3 \\ 1 & x > 3 \end{cases}$$

d $F(2.4) = \dfrac{2.4^2}{4} - \dfrac{2.4}{2} + \dfrac{1}{4} = 0.49$

$F(2.5) = \dfrac{2.5^2}{4} - \dfrac{2.5}{2} + \dfrac{1}{4} = 0.5625$

Since 0.5 lies in between the median is between 2.4 and 2.5

e Mean < median, negative skew

f 1.265 (4 s.f.)

12 a

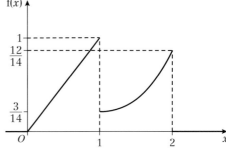

b Mode = 1   **c** $\dfrac{191}{84}$ **d** 1.14

e
$$F(x) = \begin{cases} 0 & x < 0 \\ \dfrac{x^2}{2} & 0 < x < 1 \\ \dfrac{x^3}{14} + \dfrac{3}{7} & 1 \leqslant x \leqslant 2 \\ 1 & x > 2 \end{cases}$$

f Median = 1

13 a

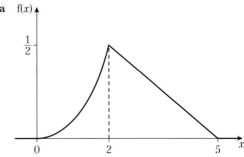

b Mode = 2

c Using the sketch, $P(X > 2)$ = area of triangle
$= \dfrac{1}{2} \times 3 \times \dfrac{1}{2} = 0.75$

d
$$F(x) = \begin{cases} 0 & x < 0 \\ \dfrac{x^4}{64} & 0 \leqslant x < 2 \\ \dfrac{10x - x^2 - 13}{12} & 2 \leqslant x \leqslant 5 \\ 1 & x > 5 \end{cases}$$

e $5 - \sqrt{6} = 2.55$ (s.f.)

14 a $f(x) = \begin{cases} \dfrac{1}{81}(-6x^2 + 30x) & 2 \leqslant x \leqslant 5 \\ 0 & \text{otherwise} \end{cases}$

b Mode = 2.5

c

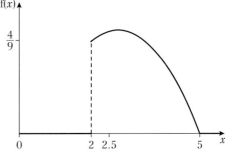

d $\dfrac{19}{6}$

e $F(\mu) = F\left(\dfrac{19}{6}\right) = \dfrac{1}{81}\left[-2\left(\dfrac{19}{6}\right)^3 + 15\left(\dfrac{19}{6}\right)^2 - 44\right]$
$= 0.5297$ (4 d.p.) > 0.5

f F(2.5) = 0.2284 so as 0.2284 < 0.5 < 0.5297, for this distribution mode < median < mean, which is positive skew

15 a F(5) = 1 ⇒ $k(35 \times 5 - 2 \times 5^2) = 1 \Rightarrow 125k = 1$

$\Rightarrow k = \frac{1}{125}$

b 2.02 (3 s.f.)

c $f(x) = \begin{cases} \dfrac{35 - 4x}{125} & 0 \leqslant x \leqslant 5 \\ 0 & \text{otherwise} \end{cases}$

d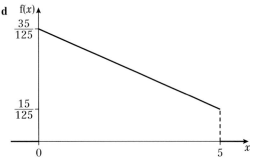

e Mode = 0

f $\frac{13}{6}$

g Positive skew as mean > median > mode

16 $a = \frac{3}{16},\ b = \frac{5}{16}$

17 a $\int_{-1}^{0} k(x + 1)^3\, dx = 1 \Rightarrow \left[\dfrac{k(x+1)^4}{4}\right]_{-1}^{0} = 1$

$\Rightarrow \dfrac{k}{4} = 1 \Rightarrow k = 4$

b −0.2

c $F(x) = \begin{cases} 0 & x < -1 \\ (x+1)^4 & -1 \leqslant x \leqslant 0 \\ 1 & x > 0 \end{cases}$

d −0.159 (3 s.f.)

18 a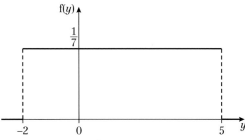

b 1.5 **c** $4\frac{1}{12}$

d $F(x) = \begin{cases} 0 & x < -2 \\ \dfrac{x + 2}{7} & -2 \leqslant x \leqslant 5 \\ 1 & x > 5 \end{cases}$

e $\frac{3}{14}$ **f** 0 **g** $\frac{1}{2}$ **h** $\frac{2}{5}$

19 a $k = 1$ **b** 0.2 **c** −1.5 **d** $\frac{25}{12}$

e $F(x) = \begin{cases} 0 & x < -4 \\ \dfrac{x + 4}{5} & -4 \leqslant x \leqslant 1 \\ 1 & x > 1 \end{cases}$

20 a $b = 5$ $a = -1$

b 0.533 (3 s.f.)

21 a Continuous uniform distribution $Y \sim U[0, 10]$

b $\frac{3}{10}$ **c** $\frac{3}{5}$

22 a Continuous uniform distribution $X \sim U[0, 20]$

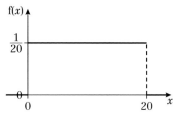

b $E(X) = 10$ $\text{Var}(X) = \frac{100}{3}$ **c** 0.6

23 a $X \sim U[-0.5, 0.5]$ **b** 0.4 **c** $\frac{1}{12}$

24 a $f(x) = \begin{cases} \dfrac{1}{13} & -3 \leqslant x \leqslant 10 \\ 0 & \text{otherwise} \end{cases}$

b 3.5 minutes

c $F(x) = \begin{cases} 0 & x < -3 \\ \dfrac{x + 3}{13} & -3 \leqslant x \leqslant 10 \\ 1 & x > 10 \end{cases}$

d $\frac{5}{13}$

25 a $X \sim U[-0.5, 0.5]$ **b** 0.4 **c** 0.064

26 a $f(x) = \begin{cases} \dfrac{1}{20} & 190 \leqslant x \leqslant 210 \\ 0 & \text{otherwise} \end{cases}$

b **i** $\frac{2}{5}$ **ii** 0 **c** 10 **d** $\frac{2}{3}$

27 a Continuous uniform distribution

b Normal distribution

28 a $F(t) = \begin{cases} 0 & t < 0 \\ 1 - \dfrac{(6 - t)^3}{216} & 0 \leqslant t \leqslant 6 \\ 1 & t > 6 \end{cases}$

b 1.24 hours (3 s.f.) **c** 1.5 hours

29 a $f(x) = \begin{cases} \dfrac{1}{4b} & b \leqslant x \leqslant 5b \\ 0 & \text{otherwise} \end{cases}$

b $3b$

c $E(X^2) = \int_{b}^{5b} \dfrac{x^2}{4b}\, dx = \dfrac{1}{4b}\left[\dfrac{x^3}{3}\right]_{b}^{5b} = \dfrac{1}{4b}\left[\dfrac{125b^3 - b^3}{3}\right]$

$= \dfrac{124b^2}{12} = \dfrac{31b^2}{3}$

d $\frac{5}{12}$

e 0.246 (3 s.f.)

30 $F(x) = \begin{cases} 0 & x < 1 \\ \dfrac{\ln(2x - 1)}{\ln 5} & 1 \leqslant x \leqslant 3 \\ 1 & x > 3 \end{cases}$

31 a $k = \pi$

b 0.5947 = 59.47%

32 a $k = \frac{2}{3}$

b $\frac{1}{3} + \frac{2}{3}\ln 2 = 0.795$ (3 s.f.)

c 0.256 (3 s.f.)

Challenge

1 a $\theta \sim U[0, 2\pi]$

b $\dfrac{2r}{\pi} = 0.6366r$ (4 d.p.)

c Spin the spinner 100 times and measure X each time. Take the mean of these observations and divide $2r$ by this value.

2 a $E(X) = \int_0^\infty x f(x)\,dx = \int_0^\infty x(\lambda e^{-\lambda x})\,dx$

$= [x(\lambda e^{-\lambda x})]_0^\infty - \int_0^\infty -e^{-\lambda x}\,dx$

$= 0 + \int_0^\infty e^{-\lambda x}\,dx = -\dfrac{1}{\lambda}(0-1) = \dfrac{1}{\lambda}$

$E(X^2) = \int_0^\infty x^2 f(x)\,dx = \int_0^\infty x^2(\lambda e^{-\lambda x})\,dx$

$= [x^2(\lambda e^{-\lambda x})]_0^\infty - \int_0^\infty -2x e^{-\lambda x}\,dx$

$= 0 + 2\int_0^\infty x e^{-\lambda x}\,dx = \dfrac{2}{\lambda^2}$

$\text{Var}(X) = E(X^2) - (E(X))^2 = \dfrac{2}{\lambda^2} - \left(\dfrac{1}{\lambda}\right)^2 = \dfrac{1}{\lambda^2}$

b $P(X > a) = 1 - P(X < a) = 1 - \int_0^a \lambda e^{-\lambda x}\,dx$

$= 1 - [-e^{-\lambda x}]_0^a = 1 - (-e^{-\lambda a} + 1)0 = e^{-\lambda a}$

Similarly, $P(X > b) = e^{-\lambda b}$ and $P(X > a + b) = e^{-\lambda(a+b)}$

$= e^{-\lambda a} \times e^{-\lambda b}$

$P(X > a + b \mid X > a) = \dfrac{P(X > a + b)}{P(X > a)}$

$= \dfrac{e^{-\lambda a} \times e^{-\lambda b}}{e^{-\lambda a}} = e^{-\lambda b} = P(X > b)$

CHAPTER 4

Prior knowledge check

1 a 0.8944 **b** 0.2902
2 a 0.6745 **b** 1.0364
3 a 14 **b** 27 **c** −41

Exercise 4A

1 a $W \sim N(130, 13)$ **b** $W \sim N(30, 13)$
2 $R \sim N(148, 18)$
3 a $T \sim N(180, 225)$ or $N(180, 15^2)$
  b $T \sim N(350, 784)$ or $N(350, 28^2)$
  c $T \sim N(530, 1009)$
  d $T \sim N(-40, 89)$
4 a $A \sim N(35, 9)$ or $N(35, 3^2)$ **b** $A \sim N(7, 6)$
  c $A \sim N(41, 41)$ **d** $A \sim N(84, 82)$
  e $A \sim N(19, 15)$
5 a 0.909 **b** 0.0512 **c** 0.319
  d 0.0614 **e** 0.857 **f** 0.855
6 a 10 **b** 9 **c** 0.136
7 a 0.7881 **b** 0.2119 **c** 0.3400 **d** 0.2882
8 a 64 **b** 148 **c** 0.0293 **d** 28
9 a 0.4497 **b** 0.0495
10 0.0385
11 0.0537
12 a i 0.268 **ii** 0.436
  b 37.1 cm (3 s.f.)
13 a 0.3768 **b** 0.6226 **c** 0.9059
14 0.732
15 a i 60 **ii** 25
  b i $R \sim N(50, 20)$ **ii** 0.9320
16 a 0.8644
  b All random variables were independent – reasonable as games were chosen at random and game size and hard drive size are unconnected.

17 0.9044
18 a 0.7390 **b** 0.4018

Challenge

$\text{Var}(X + Y) = E((X + Y)^2) - (E(X + Y))^2$
$= E(X^2 + 2XY + Y^2) - (E(X) + E(Y))^2$
$= E(X^2) + 2E(X)E(Y) + E(Y^2) - (E(X^2))^2 - 2E(X)E(Y) - (E(Y))^2$
$= E(X^2) - (E(X))^2 + E(Y^2) - (E(Y))^2$
$= \text{Var}(X) + \text{Var}(Y)$

Review exercise 1

1 a $y = -0.425 + 0.395x$ (3 s.f.)
  b $f = 0.735 + 0.395m$ (3 s.f.)
  c 93.6 litres (3 s.f.)
2 a

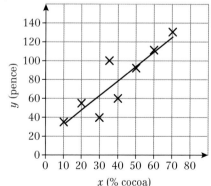

(graph: Evaporation loss (y ml) against Time (x weeks))

  b The points lie close to a straight line.
  c $a = 29.02, b = 3.90$
  d 3.90 ml of the chemicals evaporate each week.
  e i 103 ml **ii** 166 ml
  f i This estimate is reasonably reliable, since it is just outside the range of the data.
   ii This estimate is unreliable, since it is far outside the range of the data.
3 a $S_{xy} = 71.4685$, $S_{xx} = 1760.459$
  b $y = 0.324 + 0.0406x$ (3 s.f.)
  c 2461.95 mm (2 d.p.)
  d $l = 2460.324 + 0.0406t$
  e 2463.98 mm (2 d.p.)
  f This estimate is unreliable, since it is outside the range of the data.
4 a & d

(graph: y (pence) against x (% cocoa))

  b $S_{xy} = \sum xy - \dfrac{\sum x \sum y}{n} = 28\,750 - \dfrac{315 \times 620}{8} = 4337.5$
  $S_{xx} = 2821.875$
  c $a = 17.0, b = 1.54$
  e i Brand D is overpriced, since it is a long way above the regression line.
   ii 69p or 70p since this is the predicted price for a bar of chocolate with 35% cocoa.

5 a 1.749, −0.806, −1.05, −1.672, 0.5505, 1.084, −1.471, 1.7515

b A linear model is a suitable model as the residuals are randomly scattered about zero.

c 14.3 (3 s.f.)

d The first sample as the RSS is smaller.

6 a $S_{mm} = 37.9$, $S_{md} = 24.0$ (3 s.f.)

b $d = −2.48 + 0.633m$ (3 s.f.)

c 7.33 (3 s.f.)

d 3.75 (3 s.f.)

e −0.50765

f Not suitable since the residuals are not randomly scattered about zero.

7 a $y = 41.9 + 264x$ (3 s.f.)

b RSS $= 289\,771.4 \times \dfrac{895.5714^2}{3.388\,571} = 53\,100$ (3 s.f.)

c −41.5, 2.1, −41.9, 10.5, −55.1, −72.7, 202.9

d 1.8

e e.g. It could be a legitimate data point (a company that thrives despite (relatively low) spend on advertising).

f $y = −22.9 + 279x$ (3 s.f.)

g £479 000 (3 s.f.)

h This estimate is reliable as it is within the range of the data.

8 a $h = 80.6 + 1.92l$ (3 s.f.)

b 167 cm (3 s.f.)

c 2.2071

d The residuals are randomly scattered about zero so the model is suitable.

e 30.8 (3 s.f.)

f The female sample since the RSS is lower.

9 Diagram A corresponds to −0.79, since there is negative correlation.
Diagram B corresponds to 0.08, since there is very weak or no correlation.
Diagram C corresponds to 0.68, since there is positive correlation.

10 a −0.816

b Houses are cheaper the further away they are from the railway station.

c −0.816

11 a £17

b $S_{tt} = 983.6$, $S_{mm} = 1728.9$, $S_{tm} = 1191.8$

c 0.914 (3 s.f.)

d 0.914. Linear coding does not affect the correlation coefficient.

e 0.914 suggests a relationship between the time spent shopping and the money spent.
0.178 suggests that there was no such relationship.

f e.g. Shopping behaviours may be different on different days of the week.

12 a $\frac{13}{21}$ or 0.619 (3 s.f.)

b $H_0 : \rho = 0$, $H_1 : \rho > 0$
5% critical value = 0.6429
0.619 < 0.6429 so do not reject H_0. The evidence does not show positive correlation between the judges' marks, so the competitor's claim is justified.

13 a $r_s = −\frac{11}{15}$ or −0.733 (3 s.f.)

b $H_0 : \rho = 0$; $H_1 : \rho < 0$
5% critical value = −0.5636.

−0.733 < −0.5636 so reject H_0. There is evidence of a significant negative correlation between the price of an ice lolly and the distance from the pier. The further from the pier you travel, the less money you are likely to pay for an ice lolly.

14 a The variables cannot be assumed to be normally distributed.

b $r_s = \frac{5}{7}$ or 0.714 (3 s.f.)

c $H_0 : \rho = 0$; $H_1 : \rho > 0$.
5% critical value = 0.8286.
0.714 < 0.8286 so do not reject H_0. There is no evidence that the relative vulnerabilities of the different age groups are similar for the two diseases.

15 a i **ii**

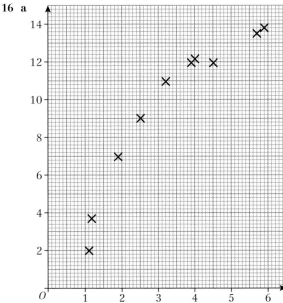

b i $r_s = \frac{23}{28}$ or 0.821 (3 s.f.)

ii $H_0 : \rho = 0$; $H_1 : \rho > 0$.
5% critical value = 0.7143.
0.821 > 0.7143 so reject H_0. There is evidence of a (positive) correlation between the ranks awarded by the judges.

16 a

b The strength of the linear link between two variables.

c $S_{tt} = 26.589$; $S_{pp} = 152.444$; $S_{tp} = 59.524$

d 0.93494...

e $H_0 : \rho = 0$, $H_1 : \rho > 0$.
5% critical value = 0.7155.
0.935 > 0.7155 so reject H_0; reactant and product are positively correlated.

f Linear correlation is significant but the scatter diagram looks non-linear. The product moment correlation coefficient should not be used here since the association/relationship is not linear.

17 a $H_0 : \rho = 0$, $H_1 : \rho < 0$.
5% critical value = −0.8929
−0.93 < −0.8929 so reject null hypothesis. There is evidence supporting the geographer's claim.

Online Full worked solutions are available in SolutionBank.

b i No effect since rank stays the same.
ii It will increase since $d = 0$ and n is bigger.

c The mean of the tied ranks is given to each and then the PMCC is used.

18 a $\int_0^2 k(4x - x^3)\,dx = 1 \Rightarrow \left[2kx^2 - \frac{k}{4}x^4\right]_0^2 = 1$

$\Rightarrow 4k = 1 \Rightarrow k = \frac{1}{4}$

b

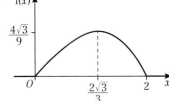

c $E(X) = \frac{16}{15}$ (or 1.07 to 3 s.f.)

d Mode $= \frac{2\sqrt{3}}{3}$ (or 1.15 to 3 s.f.)

e Median = 1.08 (3 s.f.)

f Mean (1.07) < Median (1.08) < Mode (1.15) so negative skew.

19 a $\int_2^3 kx(x-2)\,dx = 1 \Rightarrow k\left[\frac{1}{3}x^3 - x^2\right]_2^3 = 1$

$k\left((9-9) - \left(\frac{8}{3} - 4\right)\right) = 1 \Rightarrow k = \frac{3}{4}$

b $\frac{67}{1280}$

c $F(x) = \begin{cases} 0 & x < 2 \\ \frac{1}{4}(x^3 - 3x^2 + 4) & 2 \le x \le 3 \\ 1 & x > 3 \end{cases}$

d $F(2.70) = 0.453$ and $F(2.75) = 0.527$
$0.453 < 0.5 < 0.527$ so the median lies between 2.70 and 2.75.

20 a $F'(y) < 0$ for $1.625 < y < 2$ so his model cannot be a cumulative distribution function.

b $k(2^4 + 2^2 - 2) - k(1 + 1 - 2) = 1$
$\Rightarrow k(16 + 4 - 2) = 1 \Rightarrow 18k = 1 \Rightarrow k = \frac{1}{18}$
$\Rightarrow 18k = 1 \Rightarrow k = \frac{1}{18}$

c $\frac{203}{288}$ or 0.705 (3 s.f.)

d $f(y) = \begin{cases} \frac{1}{9}(2y^3 + y) & 1 \le y \le 2 \\ 0 & \text{otherwise} \end{cases}$

21 a

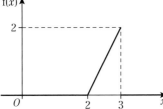

b Mode of X is 3.

c $\frac{1}{18}$

d Median of X is 2.71.

e Mean (2.67) < median (2.71) < mode (3) so negative skew.

22 a

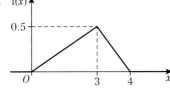

b Mode of X is 3.

c $F(x) = \begin{cases} 0 & x < 0 \\ \frac{1}{12}x^2 & 0 \le x < 3 \\ 2x - \frac{1}{4}x^2 - 3 & 3 \le x \le 4 \\ 1 & x > 4 \end{cases}$

d Median $= \sqrt{6} = 2.45$ (3 s.f.)

e 2.272 (3 d.p.)

23 a 0.847

b $F(0.59) = 0.491$ and $F(0.60) = 0.504$
$0.491 < 0.5 < 0.504$ so the median lies between 0.59 and 0.60.

c $f(x) = \begin{cases} 4x - 3x^2 & 0 \le x \le 1 \\ 0 & \text{otherwise} \end{cases}$

d $E(X) = \frac{7}{12}$ or 0.583 (3 s.f.)

e Mode $= \frac{2}{3}$ or 0.667 (3 s.f.)

f Mean (0.583) < median (0.59 – 0.6) < mode (0.667) so negative skew.

24 a $\int_0^2 k\,dx + \int_2^4 \frac{k}{x}\,dx = 1$

$\Rightarrow [kx]_0^2 + [k \ln x]_2^4 = 1$

$\Rightarrow k(2 + \ln 2) = 1$

$\Rightarrow k = \frac{1}{2 + \ln 2}$

b $\frac{4}{2 + \ln 2} (= 1.485...)$

25 a

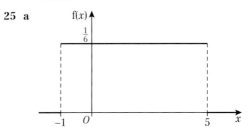

b $E(X) = 2$

c $Var(X) = 3$

d 0.6

26 a $f(x) = \begin{cases} \frac{1}{4} & 2 \le x \le 6 \\ 0 & \text{otherwise} \end{cases}$

b $E(X) = 4$

c $Var(X) = \frac{4}{3}$

d $F(x) = \begin{cases} 0 & x < 2 \\ \frac{1}{4}(x - 2) & 2 \le x \le 6 \\ 1 & x > 6 \end{cases}$

e 0.275

27 a Continuous uniform distribution

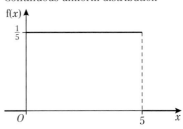

b $E(X) = 2.5$; $Var(X) = \frac{25}{12}$

c $\frac{2}{5}$

d 0

28 a $f(x) = \begin{cases} \dfrac{1}{\beta - \alpha} & \alpha \leqslant x \leqslant \beta \\ 0 & \text{otherwise} \end{cases}$

b $\alpha = -2$ $\quad \beta = 6$

29 a 75 cm **b** 43.3 (3 s.f.) **c** $\dfrac{60}{150} = \dfrac{2}{5}$

30 a 0.7586

b The durations of the two rides are independent. This is likely to be the case as two separate control panels operate each of the rides.

c $D \sim N(246, 27)$

d 0.5713

31 a 30 **b** 4.84 **c** 0.5764

32 a 0.7377 **b** 0.6858

33 a 0.9031 **b** 0.8811

34 a 0.1336 **b** 0.8413 **c** 0.1610

d All random variables are independent and normally distributed.

Challenge

1 a i Linear model: $y = -2.63 + 2.285x$

ii Quadratic model: $y = 1.04 + 0.1206x + 0.2353x^2$

iii Exponential model: $y = 1.1762e^{0.3484x}$

b Linear residuals: 1.845, −0.925, −1.21, −1.295, 0.435, 1.15

Quadratic residuals: 0.1041, −0.2195, 0.0128, −0.0255, 0.3861, −0.264

Exponential residuals: −0.16644, −0.04507, 0.560703, 0.785369, 0.321593, −2.29619

Hence quadratic model is most suitable as the residuals are smaller and are randomly scattered around zero.

2 a $\displaystyle\int_0^\infty ke^{-x}\,dx = k[-e^{-x}]_0^\infty = k(-1) \Rightarrow k = 1$

b $F(x) = \begin{cases} 0 & x < 0 \\ 1 - e^{-x} & x \geqslant 0 \end{cases}$

c $e^{-1} - e^{-4}$ or $\dfrac{e^3 - 1}{e^4}$

3 a

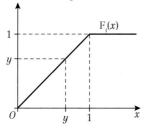

$F_Y(y) = P((X_1 \leqslant y) \cap \dots \cap (X_n \leqslant y))$
$= F_1(y) \dots F_n(y) = y^n$
So $f_y(y) = ny^{n-1}$
$\Rightarrow E(Y) = \displaystyle\int_0^1 y(ny^{n-1})\,dy = n\int_0^1 y^n\,dy = \dfrac{n}{n+1}$

b $\sqrt[n]{0.5}$

c $f_Y(z) = \begin{cases} z & 0 \leqslant z \leqslant 1 \\ 2 - z & 1 < z \leqslant 2 \\ 0 & \text{otherwise} \end{cases}$

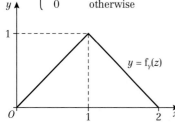

d $f_z(z) = \begin{cases} \frac{1}{2}z^2 & 0 \leqslant z \leqslant 1 \\ \frac{3}{4} - \left(z - \frac{3}{2}\right)^2 & 1 < z \leqslant 2 \\ \frac{1}{2}(z - 3)^2 & 2 < z \leqslant 3 \\ 0 & \text{otherwise} \end{cases}$

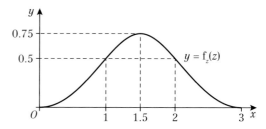

CHAPTER 5
Prior knowledge check

1 a 0.391 **b** N(25, 45)

2 Reject H_0 (p-value = 0.00143)

3 0.1870

Exercise 5A

1 i a $N(10\mu, 10\sigma^2)$ **b** $N(\mu, \frac{13}{25}\sigma^2)$

c $N(0, 10\sigma^2)$ **d** $N\left(\mu, \dfrac{\sigma^2}{10}\right)$

e $N(0, 10\sigma^2)$ **f** $N(0, 10)$

ii a, b, d, e, are statistics since they do not contain μ or σ the unknown population parameters.

2 a $\mu = E(x) = \frac{22}{5}$ or 4.4

$\sigma^2 = 11.04$ or $\frac{276}{25}$

b $\{1, 1\} \{1, 5\}^{\times 2} \{1, 10\}^{\times 2}$
$\{5, 5\} \{5, 10\}^{\times 2}$
$\{10, 10\}$

c

$\bar{x}$	1	3	5	5.5	7.5	10
$P(\bar{X} = \bar{x})$	$\frac{4}{25}$	$\frac{8}{25}$	$\frac{4}{25}$	$\frac{4}{25}$	$\frac{4}{25}$	$\frac{1}{25}$

e.g. $P(\bar{X} = 5.5) = \frac{4}{25}$

d $E(\bar{X}) = 1 \times \frac{4}{25} + 3 \times \frac{8}{25} + \dots + 10 \times \frac{1}{25} = 4.4 = \mu$

$Var(\bar{X}) = 1^2 \times \frac{4}{25} + 3^2 \times \frac{8}{25} + \dots + 10^2 \times \frac{1}{25} - 4.4^2$

$= 5.52 = \dfrac{\sigma^2}{2}$

3 a $\bar{x} = 19.3, s^2 = 3.98$ **b** $\bar{x} = 3.375, s^2 = 4.65$

c $\bar{x} = 223, s^2 = 7174$ **d** $\bar{x} = 0.5833, s^2 = 0.0269$

4 a 36.4, 29.2 (3 s.f.) **b** 9, 4

c 1.1, 0.0225 **d** 11.2, 2.24 (3 s.f.)

5 a An estimator of a population parameter that will 'on average' give the correct value.

b $\bar{x} = 236, s^2 = 7.58$

6 $\bar{x} = 205$ (3 s.f.), $s^2 = 9.22$ (3 s.f.)

7 a $\mu = \frac{20}{3}, \sigma^2 = \frac{50}{9}$

b $\{5, 5, 5\}$
$\{5, 5, 10\}^{\times 3} \{5, 10, 10\}^{\times 3}$
$\{10, 10, 10\}$

c

$\bar{x}$	5	$\frac{20}{3}$	$\frac{25}{3}$	10
$P(\bar{X} = \bar{x})$	$\frac{8}{27}$	$\frac{12}{27}$	$\frac{6}{27}$	$\frac{1}{27}$

d $E(\bar{X}) = \dfrac{20}{3} = \mu, Var(\bar{X}) = \dfrac{50}{27} = \dfrac{\sigma^2}{3}$

e

m	5	10
P($M = m$)	$\frac{20}{27}$	$\frac{7}{27}$

f E(M) = 6.296...
Var(M) = 4.80...

g Bias = 1.30 (3 s.f.)

8 a $10p$

b $\bar{X} = \dfrac{X_1 + ... + X_{25}}{25}$

$E(\bar{X}) = \dfrac{E(X_1) + ... + E(X_{25})}{25} = \dfrac{25 \times 10p}{25} = 10p$

∴ $\bar{X}$ is a biased estimator of p
so bias = $10p - p = 9p$

c $\dfrac{\bar{X}}{10}$ is an unbiased estimator of p.

9 a E(X) = 0
∴ E(X^2) = $\dfrac{\alpha^2}{3}$

b $Y = X_1^2 + X_2^2 + X_3^2$
$E(Y) = E(X_1^2) + E(X_2^2) + E(X_3^2) = \dfrac{\alpha^2}{3} \times 3 = \alpha^2$
∴ Y is an unbiased estimator of α^2

10 a $\bar{y} = 16.2$, $s_y^2 = 12.0$ (3 s.f.)

b $w = 15.92$, $s_w^2 = 10.34$

c Standard error is a measure of the statistical accuracy of an estimate.

d $\dfrac{s_x}{\sqrt{20}} = 0.632$ (3 s.f.), $\dfrac{s_y}{\sqrt{30}} = 0.633$ (3 s.f.),

$\dfrac{s_w}{\sqrt{50}} = 0.455$ (3 s.f.)

e Prefer to use w since it is based on a larger sample size and has smallest standard error.

11 a $\bar{x} = 65$, $s_x^2 = 9.74$ (3 s.f.)

b need a sample of 37 or more

c No. Because the recommendation is based on the assumed value of s^2 from the original sample OR The value of s^2 for the new sample might be different/larger.

d 65.6 (3 s.f.)

12 Need a sample of 28 (or more)

13 a 4.89

b 0.0924 (3 s.f.)

c Need $n = 35$ (or more)

14 a E(X_1) = np, E(X_2) = $2np$, Var(X_1) = $np(1-p)$,
Var(X_2) = $2np(1-p)$

b Prefer $\dfrac{X_2}{2n}$ since based on larger sample (and therefore will have smaller variance)

c $X = \dfrac{1}{2}\left(\dfrac{X_1}{n} + \dfrac{X_2}{2n}\right)$

$\Rightarrow E(X) = \dfrac{1}{2}\left(\dfrac{E(X_1)}{n} + \dfrac{E(X_2)}{2n}\right) = \dfrac{1}{2}\left(\dfrac{np}{n} + \dfrac{2np}{2n}\right)$

$= \dfrac{1}{2}(p + p) = p$

∴ X is an unbiased estimator of p

d $Y = \left(\dfrac{X_1 + X_2}{3n}\right)$

$\Rightarrow E(Y) = \dfrac{E(X_1) + E(X_2)}{3n} = \dfrac{np + 2np}{3n} = p$

∴ Y is an unbiased estimator of p

e Var(Y) is smallest so Y is the best estimator.

f $\dfrac{p}{3}$

15 a $\mu = 1$, $\sigma^2 = 0.8$ or $\frac{4}{5}$

b {0, 0, 0} {0, 0, 1}×3 {0, 0, 2}×3
{1, 1, 1} {1, 1, 0}×3 {1, 1, 2}×3
{2, 2, 2} {2, 2, 0}×3 {2, 2, 1}×3 {0, 1, 2}×3!=6

c

$\bar{x}$	0	$\frac{1}{3}$	$\frac{2}{3}$	1	$\frac{4}{3}$	$\frac{5}{3}$	2
P($\bar{X} = \bar{x}$)	$\frac{8}{125}$	$\frac{12}{125}$	$\frac{30}{125}$	$\frac{25}{125}$	$\frac{30}{125}$	$\frac{12}{125}$	$\frac{8}{125}$

d E($\bar{X}$) = 1 (= μ)

Var($\bar{X}$) = $\dfrac{4}{15} = \left(\dfrac{\sigma^2}{3}\right)$

e

n	0	1	2
P($N = n$)	$\frac{44}{125}$	$\frac{37}{125}$	$\frac{44}{125}$

f E(N) = 1
Var(N) = $0 + 1^2 \times \frac{37}{125} + 2^2 \times \frac{44}{125} - 1^2 = \frac{88}{125}$ (= σ^2)

g E(N) = 1 = μ

h $\bar{X}$ because Var($\bar{X}$) is smaller.

Challenge

a $\dfrac{1}{n-1}\sum_{i=1}^{n}(x_i - \bar{x})^2 = \dfrac{S_{xx}}{n-1} = \dfrac{n}{n-1}\left(\dfrac{\sum x^2}{n} - \bar{x}^2\right)$

$= \dfrac{1}{n-1}(\sum x^2 - n\bar{x}^2)$

b $\sigma^2 = \text{Var}(X) = E(X^2) - \mu^2$

$E(X^2) = \sigma^2 + \mu^2$ \hfill (1)

$\text{Var}(\bar{X}) = \dfrac{\sigma^2}{n}$ and E($\bar{X}$) = μ

$\dfrac{\sigma^2}{n} = E(\bar{X}^2) - \mu^2$

$E(\bar{X}^2) = \dfrac{\sigma^2}{n} + \mu^2$ \hfill (2)

$S^2 = \dfrac{1}{n-1}\left(\sum X^2 - n\bar{X}^2\right)$

$E(S^2) = \dfrac{1}{n-1} E\left(\sum X^2 - n\bar{X}^2\right)$

$= \dfrac{1}{n-1}\left(E(\sum X^2) - nE(\bar{X}^2)\right)$

$E(\sum X^2) = \sum E(X^2) = nE(X^2)$

$E(S^2) = \dfrac{1}{n-1}\left(E(X^2) - nE(\bar{X}^2)\right)$

$= \dfrac{1}{n-1}\left(n(\sigma^2 + \mu^2) - n\left(\dfrac{\sigma^2}{n} + \mu^2\right)\right)$ by (1) and (2)

$= \sigma^2$

So the statistic S^2 is an unbiased estimator of the population variance σ^2, and s^2 is an unbiased estimate for σ^2.

Exercise 5B

1 a (124, 132) **b** (123, 133)

2 a (83.7, 86.3) **b** (83.4, 86.6)

3 a Niall must assume that the underlying population is normally distributed.

b (20.6, 25.4)

4 a $n = 609$ **b** $n = 865$ **c** $n = 1493$

5 a A 95% confidence interval for a population parameter θ is one where there is a probability of 0.95 that the interval found contains θ.

b (1.76, 2.04)

6 a (304, 316) (3 s.f.)   **b** 0.0729
7 a Yes, Amy is correct. By the central limit theorem, for a large sample size, underlying population does not need to be normally distributed.
  b (73 113, 78 631) (nearest integer)
    or (73 100, 78 600) (3 s.f.)
8 a Must assume that these students form a random sample or that they are representative of the population.
  b (66.7, 70.1)
  c If $\mu = 65.3$ that is outside the C.I. so the examiner's sample was not representative. The examiner marked more better than average candidates.
9 a $E(H) = \mu$, $Var(H) = \dfrac{20^2}{12} = \dfrac{400}{12} = \dfrac{100}{3}$
  b (77.7, 79.7) (3 s.f.)
10 a (23.2, 26.8) is 95% C.I. since it is the narrower interval.
  b 0.918   **c** 25
11 a (130, 140)   **b** 85%   **c** Need $n = 189$ or more
12 (30.4, 32.4) (3 s.f.)
13 (258, 274) (3 s.f.)
14 a 0.311   **b** 0.866   **c** (21.8, 23.3) (3 s.f.)

Exercise 5C

1 H_0: $\mu_1 = \mu_2$ H_1: $\mu_1 > \mu_2$ 5% c.v. is $z = 1.6449$
  t.s. $= Z = \dfrac{(23.8 - 21.5) - 0}{\sqrt{\frac{5^2}{15} + \frac{4.8^2}{20}}} = 1.3699\ldots$
  $1.3699 < 1.6449$ so result is not significant, accept H_0.
2 H_0: $\mu_1 = \mu_2$ H_1: $\mu_1 \neq \mu_2$ 5% c.v. is $z = \pm 1.96$
  t.s. $= Z = \dfrac{(51.7 - 49.6) - 0}{\sqrt{\frac{4.2^2}{30} + \frac{3.6^2}{25}}}$
  [Choose $\bar{x}_2 - \bar{x}_1$ to get $z > 0$]
  t.s. $Z = 1.996\ldots > 1.96$ so result is significant. Reject H_0.
3 H_0: $\mu_1 = \mu_2$ H_1: $\mu_1 < \mu_2$ 1% c.v. is $z = -2.3263$
  t.s. $= Z = \dfrac{(3.62 - 4.11) - 0}{\sqrt{\frac{0.81^2}{25} + \frac{0.75^2}{36}}} = -2.3946\ldots$
  t.s. $= -2.3946\ldots < -2.3263$ so result is significant Reject H_0.
4 H_0: $\mu_1 = \mu_2$ H_1: $\mu_1 \neq \mu_2$ 1% c.v. is $z = 62.575_8$
  t.s. $= Z = \dfrac{(112.0 - 108.1) - 0}{\sqrt{\frac{8.2^2}{85} + \frac{11.3^2}{100}}} = 2.712\ldots > 2.5758$
  significant result so reject H_0.
  Central limit theorem applies since n_1, n_2 are large and enables you to assume $\bar{X}_1$ and $\bar{X}_2$ are both normally distributed.
5 H_0: $\mu_1 = \mu_2$ H_1: $\mu_1 > \mu_2$ 5% c.v. is $z = 1.96$
  t.s. $= Z = \dfrac{(72.6 - 69.5) - 0}{\sqrt{\frac{18.3^2}{100} + \frac{15.4^2}{150}}} = 1.396\ldots < 1.96$
  Result is not significant so accept H_0.
  Central limit theorem applies since n_1, n_2 are both large and enables you to assume $\bar{X}_1$ and $\bar{X}_2$ are normally distributed.
6 H_0: $\mu_1 = \mu_2$ H_1: $\mu_1 < \mu_2$ 1% c.v. is $z = -2.3263$
  t.s. $= Z = \dfrac{(0.863 - 0.868) - 0}{\sqrt{\frac{0.013^2}{120} + \frac{0.015^2}{90}}}$
  $= -2.5291\ldots , -2.3263$
  Result is significant so reject H_0.
  Central limit theorem is used to assume $\bar{X}_1$ and $\bar{X}_2$ are normally distributed since both samples are large.

7 Not significant. There is insufficient evidence to suggest that the machines are producing pipes of different lengths.
8 a H_0: $\mu_{\text{new}} - \mu_{\text{old}} = 1$; H_1: $\mu_{\text{new}} - \mu_{\text{old}} > 1$
    Test statistic $= 3.668\ldots > $ c.v. (1.6449), therefore evidence that yes, the mean yield is more than 1 tonne greater.
  b Mean yield is normally distributed; Sample size is large.
9 a H_0: $\mu_{\text{grain}} - \mu_{\text{grain/grass}} = 0$; H_1: $\mu_{\text{grain}} \neq \mu_{\text{grain/grass}}$
    Test statistic $= 2.376\ldots > $ c.v. (1.96), therefore evidence that there is a difference between the mean fat content of the milk of cows fed on these two diets.
  b Mean fat content is normally distributed.

Challenge

a $\hat{\mu} = \dfrac{\sum\limits_{1}^{n_x} x_i + \sum\limits_{1}^{n_y} y_i}{\sum n} = \dfrac{n_x \bar{x} + n_y \bar{y}}{n_x + n_y}$

b (45.76, 47.33)

Exercise 5D

1 $2.34 > 1.6449$ so the result is significant.
  There is evidence that Quickdry dries faster than Speedicover.
2 a Not significant. There is insufficient evidence to confirm that mean expenditure in the week is more than at weekends.
  b We have assumed that $s_1 = \sigma_1$ and $s_2 = \sigma_2$.
3 a Not significant. Insufficient evidence to support a change in mean mass.
  b We have assumed that $s = \sigma$ since n is large.
4 a Not significant so accept H_0.
  b t.s. $= 1.6535\ldots > 1.6449$
    Significant so reject H_0.
  c We have assumed that $s_A = \sigma_A$ and $s_B = \sigma_B$ since the samples are both large.
5 t.s. $= -1.944\ldots < -1.6449$
  Significant. There is evidence that the weights of chocolate bars are less than the stated value.
6 a Result is significant. There is evidence of difference in mean age of first-time mothers between these two dates.
  b There is no need to have to assume that both populations were normally distributed since both samples were large so the central limit theorem allows you to assume both sample means are normally distributed.
    We have assumed that $s_1 = \sigma_1$ and $s_2 = \sigma_2$.

Mixed exercise 5

1 a H_0: $\mu = 0.48$, H_1 $\mu \neq 0.48$; Significance level = 10%; 0.48 is in confidence interval so accept H_0.
  b (0.4482, 0.5158)
2 a **i** and **iii** are since they only contain known data; **ii** is not since it contains unknown population parameters.
  b μ; $\dfrac{17\sigma^2}{9}$
3 a t.s. $= 2.645 . 1.6449$
    Result is significant so reject H_0.
    There is evidence that the new bands are better.
  b (46.7, 47.6) (3 s.f.)
4 a 4.53, 0.185 (3 d.p.)   **b** (4.10, 4.96) (3 s.f.)
  c 0.3520

Online Full worked solutions are available in SolutionBank.

5 14.01, 0.04 (2 d.p.)

6 The smallest value of n is 14.

7 a (41.1, 47.3) (3 s.f.)

  b t.s. $= -1.375 > -1.6449$
Not significant so accept H_0. There is insufficient evidence to support the headteachers' claim.

8 a $\bar{X} \sim N\left(\mu, \dfrac{\sigma^2}{n}\right)$

  b Exact because X is normally distributed.

  c need $n = 28$ or more

9 a 0.75, 4.84

  b i assume that $\bar{X}$ has a normal distribution
    ii assume that the sample was random

  c (−0.346, 1.85) (3 s.f.)

  d Since 0 is in the interval it is reasonable to assume that trains do arrive on time.

10 a $\bar{X} \sim N\left(\mu, \dfrac{\sigma^2}{n}\right)$

  b 95% C.I. is an interval within which we are 95% confident m lies.

  c (3.37, 14.1) (3 s.f.)

  d (9.07, 10.35) (nearest penny)

  e t.s. $= 1.8769\ldots . 1.6449$
Significant so reject H_0. There is evidence that the mean sales of unleaded petrol in 1990 were greater than in 1989.

  f $n = 163$

11 a A 98% C.I. is an interval within which we are 98% sure the population mean will lie.

  b 0.182 (3 d.p.)

  c (9.84, 10.56) (2 d.p.)

  d $n = 13$

12 a $E(\bar{X}) = \mu$, $Var(\bar{X}) = \dfrac{\sigma^2}{n}$

  b i $\bar{X} \sim N\left(\mu, \dfrac{\sigma^2}{n}\right)$ **ii** $\bar{X} \sim N\left(\mu, \dfrac{\sigma^2}{n}\right)$

  c (17.7, 19.3) (3 s.f.)

  d $n = 278$ or more

  e t.s. $= -2.1176\ldots < -1.96$
Significant so reject H_0. There is evidence that the mean of till receipts in 2014 is different from the mean value in 2013.

13 a Approx 83 samples will have mean < 0.823

  b (0.823, 0.845) (3 s.f.)

  c Since $0.82\dot{4}$ is in the C.I. we can conclude that there is insufficient evidence of a malfunction.

14 a $H_0: \mu_d = \mu_w$; $H_1: \mu_d > \mu_w$; critical value is 2.6512 which is greater than the sig level (1.6449) so reject H_0 – cardiologist's claim is supported.

  b Assume normal distribution or assume sample sizes large enough to use the central limit theorem; Assume individual results are independent; Assume $\sigma^2 = s^2$ for both populations.

  c 58.5 (3 s.f.)

Challenge

a $T = r\bar{X} + s\bar{Y}$
$E(T) = r\mu + s\mu = (r + s)\mu$
So if T is unbiased then $r + s = 1$

b $r + s = 1 \Rightarrow s = 1 - r$
$\therefore T = r\bar{X} + (1 - r)\bar{Y}$
$Var(T) = r^2 Var(\bar{X}) + (1 - r)^2 Var(\bar{Y})$
$\quad = r^2\dfrac{\sigma^2}{n} + (1 - r)^2\dfrac{\sigma^2}{m} = \sigma^2\left(\dfrac{r^2}{n} + \dfrac{(1 - r)^2}{m}\right)$

c $\dfrac{d}{dr}Var(T) = \left(\sigma^2\dfrac{2r}{n} + \dfrac{2(1 - r)(-1)}{m}\right)$

$\dfrac{d}{dr}Var(T) = 0 \Rightarrow rm = (1 - r)n \Rightarrow r(m + n) = n \Rightarrow r = \dfrac{n}{m + n}$

d $\dfrac{n\bar{X} + m\bar{Y}}{m + n}$

CHAPTER 6
Prior knowledge check

1 a 26.4   **b** 13.8

2 1.690

3 (16.0, 19.0)

Exercise 6A

1 Confidence interval $= \left(\dfrac{(n-1)s^2}{\chi^2_{n-1}\left(\frac{\alpha}{2}\right)}, \dfrac{(n-1)s^2}{\chi^2_{n-1}\left(1-\frac{\alpha}{2}\right)}\right)$

$= \left(\dfrac{14 \times 4.8}{26.119}, \dfrac{14 \times 4.8}{5.628}\right)$

$= (2.573, 11.938)$

2 $\bar{x} = 6.62$ $s^2 = \dfrac{1}{19}\left(884.3 - \dfrac{132^2}{20}\right) = 0.4111\ldots$

Confidence interval $= \left(\dfrac{(n-1)s^2}{\chi^2_{n-1}\left(\frac{\alpha}{2}\right)}, \dfrac{(n-1)s^2}{\chi^2_{n-1}\left(1-\frac{\alpha}{2}\right)}\right)$

$= \left(\dfrac{19 \times 0.411\ldots}{30.114}, \dfrac{19 \times 0.411\ldots}{10.117}\right)$

$= (0.259, 0.772)$

3 $\bar{x} = 2.878\ldots$ $s^2 = 0.458\ldots$

Confidence interval $= \left(\dfrac{(n-1)s^2}{\chi^2_{n-1}\left(\frac{\alpha}{2}\right)}, \dfrac{(n-1)s^2}{\chi^2_{n-1}\left(1-\frac{\alpha}{2}\right)}\right)$

$= \left(\dfrac{13 \times 0.458\ldots}{24.736}, \dfrac{13 \times 0.458\ldots}{5.009}\right)$

$= (0.241, 1.191)$

4 (0.731, 16.835)

5 a (22.099, 112.45)   **b** Normal distribution

6 (148.137, 1043.704)

7 6972.0

8 (5.04, 8.18)

9 a (0.1684, 0.7625)

  b Diameters of fastenings have an underlying normal distribution.

  c Lower limit of CI > 0.15 therefore the machine needs recalibrating.

Exercise 6B

1 a $\bar{x} = 16.605$ $s^2 = \dfrac{5583.63 - 20(16.605)^2}{19} = 3.637\ldots$

  b $H_0 : \sigma^2 = 1.5$ $H_1 : \sigma^2 > 1.5$
Critical region $\geqslant 30.144$

Test statistic $= \dfrac{(n-1)s^2}{\sigma^2} = \dfrac{19 \times 3.637\ldots}{1.5} = 46.073$

The test statistic is in the critical region so reject H_0. There is evidence to suggest $\sigma^2 > 1.5$.

2 $\bar{x} = 0.337$ $s^2 = 0.00286\ldots$
$H_0 : \sigma^2 = 1.5$ $H_1 : \sigma^2 < 0.09$
Critical region $\leqslant 2.700$

Test statistic $= \dfrac{(n-1)s^2}{\sigma^2} = \dfrac{9 \times 0.00286\ldots}{0.09} = 0.287$

The test statistic is in the critical region so reject H_0. There is evidence to suggest that variance is less than 0.09.

3 $H_0 : \sigma^2 = 4.1$ $H_1 : \sigma^2 \neq 4.1$
$\bar{x} = 5.74$ $s^2 = 6.940...$
Critical region $\leqslant 2.7$ and $\geqslant 19.023$
Test statistic $= \dfrac{(n-1)s^2}{\sigma^2} = \dfrac{9 \times 6.940...}{4.1} = 15.235$
The test statistic is not in the critical region so do not reject H_0.
There is no evidence the variance does not equal 4.1.

4 $H_0 : \sigma^2 = 1.12^2$ $H_1 : \sigma^2 \neq 1.12^2$
Critical region $\leqslant 8.907$ and $\geqslant 32.852$
Test statistic $= \dfrac{(n-1)s^2}{\sigma^2} = \dfrac{19 \times 1.15...}{1.12^2} = 17.419$
The test statistic is not in the critical region so do not reject H_0.
There is no evidence the variance does not equal 1.12.

5 a $\bar{x}$ is an unbiased estimate for μ.
s^2 is an unbiased estimate for σ^2.
$\bar{x} = \dfrac{149.941}{15} = 9.996...,$
$s^2 = \dfrac{1498.83 - 15 \times 9.996...^2}{14} = 0.0006977...$

b $H_0 : \sigma^2 = 0.04$ $H_1 : \sigma^2 \neq 0.04$
Critical region $\leqslant 5.629$ and $\geqslant 26.119$
Test statistic $= \dfrac{(n-1)s^2}{\sigma^2} = \dfrac{14 \times 0.0006977}{0.04}$
$= 0.244$
The test statistic is in the critical region so reject H_0.
There is evidence that the variance is not 0.04

6 a $s^2 = 0.06125$

b $H_0 : \sigma^2 = 0.19$ $H_1 : \sigma^2 \neq 0.19$
Critical region $\leqslant 2.167$ and $\geqslant 14.067$
Test statistic $= \dfrac{(n-1)s^2}{\sigma^2} = \dfrac{7 \times 0.06125}{0.19} = 2.256$
The test statistic is not in the critical region so do not reject H_0.
There is no evidence that σ^2 does not equal 0.19.

7 a $H_0 : \sigma^2 = 110.25$ $H_1 : \sigma^2 < 110.25$
Critical region $\leqslant 10.117$
Test statistic $= \dfrac{(n-1)s^2}{\sigma^2} = \dfrac{19 \times 8.5^2}{110.25} = 12.451$
The test statistic is not in the not critical region so do not reject H_0.
There is no evidence that the variance has reduced.

b Take a larger sample before committing to the new component.

8 a 3.212, standard error = 0.0875 (3 s.f.)

b $H_0: \sigma = 0.25$, $H_1: \sigma \neq 0.25$, C.I. = (2.700, 19.023),
Test statistic = 11.03616. Hence not enough evidence to show that the standard deviation is different from 0.25.

Exercise 6C

1 a 2.34   **b** 3.36   **c** 3.37

2 a $\dfrac{1}{F_{8,6}} = 0.241$ **b** $\dfrac{1}{F_{12,25}} = 0.463$ **c** $\dfrac{1}{F_{5,5}} = 0.198$

3 a 3.37   **b** 4.20   **c** 6.06

4 a $\dfrac{1}{F_{12,3}} = 0.0370$ **b** $\dfrac{1}{F_{12,8}} = 0.176$ **c** $\dfrac{1}{F_{12,5}} = 0.101$

5 a $3.07, \dfrac{1}{F_{10,8}} = 0.299$ **b** $2.91, \dfrac{1}{F_{10,12}} = 0.364$

c $5.41, \dfrac{1}{F_{5,3}} = 0.111$

6 a $P(X < 0.5) = P(F_{40,12} < 0.5)$
$= P\left(F_{12,40} > \dfrac{1}{0.5}\right)$
$= P(F_{12,40} > 2)$
From the tables $F_{12,40}(0.05) = 2$
$\therefore P(F_{12,40} > 2) = P(F_{40,12} < 0.5) = 0.05$

7 $P(X < 3.28) = 1 - P(F_{12,8} > 3.28)$
$= 1 - 0.05 = 0.95$
$P\left(X < \dfrac{1}{2.85}\right) = P\left(F_{12,8} < \dfrac{1}{2.85}\right)$
$= P(F_{8,12} > 2.85)$
$\therefore P\left(X < \dfrac{1}{2.85}\right) = 0.05$
$P\left(\dfrac{1}{2.85} < X < 5.06\right) = P(X < 5.06) - P\left(X < \dfrac{1}{2.85}\right)$
$= 0.95 - 0.05$
$= 0.90$

8 $P(X < 9.55) = 1 - P(F_{2,7} > 9.55)$
$= 1 - 0.01$
$= 0.99$

9 a Upper tail: $P(X > 3) = P(F_{6,12} > 3) = 0.05$
So $P(X < 3) = 1 - 0.05 = 0.95$
Other tail:
$P(X < 0.25) = P(F_{6,12} < 0.25) = P(F_{12,6} > 4) = 0.05$
So $P(0.25 < X < 3) = 0.95 - 0.05 = 0.9$

b $^6C_2(0.9)^2(0.1)^4 \times 0.9 = 0.00109$

Exercise 6D

1 Critical value is $F_{10,6} = 4.06$
$F_{\text{test}} = \dfrac{7.6}{6.4} = 1.1875$
Not in critical region.
Accept H_0 – there is evidence to suggest that $\sigma_1^2 = \sigma_2^2$

2 Critical value is $F_{24,40} = 2.29$
$F_{\text{test}} = \dfrac{0.42}{0.17} = 2.4706$
In critical region.
Reject H_0 – there is evidence to suggest that $\sigma_1^2 > \sigma_2^2$

3 a Critical value is $F_{12,8} = 3.28$
$F_{\text{test}} = \dfrac{225}{63} = 3.57$
In critical region.
Reject H_0 – there is evidence to suggest that the machines differ in variability.

b Population distributions are assumed to be normal.

4 Critical value is $F_{8,12} = 2.85$
$F_{\text{test}} = \dfrac{52.6}{36.4} = 1.445$
Not in critical region.
Accept H_0 – there is evidence to suggest that $\sigma_1^2 = \sigma_2^2$

5 a $\sigma_{\text{Goodstick}}^2 = 1.363$
$\sigma_{\text{Holdtight}}^2 = 0.24167$
Critical value is $F_{4,5} = 5.19$
$F_{\text{test}} = \dfrac{1.363}{0.24167} = 5.64$
In critical region.
Reject H_0 – there is evidence to suggest that the variances are not equal.

b Holdtight as it is less variable and cheaper.

6 $\sigma_{\text{Chegrit}}^2 = 22\,143.286$
$\sigma_{\text{Dicabalk}}^2 = 6570.85238$
Critical value is $F_{6,14} = 2.85$
$F_{\text{test}} = \dfrac{22\,143.286}{6570.85238} = 3.3699$
In critical region. Reject H_0 – there is evidence to suggest that their variances differ.

7 a $\mu_1 = 1046$, $s_2^1 = 1818.11$ and $\mu_2 = 997.75$, $s_2^2 = 1200.21$

b Critical value is $F_{8,7} = 3.73$

$F_{\text{test}} = \dfrac{1818.111}{1200.21} = 1.5148$

Not in critical region.

Accept H_0 – there is evidence to suggest that $\sigma_1^2 = \sigma_2^2$

c Given that the variances appear to be equal, use present supplier who appears to have a higher mean.

Mixed exercise 6

1 a Confidence interval $= \left(\dfrac{(n-1)s^2}{\chi^2_{n-1}\left(\frac{\alpha}{2}\right)}, \dfrac{(n-1)s^2}{\chi^2_{n-1}\left(1-\frac{\alpha}{2}\right)} \right)$

$= \left(\dfrac{13 \times 1.8}{24.736}, \dfrac{13 \times 1.8}{5.009} \right)$

$= (0.946, 4.672)$

b Confidence interval $= \left(\dfrac{(n-1)s^2}{\chi^2_{n-1}\left(\frac{\alpha}{2}\right)}, \dfrac{(n-1)s^2}{\chi^2_{n-1}\left(1-\frac{\alpha}{2}\right)} \right)$

$= \left(\dfrac{13 \times 1.8}{22.362}, \dfrac{13 \times 1.8}{5.892} \right)$

$= (1.046, 3.971)$

2 $\bar{x} = \dfrac{1428}{20} = 71.4$, $\quad s^2 = \dfrac{102\,286 - 20 \times 71.4^2}{19} = 17.2$

a Confidence interval $= \left(\dfrac{(n-1)s^2}{\chi^2_{n-1}\left(\frac{\alpha}{2}\right)}, \dfrac{(n-1)s^2}{\chi^2_{n-1}\left(1-\frac{\alpha}{2}\right)} \right)$

$= \left(\dfrac{19 \times 17.2}{32.852}, \dfrac{19 \times 17.2}{8.907} \right)$

$= (9.948, 36.69)$

b $10 = 1.6449\sigma$ so $\sigma = \dfrac{10}{1.6449} = 6.079$

c $\sqrt{36.69} < 6.079$ so the supervisor should not be concerned.

3 a A confidence interval for a population parameter is a range of values defined so that there is a specific probability that the true value of the parameter lies within that range.

b $(2.98, 5.14)$

4 $H_0: \sigma = 2.7$, $H_1: \sigma \neq 2.7$

Critical regions are

$\dfrac{(n-1)s^2}{\sigma^2} \geq 14.449$ and $\dfrac{(n-1)s^2}{\sigma^2} \leq 1.237$.

Test statistic $= 6.58$

6.58 is not in the critical region, so not enough evidence to show that the standard deviation is different from 2.7.

5 a $(3.266, 8.669)$

b Times taken have an underlying normal distribution

c Lower limit of C.I. > 3.1 therefore the dosage needs changing.

6 $\bar{x} = 45.1$ $\quad s = 6.838\ldots$

$H_0: \sigma = 5$ $\quad H_1: \sigma \neq 5$

Critical region ≥ 19.023 and ≤ 2.700

Test statistic $= \dfrac{(n-1)s^2}{\sigma^2} = \dfrac{9 \times 6.838\ldots^2}{5^2} = 16.836$

There is insufficient evidence to reject H_0, therefore there is evidence that the variance has not altered.

7 $P(F_{5,10} \geq 3.33) = 0.05 \Rightarrow b = 3.33$

$P(F_{10,5} \geq 4.74) = 0.05 \Rightarrow P\left(F_{5,10} \leq \dfrac{1}{4.74}\right) = 0.05$

so $a = 0.2110$ (4 s.f.)

8 a $H_0: \sigma_1^2 = \sigma_2^2$; $H_1: \sigma_1^2 \neq \sigma_2^2$

$\dfrac{s_1^2}{s_2^2} = \dfrac{14^2}{8^2} = 3.0625$

C.V.: $F_{12,7} = 3.57$

Since 3.0625 is not in the critical region, there is insufficient evidence to reject H_0. There is insufficient evidence of a difference in the variances of the lengths of the fence posts.

b The distribution of the population of lengths of fence posts is normally distributed.

9 $H_0: \sigma_F^2 = \sigma_M^2$ $\quad H_1: \sigma_F^2 \neq \sigma_M^2$

$s_F^2 = \dfrac{1}{6}(17\,956.5 - 7 \times 50.6^2) = \dfrac{33.98}{6} = 5.66333\ldots$

$s_M^2 = \dfrac{1}{9}(28\,335.1 - 10 \times 53.2^2) = \dfrac{32.7}{9} = 3.63333\ldots$

$\dfrac{s_F^2}{s_M^2} = 1.5587\ldots$

$F_{6,9} = 3.37$

Not in critical region. There is no reason to doubt the variances of the two distributions are the same.

Challenge

a $\dfrac{(n-1)S^2}{\sigma^2} \sim \chi^2_{n-1}$

$\text{Var} = \left(\chi^2_{n-1}\right) = \text{Var}\left(\dfrac{(n-1)S^2}{\sigma^2}\right) = 2(n-1)$

Take out factor from variance and square

$\dfrac{(n-1)^2 \text{Var}(S^2)}{(\sigma^4)} = 2(n-1)$

$\Rightarrow \text{Var}(S^2) = \dfrac{2\sigma^4}{n-1}$

b Variance decreases as n increases.

CHAPTER 7
Prior knowledge check

1 [15.12, 16.88]

2 $z = 1.953\ldots$ so no evidence ($z < 1.96$).

Exercise 7A

1 a $P(X > t) = 0.025$ when $t = 2.179$ so $P(X < t) = 0.025$ when $t = -2.179$

b $P(X > t) = 0.025$ when $t = 1.782$

c $P(X > t) = 0.025$ when $t = 2.179$

$P(|X| > t) = 0.95$ when $|t| = 2.179$

2 a 2.479 $\quad\quad$ **b** 1.706

3 a $P(Y > t) = 0.05$ when $t = 1.812$ so $P(Y < t) = 0.95$ when $t = 1.812$

b $P(Y > t) = 0.005$ when $t = 2.738$ so $P(Y > t) = 0.005$ when $t = -2.738$

c $P(Y > t) = 0.025$ when $t = 2.571$ so $P(Y < t) = 0.025$ when $t = -2.571$

d $P(Y > t) = 0.01$, when $t = 2.583$, and $P(Y < t) = 0.01$ when $t = -2.583$

so $P(|Y| < t) = 0.98$ when $|t| = 2.583$

e $P(Y > t) = 0.05$ when $t = 1.734$ and $P(Y < t) = 0.05$ when $t = -1.734$

so $P(|Y| > t) = 0.10$ when $|t| = 1.734$

4 $\bar{x} = 20.95$ $s = 3.4719\ldots$ $n = 8$ $v = 7$

confidence limits $= x \pm t_{(n-1)}\left(\frac{\alpha}{2}\right) \times \frac{s}{\sqrt{n}}$

$$= 20.95 \pm 1.895 \times \frac{3.4719\ldots}{\sqrt{8}}$$

$$= 18.624 \text{ and } 23.276$$

Confidence interval = (18.624, 23.276)

5 $\bar{x} = 12.4$ $s = \sqrt{21}$ $n = 16$ $v = 15$

confidence limits $= \bar{x} \pm t_{(n-1)}\left(\frac{\alpha}{2}\right) \times \frac{s}{\sqrt{n}}$

$$= 12.4 \pm 2.131 \times \frac{\sqrt{21}}{\sqrt{16}}$$

$$= 9.9586 \text{ and } 14.8413\ldots$$

Confidence interval = (9.959, 14.841)

6 a $\bar{x} = 179.333333$ $s = 5.5015\ldots$ $n = 6$ $v = 5$

confidence limits $= \bar{x} \pm t_{(n-1)}\left(\frac{\alpha}{2}\right) \times \frac{s}{\sqrt{n}}$

$$= 179.333 \pm 2.015 \times \frac{5.5015\ldots}{\sqrt{6}}$$

$$= 174.808 \text{ and } 183.859$$

Confidence interval = (174.808, 183.859)

b confidence limits $= \bar{x} \pm t_{(n-1)}\left(\frac{\alpha}{2}\right) \times \frac{s}{\sqrt{n}}$

$$= 179.333 \pm 2.571 \times \frac{5.5015\ldots}{\sqrt{6}}$$

$$= 173.559 \text{ and } 185.108$$

Confidence interval = (173.559, 185.108)

7 a $\bar{x} = 10.36$ $s = 0.73363\ldots$ $n = 10$ $v = 9$

confidence limits $= \bar{x} \pm t_{(n-1)}\left(\frac{\alpha}{2}\right) \times \frac{s}{\sqrt{n}}$

$$= 10.36 \pm 2.821 \times \frac{0.73363\ldots}{\sqrt{10}}$$

$$= 9.706 \text{ and } 11.014$$

Confidence interval = (9.706, 11.014)

b Masses are normally distributed.

8 $\bar{x} = \frac{224.1}{8} = 28.0125$ $s^2 = \frac{1}{7}\left(6337.39 - \frac{224.1^2}{8}\right)$

$$= 8.54125$$

$s = 2.92254\ldots$ $n = 8$ $v = 7$

confidence limits $= \bar{x} \pm t_{(n-1)}\left(\frac{\alpha}{2}\right) \times \frac{s}{\sqrt{n}}$

$$= 28.0125 \pm 3.499 \times \frac{2.92254\ldots}{\sqrt{8}}$$

$$= 24.397 \text{ and } 31.628$$

Confidence interval = (24.397, 31.628)

9 $\bar{x} = 122$ $s = \sqrt{225} = 25$ $v = 25$

confidence limits $= \bar{x} \pm t_{(n-1)}\left(\frac{\alpha}{2}\right) \times \frac{s}{\sqrt{n}}$

$$= 122 \pm 2.060 \times \frac{\sqrt{225}}{\sqrt{26}}$$

$$= 115.940 \text{ and } 128.060$$

Confidence interval = (115.94, 128.06)

10

	Normal	χ^2	t
For the population mean, using a sample of size 50 from a population of unknown variance			✓
For the population mean, using a sample of size 6 from a population of known variance	✓		
For the population variance, using a sample of size 20		✓	

11 a Population variance is unknown
b (468.7, 509.7)
c The lifespan is normally distributed
d (24.19, 112.2)

Exercise 7B

1 $\bar{x} = 11.4$ $s = 1.816\ldots$

$H_0 : \mu = 11$ $H_0 : \mu > 11$

Critical region $t > 2.132$

Test statistic $t = \dfrac{\bar{x} - \mu}{\frac{s}{\sqrt{n}}} = \dfrac{11.4 - 11.0}{\frac{1.816\ldots}{\sqrt{5}}} = 0.492$

The result is not in critical region.
No evidence that μ is not 11

2 $\bar{x} = 17.1$ $s = 2$

$H_0 : \mu = 19$ $H_1 : \mu < 19$

Critical region $t < -2.473$

Test statistic $t = \dfrac{\bar{x} - \mu}{\frac{s}{\sqrt{n}}} = \dfrac{17.1 - 19}{\frac{2}{\sqrt{28}}} = -5.027$

The result is in the critical region.
There is evidence that μ is < 19

3 $\bar{x} = 3.26$ $s = 0.8$

$H_0 : \mu = 3$ $H_1 : \mu \neq 3$

Critical values 6 2.179

Critical region $t < -2.179$ or $t > 2.179$

Test statistic $t = \dfrac{\bar{x} - \mu}{\frac{s}{\sqrt{n}}} = \dfrac{3.25 - 3}{\frac{0.8}{\sqrt{13}}} = 1.172$

The result is not in the critical region.
There is no evidence that μ is not 3

4 a Population variance is unknown.

b $\bar{x} = 98.2$ $s = 15.744\ldots$

$H_0 : \mu = 100$ $H_1 : \mu \neq 100$

Critical region < -2.145 or > 2.145

Test statistic $t = \dfrac{\bar{x} - \mu}{\frac{s}{\sqrt{n}}} = \dfrac{98.2 - 100}{\frac{15.74\ldots}{\sqrt{15}}} = -0.443$

The result is not in the critical region.
There is no evidence that μ is not 100

5 $\bar{x} = 1048.75$ $s = 95.2346\ldots$

$H_0 : \mu = 1000$ $H_1 : \mu > 1000$

Critical region $t > 1.895$

Test statistic $t = \dfrac{\bar{x} - \mu}{\frac{s}{\sqrt{n}}} = \dfrac{1048.75 - 1000}{\frac{95.234\ldots}{\sqrt{8}}} = 1.448$

The result is not in the critical region.
There is no evidence that μ is not 1000

6 $\bar{x} = 6.4857\ldots$ $s^2 = 0.853626\ldots$ $s = 0.923919\ldots$

$H_0 : \mu = 6$ $H_1 : \mu > 6$

Critical region $t > 2.160$

Test statistic $t = \dfrac{\bar{x} - \mu}{\frac{s}{\sqrt{n}}} = \dfrac{6.4857142 - 6}{\frac{0.923919\ldots}{\sqrt{14}}} = 1.967$

The result is not in the critical region.
There is no evidence supporting manufacturer's claim.

7 a $\bar{x} = 1.085$ $s^2 = \dfrac{28.4 - 20 \times 1.085^2}{19} = 0.2555\ldots$

$s = 0.5055\ldots$

$H_0 : \mu = 1.00$ $H_1 : \mu < 1.00$

Critical values $t < -1.328$

Online Full worked solutions are available in SolutionBank.

Test statistic $t = \dfrac{\bar{x} - \mu}{\frac{s}{\sqrt{n}}} = \dfrac{1.085 - 1}{\frac{0.5055\ldots}{\sqrt{20}}} = 0.752$

The result is not in the critical region.
There is no evidence that μ is not 1.00

b Amounts of radiation are normally distributed.

8 a H_0: $\mu = 100$; H_1, $\mu > 100$, Test statistic is 2.98... c.f. critical value of 1.729 therefore reject H_0 – there is evidence that the training improves IQ.

b H_0: $\sigma = 12$, H_1 $\sigma \neq 12$, CI is (11.91, 20.56) so since stated value is in CI, accept H_0.

Exercise 7C

1 a i $H_0 : \mu_D = 0$ $H_1 : \mu_D \neq 0$
  ii $H_0 : \mu_D = 0$ $H_1 : \mu_D > 0$

b $\sum d = 30$ $\sum d^2 = 238$
$\bar{d} = 5$
$s^2 = \dfrac{238 - 6(5)^2}{5} = 17.6$
$s = 4.195$
Critical value $t_5(5\%) = 2.015$
The critical region is $t > 2.015$
$t = \dfrac{5 - 0}{\frac{4.195}{\sqrt{6}}}$
$\quad = 2.919$
In the critical region, reject H_0.
There is evidence to suggest that there has been an increase in shorthand speed.

2 $H_0 : \mu_D = 0$ $H_1 : \mu_D > 0$
$\sum d = 5$ $\sum d^2 = 59$
$\bar{d} = 0.5$
$s^2 = \dfrac{59 - 10(0.5)^2}{9} = 6.278$
$s = 2.50555$
Critical value $t_9(1\%) = 2.821$
The critical region is $t > 2.821$
$t = \dfrac{0.5 - 0}{\frac{2.50555}{\sqrt{10}}}$
$\quad = 0.631$
Not in the critical region. Do not reject H_0.
There is evidence to suggest that paper 2 is easier than paper 1.

3 a $H_0 : \mu_D = 0$ $H_1 : \mu_D > 0$
$\sum d = 47$ $\sum d^2 = 315$
$\bar{d} = 4.7$
$s^2 = \dfrac{315 - 10(4.7)^2}{9} = 10.456$
$s = 3.234$
Critical value $t_9(5\%) = 1.833$
The critical region is $t > 1.833$
$t = \dfrac{4.7 - 0}{\frac{3.234}{\sqrt{10}}}$
$\quad = 4.596$
In the critical region. Reject H_0.
There is evidence to suggest that chewing the gum does not reduce the craving for cigarettes.

b The differences are normally distributed.

4 $H_0 : \mu_D = 0$ $H_1 : \mu_D > 0$
$\sum d = 46$ $\sum d^2 = 336$
$\bar{d} = 4.6$
$s^2 = \dfrac{336 - 10(4.6)^2}{9} = 13.8222$
$s = 3.7178$
Critical value $t_9(5\%) = 1.833$
The critical region is $t > 1.833$.
$t = \dfrac{4.6 - 0}{\frac{3.7178}{\sqrt{10}}} = 3.913$
In the critical region. Reject H_0. There is evidence to suggest that the journey times have decreased.

5 a $H_0 : \mu_D = 0$ $H_1 : \mu_D \neq 0$
$\sum d = 20$ $\sum d^2 = 604$
$\bar{d} = 2.5$
$s^2 = \dfrac{604 - 8(2.5)^2}{7} = 79.1429$
$s = 8.896$
Critical value $t_7(5\%) = 1.895$
The critical regions are $t > -1.895$ and $t > 1.895$
$t = \dfrac{2.5 - 0}{\frac{8.896}{\sqrt{8}}} = 0.795$
Not in the critical region. Do not reject H_0.
The mock examination is a good predictor.

b The differences are normally distributed.

6 a Different people will have different productivity rates. Need a common link if want to compare before and after. This reduces experimental error due to differences between individuals so that, if a difference does exist, it is more likely to be detected.

b $H_0 : \mu_D = 0$ $H_1 : \mu_D > 0$
$\sum d = 65$ $\sum d^2 = 569$
$\bar{d} = 6.5$
$s^2 = \dfrac{569 - 10(6.5)^2}{9} = 16.278$
$s = 4.0346$
Critical value $t_9(5\%) = 1.833$
The critical region is $t > 1.833$
$t = \dfrac{6.5 - 0}{\frac{4.0346}{\sqrt{10}}} = 5.095$
In the critical region. Reject H_0.
There is evidence to suggest a tea break increases the number of garments made.

7 $H_0 : \mu_D = 0$ $H_1 : \mu_D > 0$
$\sum d = 8.6$ $\sum d^2 = 20.78$
$\bar{d} = 1.075$
$s^2 = \dfrac{20.78 - 8(1.075)^2}{7} = 1.64786$
$s = 1.2837$
Critical value $t_7(1\%) = 2.998$
The critical region is $t > 2.998$
$t = \dfrac{1.075 - 0}{\frac{1.2827}{\sqrt{8}}} = 2.3686$
Not in the critical region. Do not reject H_0.
There is evidence to suggest that the drug increases the mean number of hours sleep per night.

Exercise 7D

1 a $s_p^2 = \dfrac{(9 \times 4) + (14 \times 5.3)}{10 + 15 - 2} = 4.7913$

$s_p = 2.189$

$t_{23}(2.5\%) = 2.069$

$(25 - 22) \pm 2.069 \times 2.189\sqrt{\frac{1}{15} + \frac{1}{10}} = 3 \pm 1.849$

$= (1.151, 4.849)$

b Independent random samples, normal distributions, common variance

2 a $\bar{x}_s = 8.9125 \quad s_s^2 = 0.58125$

$\bar{x}_{ns} = 12.04 \quad s_{ns}^2 = 0.84933$

$s_p^2 = \dfrac{(7 \times 0.58125) + (9 \times 0.849)}{10 + 8 - 2} = 0.7319$

$s_p = 0.8555$

$t_{16}(5\%) = 1.746$

$(12.04 - 8.9125) \pm 1.746 \times 0.855\sqrt{\frac{1}{10} + \frac{1}{8}}$

$= 3.1275 \pm 0.7081 = (2.419, 3.836)$

b Population variances are equal – reasonable since the compost is designed to increase the *amount* of growth, not the variability.

3 a $s_p^2 = \dfrac{(19 \times 6.12) + (19 \times 5.22)}{20 + 20 - 2} = 5.67$

$s_p = 2.381...$

$t_{38}(0.5\%) = 2.712$

$(38.2 - 32.7) \pm 2.712 \times 2.381...\sqrt{\frac{1}{20} + \frac{1}{20}}$

$= 5.5 \pm 2.042 = (3.46, 7.54)$

b normality and equal variances

c zero not in interval $\Rightarrow$ method B seems better than method A

4 Assume same variances,

$s_p^2 = \dfrac{(9 \times 32.488) + (9 \times 33.344)}{10 + 10 - 2} = 32.916$

$s_p = 5.73725$

$t_{18}(5\%) = 1.734$

$(18.6 - 14.3) \pm 1.734 \times 5.73725\sqrt{\frac{1}{10} + \frac{1}{10}}$

$= 4.3 \pm 4.44905 = (-0.149, 8.749)$

5 a $H_0: \sigma_A^2 = \sigma_B^2$, $H_1: \sigma_A^2 \neq \sigma_B^2$, c.v. = 3.87, test stat = 1.33, therefore no evidence that there is a difference in variability.
Assumption: samples are taken from normally distributed populations.

b Can assume population variances are equal – this is one of the fundamental requirements to use t-distribution.

c (0.752, 3.85)

Exercise 7E

1 a $s_p^2 = \dfrac{(19 \times 12) + (10 \times 12)}{20 + 11 - 2} = 12$

b $H_0: \mu_{1st} = \mu_{2nd} \quad H_1: \mu_{1st} \neq \mu_{2nd}$
critical value $t_{29}(0.025) = 2.045$
critical region is $t \leqslant -2.045$ and $t \geqslant 2.045$

$t = \dfrac{(16 - 14) - 0}{3.464\sqrt{\frac{1}{20} + \frac{1}{11}}} = 1.538$

Not in critical region – do not reject H_0.
There is evidence to suggest that the populations have the same mean.

2 $H_0: \mu_F = \mu_c \quad H_1: \mu_F > \mu_c$

$\bar{x}_w = 38.67, \quad s_w^2 = 5.5827$

$\bar{x}_c = 41.625, \quad s_c^2 = 1.5625$

$s_p^2 = \dfrac{(5 \times 5.5827) + (3 \times 1.5625)}{6 + 4 - 2} = 4.075$

critical value $t_8(0.05) = 1.86$
critical region is $t \geqslant 1.860$

$t = \dfrac{(41.625 - 38.67) - 0}{2.0187\sqrt{\frac{1}{6} + \frac{1}{4}}} = 2.270$

In the critical region – reject H_0.
There is evidence to suggest that the salmon are wild.

3 a $\bar{x}_t = 0.1185 \quad s_t^2 = 0.0005227$

$\bar{x}_e = 0.1425 \quad s_e^2 = 0.0011319$

$s_p^2 = \dfrac{(5 \times 0.0005227) + (5 \times 0.0011319)}{6 + 6 - 2} = 0.000827$

$H_0: \mu_t = \mu_e \quad H_1: \mu_t \neq \mu_e$
critical value $t_{10}(0.025) = 2.228$
critical region $t \leqslant 2.228$ and $t \geqslant 2.228$

$t = \dfrac{(0.1425 - 0.1185) - 0}{0.02876\sqrt{\frac{1}{6} + \frac{1}{6}}} = 1.445$

Not in the critical region – accept H_0.
There is evidence to suggest that Tetracycline and Erythromycin are equally as effective.

b $\bar{x}_s = 0.2387 \quad s_s^2 = 0.0004959$

$\bar{x}_2 = 0.1305 \quad s_2^2 = 0.000909$

$s_p^2 = \dfrac{(11 \times 0.000909) + (5 \times 0.0004959)}{12 + 6 - 2} = 0.000780$

$H_0: \mu_s = \mu_2 \quad H_1: \mu_s \neq \mu_2$
critical value $t_{16}(0.05) = 1.746$
critical region $t \geqslant 1.746$

$t = \dfrac{(0.2387 - 0.1305) - 0}{0.0279\sqrt{\frac{1}{12} + \frac{1}{6}}} = 7.75$

In the critical region – reject H_0.
There is evidence to suggest that Streptomycin is more effective than the others.

4 a $H_0: \mu_{old} = \mu_{new} \quad H_1: \mu_{old} > \mu_{new}$

b $\bar{x}_{old} = 7.911 \quad s_{old}^2 = 5.206$

$\bar{x}_{new} = 5.9 \quad s_{new}^2 = 3.98$

$s_p^2 = \dfrac{(6 \times 3.98) + (8 \times 5.206)}{7 + 9 - 2} = 4.6806$

critical value $t_{14} = 1.761$
critical region $t \geqslant 1.761$

Test statistic $t = \dfrac{(7.911 - 5.9) - 0}{2.1635\sqrt{\frac{1}{9} + \frac{1}{7}}} = 1.8446$

Significant – there is evidence to suggest that new language does improve time.

c Once task is solved the programmer should be quicker next time with either language.

5 a $27 \times 34 - 384 = 534 = \sum x$

$23 = \dfrac{\sum y^2}{27 - 1} - \dfrac{918^2}{27(27 - 1)}$

$\sum y^2 = 31810$

$31810 - 12480 = 19330 = \sum x^2$

b $\bar{x}_v = 32 \quad s_v^2 = 17.45$

$\bar{x}_s = 35.6 \quad s_s^2 = 22.829$

$s_p^2 = \dfrac{(14 \times 22.829) + (11 \times 17.45)}{12 + 15 - 2} = 20.464$

Online Full worked solutions are available in SolutionBank.

c $H_0 : \mu_v = \mu_s$ $H_1 : \mu_v \neq \mu_s$
critical value $t_{25}(0.025) = 2.060$
critical region $t \leqslant -2.060$ and $t \leqslant 2.060$

$t = \dfrac{(35.6 - 32) - 0}{4.524\sqrt{\frac{1}{15} + \frac{1}{12}}} = 2.0547 -$ accept H_0

– no evidence to suggest difference in means

d normality

e same types of driving, roads and weather.

6 a $H_0 : \sigma_A{}^2 = \sigma_B{}^2$, $H_1 : \sigma_A{}^2 \neq \sigma_B{}^2$, c.v. $= 2.91$, test stat $= 1.57$, therefore no evidence that there is a difference in variability. Assumption: samples are taken from normally distributed populations.

b Can assume population variances are equal – this is one of the fundamental requirements to use two-sample t-test.

c $H_0 : \mu_A = \mu_B$, $H_1 : \mu_A \neq \mu_B$, c.v. $= 2.074$, test stat $= 6.177$, therefore evidence that there is a difference in the average weight of potatoes.

Challenge

$E[(n_x - 1)\,s_x^2 + (n_y - 1)\,s_y^2] = E\,[(n_x - 1)\,s_x^2] + E[(n_y - 1)\,s_y^2]$
$= (n_x - 1)\,s^2 + (n_y - 1)\,s^2$
$= [(n_x - 1) + (n_y - 1)]\,s^2$

so $E(S_p^2) = E\,[\dfrac{(n_x - 1)\,s_x^2 + (n_y - 1)\,s_y^2}{(n_x - 1) + (n_y - 1)}]$

$= \dfrac{E[(n_x - 1)\,s_x^2 + (n_y - 1)\,s_y^2]}{[(n_x - 1) + (n_y - 1)]}$

$= \dfrac{[(n_x - 1) + (n_y - 1)]\sigma^2}{[(n_x - 1) + (n_y - 1)]}$

$= \sigma$.

So S_p^2 is an unbiased estimator

Mixed exercise 7

1 $H_0 : \mu = 28$ $H_1 : \mu \neq 28$
Critical region < -2.160 or > 2.160

Test statistic $= \dfrac{\bar{x} - \mu}{\frac{s}{\sqrt{n}}} = \dfrac{30.4 - 28}{\frac{6}{\sqrt{14}}} = 1.4967$

The test statistic is not in the critical region.
There is no evidence to suggest that μ does not $= 28$

2 $H_0 : \mu = 10$ $H_1 : \mu > 10$
Critical region > 1.895

$\bar{x} = \dfrac{85}{8} = 10.625$

$s^2 = \dfrac{\sum x^2 - n\bar{x}^2}{n - 1} = \dfrac{970.25 - 8 \times 10.625^2}{7} = 9.589...$

Test statistic $= \dfrac{\bar{x} - \mu}{\frac{s}{\sqrt{n}}} = \dfrac{10.625 - 10}{\sqrt{\frac{9.589...}{8}}}$

$= 0.571$ – not critical – no evidence to suggest that $\mu > 10$

3 $\bar{x} = 52.833...$ $s = 1.722....$

a Confidence interval

$= \left(\bar{x} - t_{n-1}\left(\dfrac{\alpha}{2}\right) \times \dfrac{s}{\sqrt{n}}, \bar{x} + t_{n-1}\left(\dfrac{\alpha}{2}\right) \times \dfrac{s}{\sqrt{n}}\right)$

$= \left(52.833... - 2.571 \times \dfrac{1.722...}{\sqrt{6}}, 52.833... \right.$

$\left. + 2.571 \times \dfrac{1.722...}{\sqrt{6}}\right)$

$= (51.025, 54.641)$

b Confidence interval $= \left(\dfrac{(n - 1)s^2}{\chi_{n-1}^2\left(\frac{\alpha}{2}\right)}, \dfrac{(n - 1)s^2}{\chi_{n-1}^2\left(1 - \frac{\alpha}{2}\right)}\right)$

$= \left(\dfrac{5 \times 1.722...^2}{12.832}, \dfrac{5 \times 1.722...^2}{0.831}\right)$

$= (1.156, 17.850)$

c They are normally distributed.

4 a Confidence interval

$= \left(\bar{x} - t_{n-1}\left(\dfrac{\alpha}{2}\right) \times \dfrac{s}{\sqrt{n}}, \bar{x} + t_{n-1}\left(\dfrac{\alpha}{2}\right) \times \dfrac{s}{\sqrt{n}}\right)$

$= \left(9.8 - 2.110 \times \dfrac{0.7}{\sqrt{18}}, 9.8 + 2.110 \times \dfrac{0.7}{\sqrt{18}}\right)$

$= (9.451, 10.148)$

b Confidence interval $= \left(\dfrac{(n - 1)s^2}{\chi_{n-1}^2\left(\frac{\alpha}{2}\right)}, \dfrac{(n - 1)s^2}{\chi_{n-1}^2\left(1 - \frac{\alpha}{2}\right)}\right)$

$= \left(\dfrac{17 \times 0.49}{30.191}, \dfrac{17 \times 0.49}{7.564}\right)$

$= (0.276, 1.101)$

5 $\bar{x} = 20.95$ $s = 2.674...$
$H_0 : \mu = 21.5$ $H_1 : \mu < 21.5$
critical region < -1.895,

Test statistic $t = \dfrac{\bar{x} - \mu}{\frac{s}{\sqrt{n}}} = \dfrac{20.95 - 21.5...}{\frac{2.674...}{\sqrt{8}}} = -0.5817$.

The test statistic is not in the critical region.
There is no evidence to reject claim.

6 $\bar{x} = 6.1916....$ $s = 0.7549...$ $s^2 = 0.5699...$

a Confidence interval

$= \left(\bar{x} - t_{n-1}\left(\dfrac{\alpha}{2}\right) \times \dfrac{s}{\sqrt{n}}, \bar{x} + t_{n-1}\left(\dfrac{\alpha}{2}\right) \times \dfrac{s}{\sqrt{n}}\right)$

$= \left(6.1916... - 2.201 \times \dfrac{0.7549...}{\sqrt{12}}, 6.1916... \right.$

$\left. + 2.201 \times \dfrac{0.7549...}{\sqrt{12}}\right)$

$= (5.712, 6.671)$

b Confidence interval Var. $= \left(\dfrac{(n - 1)s^2}{\chi_{n-1}^2\left(\frac{\alpha}{2}\right)}, \dfrac{(n - 1)s^2}{\chi_{n-1}^2\left(1 - \frac{\alpha}{2}\right)}\right)$

$= \left(\dfrac{11 \times 0.5699...}{21.920}, \dfrac{11 \times 0.5699...}{3.816}\right)$

$= (0.286, 1.643)$

Confidence interval s.d. $= (0.535, 1.282)$

c He should measure his blood glucose at the same time each day.

7 $\bar{x} = 11.5$ $s = 2.073...$

a Confidence interval

$= \left(\bar{x} - t_{n-1}\left(\dfrac{\alpha}{2}\right) \times \dfrac{s}{\sqrt{n}}, \bar{x} + t_{n-1}\left(\dfrac{\alpha}{2}\right) \times \dfrac{s}{\sqrt{n}}\right)$

$= \left(11.5 - 2.571 \times \dfrac{2.073...}{\sqrt{6}}, 11.5 + 2.571 \times \dfrac{2.073...}{\sqrt{6}}\right)$

$= (9.324, 13.675)$

b Confidence interval $= \left(\dfrac{(n - 1)s^2}{\chi_{n-1}^2\left(\frac{\alpha}{2}\right)}, \dfrac{(n - 1)s^2}{\chi_{n-1}^2\left(1 - \frac{\alpha}{2}\right)}\right)$

$= \left(\dfrac{5 \times 2.073...^2}{12.832}, \dfrac{5 \times 2.073...^2}{0.831}\right)$

$= (1.675, 25.872)$

8 a $H_0 : \sigma = 4$ $H_1 : \sigma > 4$
Critical region $\chi^2 > 16.919$

Test statistic $= \dfrac{(n - 1)s^2}{\sigma^2} = \dfrac{9 \times 5.2^2}{4^2} = 15.21$

The test statistic is not in the critical region.
standard deviation $= 4$ months

b $H_0 : \mu = 24$ $H_1 : \mu > 24$
Critical region $t > 1.833$
Test statistic $t = \dfrac{\bar{x} - \mu}{\frac{s}{\sqrt{n}}} = \dfrac{27.2 - 24}{\frac{5.2}{\sqrt{10}}} = 1.946$
There is evidence to reject H_0
The mean battery life >24

c Lifetime is normally distributed.

9 $\bar{x} = 721.5$ $s = 10.399....$

a Confidence interval
$= \left(\bar{x} - t_{n-1}\left(\frac{\alpha}{2}\right) \times \frac{s}{\sqrt{n}}, \bar{x} + t_{n-1}\left(\frac{\alpha}{2}\right) \times \frac{s}{\sqrt{n}} \right)$
$= \left(721.5 - 2.093 \times \dfrac{10.399...}{\sqrt{20}}, \right.$
$\left. 721.5 + 2.093 \times \dfrac{10.399...}{\sqrt{20}} \right)$
$= (717, 726)$

b Confidence interval variance
$= \left(\dfrac{(n-1)s^2}{\chi^2_{n-1}\left(\frac{\alpha}{2}\right)}, \dfrac{(n-1)s^2}{\chi^2_{n-1}\left(1 - \frac{\alpha}{2}\right)} \right)$
$= \left(\dfrac{19 \times 10.399...^2}{32.852}, \dfrac{19 \times 10.399...^2}{8.907} \right)$
$= (62.553, 230.717)$
Confidence interval standard deviation =
(7.909, 15.189)

c 725 within confidence interval,
There is no evidence to reject this hypothesis.

10 a $\bar{x} = \dfrac{34.2}{10} = 3.42$
$s^2 = \dfrac{\sum x^2 - n\bar{x}^2}{n - 1} = \dfrac{121.6 - 10 \times 3.42^2}{9} = 0.5151...$

b i Confidence interval mean
$= \left(\bar{x} - t_{n-1}\left(\frac{\alpha}{2}\right) \times \frac{s}{\sqrt{n}}, \bar{x} + t_{n-1}\left(\frac{\alpha}{2}\right) \times \frac{s}{\sqrt{n}} \right)$
$= \left(3.42 - 2.262 \times \dfrac{0.7177...}{\sqrt{10}}, \right.$
$\left. 3.42 + 2.262 \times \dfrac{0.7177...}{\sqrt{10}} \right)$
$= (2.906, 3.933)$

ii Confidence interval variance
$= \left(\dfrac{(n-1)s^2}{\chi^2_9(0.025)}, \dfrac{(n-1)s^2}{\chi^2_9(0.975)} \right)$
$= \left(\dfrac{9 \times 0.515...}{19.023}, \dfrac{9 \times 0.515...}{2.700} \right)$
$= (0.244, 1.717)$
Confidence interval standard deviation =
(0.4937, 1.3103)

c 3.5 hours is inside the confidence interval on the mean, so there is no evidence of a change in the mean time.
0.5 hours is inside the confidence interval on the standard deviation so there is no evidence of a change in the variability of the time.
There is no reason to change the repair method.

d Use a 'matched pairs' experiment, getting each engineer to carry out a similar repair using the old method and the new method and use a paired t-test.

11 a Normal approximation (n is large), C.I. = (23.69, 24.31);

b Sample size is small

c $H_0 : \mu = 25$, $H_1 : \mu > 25$, c.v. = 2.015, test stat = 2.981, hence evidence that male raccoons have an average length greater than 25 cm.

12 d: 5, 13, −8, 2, −3, 4, 11, −1
$\left(\sum d = 23, \sum d^2 = 409 \right)$ $\bar{d} = 2.875$, sd = 6.9987 ($\approx$7.00)
$H_0 : \mu_d = 0$ $H_1 : \mu_d > 0$
$t = \dfrac{(2.875 - 0)}{\frac{6.9987}{\sqrt{8}}} = 1.1618...$ ($\approx$1.16)
Critical value $t_7(10\%) = 1.415$ (one-tailed)
Critical region is $t > 1.415$
Not significant
Insufficient evidence to support the chemist's claim

13 a The data were not collected in pairs.

b Use data from twin lambs.

c Age, weight, gender

d $d = B - A$
d: 2, 1.2, 1, 1.8, −1, 2.2, 2, −1.2, 1.1, 2.8
$\sum d = 11.9$; $\sum d^2 = 30.01$
$\therefore \bar{d} = 1.19$; $s^2 = 1.761$ ($s = 1.327$)
$H_0 : d = 0$ $H_1 : d \neq 0$ Allow μ_D for d
$t = \dfrac{1.19 - 0}{\sqrt{\frac{1.761}{10}}} = 2.83574...$
$\mu = 9$; C.V.: $t = 2.262$
Critical regions $t < -2.262$ or $t > 2.262$
Since 2.8357... is in the critical region ($t = 2.262$) there is evidence to reject H_0. The (mean) weight gained by the lambs is different for each diet.

e Diet B – it has the higher mean.

14 a d: 14, 2, 18, 25, 0, −8, 4, 4, 12, 20
$\left(\sum d = 91; \sum d^2 = 1789 \right)$
$\therefore \bar{d} = 9.1$ $s = \sqrt{106.7} = 10.332...$
$H_0 : \mu_d = 0$ $H_1 : \mu_d \neq 0$
$t = \dfrac{(9.1 - 0)}{\frac{10.332}{\sqrt{10}}} = 2.785$
Critical value $t_9 = \pm 1.833$
critical regions $t \leqslant -1.833$ or $t \geqslant 1.833$
Significant. There is a difference between *blood pressure* measured by arm cuff and finger monitor.

b The *difference in measurements* of blood pressure is *normally* distributed.

15 Differences: 2.1, −0.7, 2.6, −1.7, 3.3, 1.6, 1.7, 1.2, 1.6, 2.4
$\sum d = 14.1$ $\sum d^2 = 40.65$ $d = 1.41$
$H_0 : \mu_d = 0$ $H_1 : \mu_d > 0$
$s = \sqrt{\dfrac{40.65 - 10 \times 1.41^2}{9}} = 1.5191...$
$t = \dfrac{1.41}{\left(\frac{1.519...}{\sqrt{10}}\right)} = 2.935$
$t_9(1\%) = 2.821$
so critical region $t > 2.821$
2.935... > 2.821 Evidence to reject H_0.
There has been an increase in the mean weight of the mice.

16 a $s_o^2 = \dfrac{5136.3}{9} - \dfrac{225^2}{10(10-1)} = 8.2$
$s_n^2 = \dfrac{6200}{8} - \dfrac{234^2}{9(9-1)} = 14.5$

Online Full worked solutions are available in SolutionBank.

b $H_0 : \sigma_o^2 = \sigma_n^2$ $H_1 : \sigma_o^2 < \sigma_n^2$
critical value is $F_{8,9} = 3.23$
so critical region, $F \geq 3.23$

$F_{\text{test}} = \dfrac{14.5}{8.2} = 1.768$

not in critical region
accept H_0 – there is evidence to suggest that $\sigma_o^2 = \sigma_n^2$

c $s_p^2 = \dfrac{(9 \times 8.2) + (8 \times 14.5)}{10 + 9 - 2} = 11.1647$

$H_0 : \mu_o = \mu_n$ $H_1 : \mu_o \neq \mu_n$
critical value $t_{17}(0.01) = 2.567$
critical region $t \leq -2.567$ and $t \geq 2.567$

$t = \dfrac{(26 - 22.5) - 0}{\sqrt{11.1647}\,\sqrt{\frac{1}{10} + \frac{1}{9}}} = 2.2798$

Not in the critical region – do not reject H_0.
There is evidence to suggest that there is no difference in mean times between the old and new equipment.

d $t_{17}(2.5\%) = 2.110$
$(26 - 22.5) \pm 2.110 \times \sqrt{11.1647} \times \sqrt{\frac{1}{10} + \frac{1}{9}}$
$= 3.5 \pm 3.2394 = (0.261, 6.739)$

e Need to learn new equipment

f Gather data on new equipment after it has been mastered.

17 a $H_0 : \sigma_A^2 = \sigma_B^2$, $H_1 : \sigma_A^2 \neq \sigma_B^2$, c.v. = 2.15,
test stat = 1.53, therefore no evidence that there is a difference in variability. Assumption: samples are taken from normally distributed populations.

b Can assume population variances are equal – this is one of the fundamental requirements to use t-distribution.

c (0.186, 2.01)

Challenge

a $S_p^2 = \dfrac{(n_x - 1)S_x^2 + (n_y - 1)S_y^2 + (n_z - 1)S_z^2}{n_x + n_y + n_z - 3}$

b $E(S_p^2) = E\dfrac{(n_x - 1)S_x^2 + (n_y - 1)S_y^2 + (n_z - 1)S_z^2}{n_x + n_y + n_z - 3}$

$= \dfrac{E((n_x - 1)S_x^2 + (n_y - 1)S_y^2 + (n_z - 1))S_z^2}{n_x + n_y + n_z - 3}$

$= \dfrac{((n_x - 1) + (n_y - 1) + (n_z - 1))\sigma^2}{n_x + n_y + n_z - 3}$

$= \sigma^2$

So S_p^2 is an unbiased estimator.

Review exercise 2

1 $H_0 : \mu = 18$; $H_1 : \mu < 18$.
$-1.9364... < -1.6449$ so reject H_0. There is evidence that the (mean) time to complete the puzzles has reduced.

2 a $\bar{x} = 4.52$, $s^2 = 1.51$ (3 s.f.)

b $H_0 : \mu_A = \mu_B$; $H_1 : \mu_A > \mu_B$.
$1.868... > 1.6449$ so reject H_0. There is evidence that diet A is better than diet B or evidence that (mean) weight loss in the first week using diet A is more than with diet B.

c The central limit theorem enables you to assume that $\bar{A}$ and $\bar{B}$ are both normally distributed since both samples are large.

d Assumed $\sigma_A^2 = s_A^2$ and $\sigma_B^2 = s_B^2$

3 (127, 151) to 3 s.f.

4 a $\bar{x} = 50$, $s^2 = 0.193$ (3 s.f.)
b (49.7, 50.3)
c (49.6, 50.4)

5 a $\bar{x} = 110.5$, $s^2 = 672$ (3 s.f.)
b (95.0, 126) to 3.s.f.
c 0.4633 (4 d.p.)

6 a $\bar{x} = 168$, $s^2 = 27.0$ (3 s.f.)
b (166, 170) to 3 s.f.

7 a $H_0 : \mu_F = \mu_M$; $H_1 : \mu_F \neq \mu_M$.
$2.860... > 1.96$ so reject H_0. There is evidence of a difference in the (mean) amount spent on junk food by male and female teenagers.

b The central limit theorem enables you to assume that $\bar{F}$ and $\bar{M}$ are both normally distributed since both samples are large.

8 a $\bar{x} = 287$, $s^2 = 7682.5$
b Sample size $(\geq)$ 97 required.

9 a (3.28, 11.8) to 3.s.f.
b $\sigma = 3$ so $\sigma^2 = 9$. Since 9 lies in the interval, the confidence interval supports the assertion.

10 a (0.00164, 0.00719) to 3 s.f.
b $0.07^2 = 0.0049$
0.0049 is within the 95% confidence interval. There is no evidence to reject the idea that the standard deviation of the volumes is not 0.07, or the machine is working well.

11 a $H_0 : \sigma = 4$; $H_1 : \sigma > 4$.
$\nu = 19$, $\chi_{19}^2(0.05) = 30.144$

$\dfrac{(n - 1)S^2}{\sigma^2} = \dfrac{19 \times 6.25}{4} = 29.6875$

Since $29.6875 < 30.144$ there is insufficient evidence to reject H_0. There is insufficient evidence to suggest that the standard deviation is greater than 2.

b Times are normally distributed.

12 $F_{10,12}(5\%) = 2.75$ so $b = 2.75$

$a = \dfrac{1}{2.91} = 0.344$

13 $P(X > 2.85) = 0.05$

$P\left(X < \dfrac{1}{5.67}\right) = 0.01$

So $P\left(\dfrac{1}{5.67} < X < 2.85\right) = 1 - 0.05 - 0.01 = 0.94$

14 a $H_0 : \sigma_A^2 = \sigma_B^2$; $H_1 : \sigma_A^2 \neq \sigma_B^2$.
Critical value $F_{24,25} = 1.96$

$\dfrac{s_B^2}{s_A^2} = 2.10$

Since 2.10 is in the critical region, we reject H_0 and conclude there is evidence that the two variances are different.

b The lengths of pebbles are normally distributed.

15 a (10.4, 449) to 3 s.f.
b $H_0 : \sigma_M^2 = \sigma_S^2$; $H_1 : \sigma_M^2 > \sigma_S^2$

$\dfrac{s_M^2}{s_S^2} = \dfrac{318.8}{32.3} = 9.859...$

$F_{6,3}(0.01) = 27.91$
$9.859... < 27.91$ so do not reject H_0. There is insufficient evidence of an increase in variance.

16 a (1.13, 2.23) to 3 s.f.
b (1.09, 3.46) to 3 s.f.

c Require $P(X > 2.5) = P\left(Z > \dfrac{2.5 - \mu}{\sigma}\right)$ to be as small

as possible OR $\dfrac{2.5 - \mu}{\sigma}$ to be as large as possible;

so both both σ and μ must be as small as possible.
Lowest estimate for proportion with high
cholesterol

$= P\left(Z > \dfrac{2.5 - 1.13}{\sqrt{1.09}}\right) = P(Z > 1.31)$

$= 1 - 0.9049 = 0.0951 = 9.51\%$

17 a (3.44, 4.58) to 3.s.f.

   b (0.302, 2.13) to 3 s.f.

   c Require $P(X > 7) = P\left(Z > \dfrac{7 - \mu}{\sigma}\right)$ to be as high as

possible; therefore both σ and μ must be as large as
possible.
Highest estimate for proportion with high blood
glucose

$= P\left(Z > \dfrac{7 - 4.581}{\sqrt{2.13}}\right)$

$= 1 - 0.9515 = 0.0485 = 4.85\%$

18 a i (119, 127) to 3 s.f.

   ii (16.3, 115) to 3 s.f.

   b 130 is just above the confidence interval. 16 is just
below the confidence interval. Thus the supervisor
should be concerned about the speed to the new
typist since both their average speed is too slow and
their variability of time is too large.

19 a $H_0 : \sigma^2 = 0.9$ $H_1 : \sigma^2 \neq 0.9$

   $v = 19$

   CR (Lower tail 10.117)
        Upper tail 30.144

   Test statistic $= \dfrac{19 \times 1.5}{0.9} = 31.6666$, significant.

   There is sufficient evidence that the variance of the
length of spring is different to 0.9.

   b $H_0 : \mu = 100$ $H_1 : \mu > 100$

   $t_{19} = 1.328$ is the critical value

   $t = \dfrac{100.6 - 100}{\sqrt{\dfrac{1.5}{20}}} = 2.19$

   Significant. The mean length of spring is greater
than 100.

20 a $s_A^2 = \dfrac{1}{10}\left(3\,960\,540 - \dfrac{6600^2}{11}\right) = 54.0$

   $s_B^2 = \dfrac{1}{12}\left(7\,410\,579 - \dfrac{9815^2}{13}\right) = 21.16$

   $H_0 : \sigma_A^2 = \sigma_B^2$, $H_1 : \sigma_A^2 \neq \sigma_B^2$
   Critical region: $F_{10,12} > 2.75$

   $\dfrac{s_A^2}{s_B^2} = \dfrac{54.0}{21.16} = 2.55118...$

   Since 2.55118... is not in the critical region, we can
assume that the variances are equal.

   b $H_0 : \mu_B = \mu_A + 150$; $H_1 : \mu_B > \mu_A + 150$
   C.R.: $t_{22}(0.05) > 1.717$

   $s_p^2 = \dfrac{10 \times 54.0 + 12 \times 21.16}{22} = 36.0909$

   $t = \dfrac{755 - 600 - 150}{\sqrt{36.0909...\left(\dfrac{1}{11} + \dfrac{1}{13}\right)}} = 2.03157$

   Since 2.03... is in the critical region we reject H_0
and conclude that the mean weight of cauliflowers
from B exceeds that from A by at least 50 g.

c Samples from normal populations
   Equal variances
   Independent samples

21 a i $s_C^2 = 32.49$ $s_N^2 = 12.25$
   $H_0 : \sigma_C^2 = \sigma_N^2$, $H_1 : \sigma_C^2 > \sigma_N^2$
   Critical value: $F_{8,9} = 3.23$

   $\dfrac{s_C^2}{s_N^2} = \dfrac{32.49}{12.25} = 2.652...$

   $2.652... < 3.23$ so do not reject H_0 and assume
   the variances are equal.

   ii $H_0 : \bar{x}_C = \bar{x}_N$, $H_1 : \bar{x}_C > \bar{x}_N$
   $\nu = 17$
   Critical value $t_{17}(0.05) = 1.740$

   $s_P^2 = \dfrac{(9 \times 3.5^2) + (8 \times 5.7^2)}{17} = 21.774...$

   $t = \dfrac{82.3 - 78.2}{\sqrt{21.774...\left(\dfrac{1}{10} + \dfrac{1}{9}\right)}} = 1.912...$

   $1.912... > 1.74$ so reject H_0, there is evidence the
   mean scores differ.

   b Independent and samples from normal population.

   c (12.4, 47.6)

22 a $H_1 : \sigma_A^2 = \sigma_B^2$, $H_0 : \sigma_A^2 \neq \sigma_B^2$
   $s_A^2 = 22.5 > s_B^2 = 21.6$

   $\dfrac{s_A^2}{s_B^2} = 1.04$

   $F_{(8,6)} = 4.15$
   $1.04 < 4.15$ do not reject H_0. The variances are the
   same.

   b Assume the samples are selected at random
   (independent)

   c $s_p^2 = \dfrac{8(22.5) + 6(21.62)}{14} = 22.12$

   $H_0 : \mu_A = \mu_B$, $H_1 : \mu_A \neq \mu_B$

   $t = \dfrac{40.667 - 39.57}{\sqrt{22.12\left(\dfrac{1}{9} + \dfrac{1}{7}\right)}} = 0.462$

   Critical value $= t_{14}(2.5\%) = 2.145$
   $0.462 < 2.145$ No evidence to reject H_0.
   The means are the same.

   d Music has no effect on performance.

23 a $H_0 : \sigma_R^2 = \sigma_E^2$ $H_1 : \sigma_R^2 \neq \sigma_E^2$

   $F_{6,12}(5\%)_{1\,\text{tail}}$ c.v. $= 3.00$, $\dfrac{s_E^2}{s_R^2} = \dfrac{35.79}{14.48} = 2.4716...$

   Not significant so do not reject H_0.
   Insufficient evidence to suspect $\sigma_R^2 \neq \sigma_E^2$

   b $H_0 : \mu_R = \mu_E$, $H_1 : \mu_R \neq \mu_E$

   $s^2 = \dfrac{6 \times 35.79 + 12 \times 14.48}{18} = 21.58\dot{3}$

   $t = \dfrac{32.31 - 28.43}{s\sqrt{\dfrac{1}{13} + \dfrac{1}{7}}} = 1.78146...$

   $t_{18}(5\%)_{2\,\text{tail}}$ c.v. $= 2.101$
   $\therefore$ Not significant.
   Insufficient evidence of difference in mean
   performance.

   c Test in part **b** requires $\sigma_1^2 = \sigma_2^2$

   d e.g. same: type of driving
                    roads and journey length
                    weather
                    driver

a $\bar{x} = 668.125$ $s = 84.425$

$t_7(5\%) = 1.895$

Confidence limits $= 668.125 \pm \dfrac{1.895 \times 84.425}{\sqrt{8}}$

$= 611.6$ and 724.7

Confidence interval $= (612, 725)$

b Normal distribution

c £650 is within the confidence interval.
No need to worry.

5 $H_0 : \mu = 1012$ $H_1 : \mu \neq 1012$

$\bar{x} = \dfrac{13\,700}{14} (= 978.57...)$

$S_x^2 = \dfrac{13\,448\,750 - 14\bar{x}^2}{13} (= 3255.49)$

$t_{13} = \dfrac{\bar{x} - \mu}{\frac{s}{\sqrt{n}}} = \dfrac{978.6 - 1012}{\frac{57.06}{\sqrt{14}}} = -2.19...$

$t_{13}(5\%)$ two-tail critical value $= -2.160$

Significant result – there is evidence of a change in mean weight of squirrels.

26 Let x represent weight of flour

$\sum x = 8055$ $\therefore \bar{x} = 1006.875$

$\sum x^2 = 8\,110\,611$ $\therefore s^2 = \dfrac{1}{7}\left(8\,110\,611 - \dfrac{8055^2}{8}\right)$

$= 33.26785...$

$\therefore s = 5.767825$

$H_0 : \mu = 1010$; $H_1 : \mu < 1010$

Critical value: $t = -1.895$ so critical region $t < -1.895$

$t = \dfrac{(1006.875 - 1010)}{\left(\frac{5.7678}{\sqrt{8}}\right)} = -1.5324$

Since -1.53 is not in the critical region there is insufficient evidence to reject H_0.
The mean weight of flour delivered by the machine is $1010\,g$.

27 d: 7, 2, –3, 1, –1, –2, 10, 5

$\sum d = 19; \sum d^2 = 193$

$\therefore \bar{d} = \dfrac{19}{8} = 2.375; s_d^2 = \dfrac{1}{7}\left\{193 - \dfrac{19^2}{8}\right\} = 21.125$

$H_0 : \mu_D = 0$; $H_1 : \mu_D > 0$

$t = \dfrac{2.375 - 0}{\sqrt{\frac{21.125}{8}}} = 1.4615...$

$v = 7 \Rightarrow$ critical region: $t > 1.895$

Since $1.4195...$ is *not* in the critical region there is insufficient evidence to reject H_0 and we conclude that there is insufficient evidence to support the doctors' belief.

28 **a** d = coursework – written: 4, –3, –3, 4, 6, 3, –4, 17, 7, 7

$\bar{d} = \dfrac{38}{10} = 3.8, s_d^2 = \dfrac{498 - 10\bar{d}^2}{9} = 39.2\dot{8}$

test statistic: $t = \dfrac{3.8}{\frac{s_d}{\sqrt{10}}} = 1.917...$

$H_0 : \mu_d = 0$, $H_1 : \mu_d > 0$

$t_9(5\%)$ c.v. is 1.833

$\therefore$ Significant – there is evidence coursework marks are higher.

b The difference between the marks follows a normal distribution.

29 D = dry – wet $H_0 : \mu_D = 0$, $H_1 : \mu_D \neq 0$

d: 0.6, –1, –1.9, –1.4, –1.3, 0.5, –1.6, –0.6, –1.8

$\bar{d}: -\dfrac{8.5}{9} = -0.9\dot{4}, s_d^2 = \dfrac{15.03 - 9 \times (\bar{d})^2}{8} = 0.87527...$

$t = \dfrac{-0.9\dot{4}}{\frac{s_d}{\sqrt{9}}} =$ awrt -3.03

t_8 two-tailed 1% critical value $= 3.355$

Not significant – insufficient evidence of a difference between mean strength.

30 **a** $H_0 : \mu_d = 0$ ⎫   where $d =$
$H_1 : \mu_d > 0$ ⎭   without solar heating – with solar heating

d = 6, –3, 7, –2, –8, 6, 5, 11, 5

$\bar{d} = 3$

$s_d = 6$

$n_d = 9$

t test statistic $= \dfrac{(3.0)}{\left(\frac{6}{\sqrt{9}}\right)}$

t.s. $= 1.5$

Critical value $= t_8(5\%) = 1.860$

so critical region: $t > 1.860$

Test statistic not in critical region so accept H_0.
Conclude there is insufficient evidence that solar heating reduces mean weekly fuel consumption.

b The differences are normally distributed.

31 **a** $s_p^2 = \dfrac{7 \times 7.84 + 7 \times 4}{7 + 7} = 5.92$

$s_p = 2.433105$

$H_0 : \mu_A = \mu_B$, $H_1 : \mu_A > \mu_B$

$t = \dfrac{26.125 - 25}{2.43\sqrt{\frac{1}{8} + \frac{1}{8}}} = 0.92474$

$t_{14}(2.5\%) = 2.145$

Insufficient evidence to reject H_0.
Conclude that there is no difference in the means.

b d = 2, 5, –2, 1, 3, –4, 1, 3

$\bar{d} = \dfrac{9}{8} = 1.125$

$s_d^2 = \dfrac{69 - 8 \times 1.125^2}{7} = 8.410714$

$H_0 : \delta = 0$, $H_1 : \delta \neq 0$

$t = \dfrac{1.125}{\sqrt{\frac{8.41}{8}}} = 1.0972$

$t_7(2.5\%) = 2.365$

There is no significant evidence of a difference between method A and method B.

c Paired sample as they are two measurements on the same orange.

32 **a** Confidence interval is given by

$\bar{x} \pm t_{19} \times \dfrac{s}{\sqrt{n}}$

i.e.: $207.1 \pm 2.539 \times \sqrt{\dfrac{3.2}{20}}$

i.e.: 207.1 ± 1.0156

i.e.: $(206.08..., 208.1156)$

b $s_G^2 = \dfrac{418\,785.4 - \dfrac{2046.2^2}{10}}{9}$

$s_G^2 = 10.217\dot{3}$

$\bar{x}_G = \dfrac{2046.2}{10} = 204.62$

$s_p^2 = \dfrac{19 \times 3.2 + 9 \times 10.2173}{28}$

$= 5.45557\ldots$

Confidence interval is given by

$\bar{x}_B - \bar{x}_G \pm t_{28} \times \sqrt{5.45557\left(\dfrac{1}{20} + \dfrac{1}{10}\right)}$

i.e.: $(207.1 - 204.62) \pm 1.701\sqrt{5.45557\left(\dfrac{1}{20} + \dfrac{1}{10}\right)}$

i.e.: 2.48 ± 1.53875

i.e.: $(0.94125, 4.0187)$

Challenge

1 a $E\left(\dfrac{2}{3}X_1 - \dfrac{1}{2}X_2 + \dfrac{5}{6}X_3\right) = \dfrac{2}{3}\mu - \dfrac{1}{2}\mu + \dfrac{5}{6}\mu$

$E(Y) = \mu \Rightarrow$ unbiased

b $E(aX_1 + bX_2) = a\mu + b\mu = \mu$

$a + b = 1$

$\text{Var}(aX_1 + bX_2) = a^2\sigma^2 + b^2\sigma^2$

$= a^2\sigma^2 + (1-a)^2\sigma^2 = (2a^2 - 2a + 1)\sigma^2$

c Minimum value when $(4a - 2)\sigma^2 = 0$

$\Rightarrow 4a - 2 = 0$

$a = \dfrac{1}{2}, b = \dfrac{1}{2}$

2 a Consider cumulative distribution functions:
X has c.d.f. $F(x) = x$ in the interval $0 \le x \le 1$.
In order for $M < x$, you must have no more than 1
of the $X_i > x$.
So either all three $X_i < x$ with probability $F(x)^3 = x^3$.
Or exactly one of the $X_i > x$, with probability
$3 \times F(x)^2 \times (1 - F(x)) = 3x^2(1 - x)$ since it can happen
in 3 different ways.
So $P(M < x) = x^3 + 3x^2(1 - x) = 3x^2 - 2x^3$
So M has c.d.f.

$F(x) = \begin{cases} 0 & x < 0 \\ 3x^2 - 2x^3 & 0 \le x \le 1 \\ 1 & x > 1 \end{cases}$

Differentiating, M has p.d.f.

$h(x) = \begin{cases} 6x - 6x^2 & 0 \le x \le 1 \\ 0 & \text{otherwise} \end{cases}$

b $E(M) = \displaystyle\int_0^1 x(6x - 6x^2)\,dx = \left[2x^3 - \dfrac{3}{2}x^4\right]_0^1 = 0.5 = E(X)$

c $E(M^2) = \displaystyle\int_0^1 x^2(6x - 6x^2)\,dx = \left[\dfrac{3}{2}x^4 - \dfrac{6}{5}x^5\right]_0^1 = 0.3$

$\text{Var}(M) = E(M^2) - (E(M))^2 = 0.3 - 0.5^2 = 0.05$

Standard error of $M = \sqrt{0.05} = 0.224$ (3 s.f.)

Exam-style practice: AS level

1 a
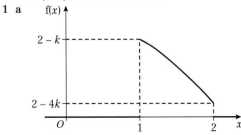

b Mode is 1

c $\displaystyle\int_1^2 (2 - kx^2)\,dx = \left[2x - \dfrac{k}{3}x^3\right]_1^2 = 1 \Rightarrow k = \dfrac{3}{7}$

d $\dfrac{39}{28}(= 1.393)$

e $F(x) = \begin{cases} 0 & x < 1 \\ 2x - \dfrac{1}{7}x^3 - \dfrac{13}{7} & 1 \le x \le 2 \\ 1 & x > 2 \end{cases}$

f $F(m) = 0.5$

$\Rightarrow 2m - \dfrac{1}{7}m^3 - \dfrac{13}{7} = 0.5$

$\Rightarrow 2m^3 - 28m + 33 = 0$

g Mean > Median > Mode therefore POSITIVE skew.

2 a $f(x) = \begin{cases} \dfrac{1}{40} & 0 \le x \le 40 \\ 0 & \text{otherwise} \end{cases}$

b 20

c $\dfrac{400}{3}$

d $\dfrac{3}{8}$

e $\dfrac{2}{5}$

3 a 0.455 (3 s.f.)

b H_0: There is no association between French and
Spanish scores. H_1: There is an association between
French and Spanish scores. Critical value is 0.6485
(two-tailed test) and $0.455 < 0.6485$ so accept
the null hypothesis – there is no evidence of an
association between French and Spanish scores.

c H_0: $\rho = 0$; H_1: $\rho > 0$; critical value is 0.5494 (one-
tailed test) and $0.568 > 0.5494$ so reject the
null hypothesis – there is evidence of a (positive)
correlation between French and Spanish scores.

d Spearman's rank does not use the actual data, just
the ranks.

4 $S_{tc} = \displaystyle\sum tc - \dfrac{\sum t \sum c}{n} = 561.3 - \dfrac{88 \times 32.3}{5} = -7.18$

$\text{RSS} = S_{cc} - \dfrac{(S_{tc})^2}{S_{tt}} = 0.852 - \dfrac{(-7.18)^2}{65.2} = 0.0613$ (3 s.f.)

Thus since $0.0524 < 0.0613$, we conclude that ice-
cream sales are more likely to have a linear model.

Exam-style practice: A level

1 a 0.665

b 0.312

2 a Let σ_f^2 be variance with fertiliser;

H_0: $\sigma^2 = \sigma_f^2$; H_1: $\sigma^2 \ne \sigma_f^2$

$F = \dfrac{6.84^2}{5.29^2} = 1.671\ldots$

$F_{12,9}$ critical value = 3.07
So the test is not significant – the variances can be
assumed equal.

b Let μ_f be mean with fertiliser; H_0: $\mu = \mu_f$; H_1: $\mu \ne \mu_f$

$s_p^2 = \dfrac{9 \times 5.29^2 + 12 \times 6.84^2}{10 + 13 - 2} = 38.72\ldots$

So $t = \dfrac{23.36 - 19.96 - 0}{\sqrt{38.72\ldots} \times \sqrt{\dfrac{1}{10} + \dfrac{1}{13}}} = 1.298\ldots$

t_{21} 5% two-tail c.v. is 2.080.
Test is not significant – there is insufficient evidence
of a difference in mean heights. No evidence to
support the idea that the fertiliser increases the
average height gained.

Online Full worked solutions are available in SolutionBank.

c The test in part **b** requires that both the variances are equal. The test in part **a** established that this was reasonable.

3 a 95% C.I. uses t value of 2.571;

$$\frac{\hat{\sigma}}{\sqrt{6}} \times 2.571 = \tfrac{1}{2}(223.5 - 206.2) \Rightarrow \hat{\sigma} = 8.241...$$
$$\Rightarrow \hat{\sigma}^2 = 67.9 \text{ (3 s.f.)}$$

b (5.14, 20.2)

c Let S = span of an adult male's hand

$$P(S > 230) = P\left(Z > \frac{230 - \mu}{\sigma}\right)$$

For a maximum value, you need largest μ and largest σ

$$P\left(Z > \frac{230 - 223.5}{20.2}\right) = P(Z > 0.3218)$$
$$= 1 - 0.626$$
$$= 0.374$$

So highest estimate of the proportion is 0.37 (2 d.p.)

4 a $S_{hh} = 272\,094 - \dfrac{1562^2}{9} = 1000.\dot{2}$

$S_{cc} = 2\,878\,966 - \dfrac{5088^2}{9} = 2550$

$S_{hc} = 884\,484 - \dfrac{1562 \times 5088}{9} = 1433.\dot{3}$

b $b = \dfrac{1433.3}{1000.2} = 1.433015$

$a = \dfrac{5088}{9} - b \times \dfrac{1562}{9} = 316.6256$

$c = 317 + 1.43h$

c For every 1 cm increase in height, the confidence measure increases by 1.43.

d a is the value of c when $h = 0$ which makes no sense since this implies a person could be 0 cm tall.

e i −3.97, 2.33, −5.41, 6.62, −8.66, 5.01, −5.23, −5.11, 15.76
ii 573 since residual is far greater than the others.

f $c = 297 + 1.534h$

g 561

5 a 0.810 (3 s.f.)

b $H_0: \rho = 0$, $H_1: \rho > 0$. Critical value is $0.6429 < 0.810$, so at the 5% significance level there is evidence to suggest a positive correlation between qualifying lap-times and race results.

c Race ranks are not measurable on a continuous scale.

d Data will have 4 values with tied rank. Assign a rank equal to the mean of the tied ranks. Calculate the PMCC directly from the ranked data rather than using the formula.

6 a $\frac{31}{40}$ and $\frac{157}{4800}$

b 1 since f(x) is increasing function for $0 \le x \le 1$.

c Mode > mean therefore negative skew

d Substitute $3k$ and k into F(x) and subtract before simplifying to get the required result.

7 $3\,\text{m}^2$